RISK

Negotiating Safety in American Society

Arwen P. Mohun

The Johns Hopkins University Press
Baltimore

Printed in the United States of America on acid-free paper
2 4 6 8 9 7 5 3 1

The Johns Hopkins University Press
2715 North Charles Street
Baltimore, Maryland 21218-4363
www.press.jhu.edu

Library of Congress Cataloging-in-Publication Data

Mohun, Arwen, 1961–
Risk : negotiating safety in American society / Arwen P. Mohun.
p. cm.
Includes bibliographical references and index.
ISBN 978-1-4214-0790-6 (hdbk. : alk. paper) —
ISBN 978-1-4214-0825-5 (electronic) — ISBN 1-4214-0790-6 (hdbk. : alk.
paper) — ISBN 1-4214-0825-2 (electronic)
1. Public safety—United States. 2. Risk assessment—United
States—Public opinion. 3. Emergency management—United States.
4. Accidents—Social aspects—United States. I. Title.
HD4605.M63 2013
363.190973—dc23 2012017649

A catalog record for this book is available from the British Library.

Special discounts are available for bulk purchases of this book.
For more information, please contact Special Sales at 410-516-6936 or
specialsales@press.jhu.edu.

The Johns Hopkins University Press uses environmentally friendly
book materials, including recycled text paper that is composed of at least
30 percent post-consumer waste, whenever possible.

CONTENTS

ACKNOWLEDGMENTS

Writing a book is, in its own way, a distinctly risky endeavor—particularly a book about a huge and amorphous topic such as this one. I owe a debt of gratitude to many people who have helped and encouraged me along the way.

In 1993, Julie Johnson and I organized a conference at the Hagley Museum and Library entitled "Danger, Risk, and Safety." Since then, Hagley has been both a resource and a refuge. Roger Horowitz, Julie's successor at Hagley's Center for Business, Technology, and Society, found me an office and a stack pass. Roger and Philip Scranton also invited me to present a version of the gun chapter to the Center's Research Seminar. Comments from former staff member and firearms expert Rob Howard helped save me from a number of significant factual errors. Hagley's first-rate staff continues to be a pleasure to work with. Special thanks to Lyn Cantanese and Chris Baer for help with the railroad collections. In Imprints, Linda Gross worked her magic with interlibrary loan and is responsible for finding the Jacques Collection at Penn State.

The Smithsonian's National Museum of American History provided a senior research fellowship and intellectual home at a critical point in developing this project. Thanks to Steve Lubar and Peter Liebhold for your support, enthusiasm, humor, and good ideas. Someday I'll write that piece about banana peels and risk in American culture. Pete Daniel made me feel welcome, listened to my ideas, and showed me the museum's lawn mower collection. Various staff members in the museum's library and archives center also provided invaluable help.

Librarians and archivists in a number of other institutions also went out of their way to be helpful. Special thanks are due to the first-rate reference staff at the University of Delaware, especially David Langenberg, Rebecca Knight, and Tom Melvin. I am lucky to live only a short train ride away from the Library of Congress, one of America's great cultural treasures. The staff in the law library was particularly gracious in helping a nonlawyer understand the mysteries of legal research. Paul Dzyak and Jackie Esposito were unfailingly helpful during my visit to Penn State Special Collections. Erika Piola at the Library Company enlightened me about early American images.

A number of other organizations also provided research support and opportunities to present and publish my work. A National Science Foundation Scholars Award funded much-needed leave time. Thanks are due to Keith Benson, then a program officer, for understanding the significance of my project and giving it his support. An invitation by David Rhees, Oliver Hochadel, and Peter Heering to attend a lightning rod conference at the Bakken Institute helped inspire the weather chapter. Thanks also to faculty and graduate students at Lehigh University, Stevens Institute, and the University of Pennsylvania for inviting me to speak and showering me with hospitality and useful feedback.

A number of friends, students, and colleagues brought their own expertise and analytical skills to reading and commenting on parts of the manuscript, including Kevin Borg, Jim Brophy, Ann Greene, Kasey Grier, Elizabeth Higginbotham, Roger Horowitz, Cathy Matson, and Susan Strasser. Carole Haber took time out from her busy schedule to read and comment on an entire draft, offering many valuable suggestions. Thanks also to the Mohun-Strasser dissertation group, especially Andy Bozanic and Cristina Turdean for reading an earlier draft. Without Dave McClemens, retired DuPont chemist and aspiring historian, I could never have written the railroad chapter. Ben Schwantes and Cristina Turdean patiently found and copied many articles and government documents. Angie Hoseth helped digitize many of the illustrations. Late in the game, Audra Wolf contributed her editorial skills to making this a much better book.

A few other people deserve thanks for help that is not so neatly categorized. Alan Meyer brought a small arsenal in the trunk of his Toyota up from Washington and took me shooting. Dave Amber initiated a day of roller coaster research at King's Dominion, which left me with whiplash and a much better understanding of risk. Bob Brugger of the Johns Hopkins University Press continued to believe in the project despite significant setbacks.

Finally, I express my heartfelt gratitude to Erik Rau. I'm fortunate to have a spouse who is not only a generous editor and sounding board but also an enthusiastic participant in more experiential forms of risk research. During the process of writing this book, we searched for nineteenth-century lightning rods, toured factories, galloped horses across Botswana and the Maryland countryside, toured the Harness Racing Hall of Fame, went target shooting, visited amusement parks, and had many more adventures. Erik, the front seat of the coaster belongs to you.

RISK

Introduction

The camera caught them arrayed like spiders across a vast, man-made web: workmen posing against the cables of the Brooklyn Bridge for a city-employed photographer on a foggy October morning.[1] More than a hundred feet above the water, they recline casually, as if leaning on the polished counter of a saloon. A century later, this photograph clearly dates from another time, its antiquity indicated not just by the men's old-fashioned clothing or the unfamiliar profile of Manhattan's skyline, but also by their cavalier attitude toward the dangers of their profession. People still work on bridges, high above the ground, but they rarely flaunt a relaxed attitude toward risk for official photographers. And not a single piece of safety equipment is in sight—no hard hats, no safety harnesses—nothing but wits, experience, and luck between the bridge painters and a hard fall into the East River.

Eugene de Salignac's 1914 photograph opens a window onto a society that understood and managed risk differently from our own. Predominantly Norwegian and Swedish immigrants, many bridge painters once worked as sailors, tutored in a studied disregard for heights while clinging to the rigging of tall ships.[2] They negotiated the uncertainties of their profession (and the dangerous city below) through what this book will call a vernacular risk culture.[3] This set of rules, customs, and beliefs was generated in the course of everyday activities. It passed informally from person to person, reinforced through the authority of experience and social status. In the bridge painters' worldview, skill, carefulness, and attention to duty were the only means of avoiding accidents.

Until 1918, the painters, their supervisors, and the public viewed falling from

Eugene de Salignac, Painters on Brooklyn Bridge, 1914. *Courtesy, NYC Municipal Archives.*

the bridge as unfortunate and avoidable, but ultimately a matter of personal responsibility. Bridge painting was a dangerous job, so accidents were bound to happen. In the previous year, however, New York's mayor, John Mitchell, ordered supervisors to collect statistics on workplace deaths among city employees. Until then, no one knew how many workers died or suffered injuries each year because no one counted. In response to Mitchell's edict, the Department of Bridges began promoting worker safety as an organizational priority. Eventually, new work rules required helmets and other gear and specified various safety protocols taken for granted in our own times.[4]

Physical risk is a perpetual part of the human condition. But the kinds of risks most people face in their everyday lives, how they distinguish between acceptable and unacceptable risks, and their options for limiting risk have under gone an extraordinary transformation over the past three centuries. The character of modern technology conditions how Americans experience risk. In the twenty-first century, most accidents and fatalities among young people result from encounters with machines and the built environment (including bridges) rather than from the natural processes that most often injured and killed our ancestors.[5] New ways of knowing about risk and new kinds of expertise play equally important roles in determining how risk is assessed and managed. We are bombarded by accident statistics, probability assessments, and advice from experts—all of which barely existed a hundred years ago. Vernacular risk cultures still shape the way we address everyday risks, but they no longer represent the unquestioned norm.

The evolution of industrial and consumer capitalism has also transformed the social relationships through which the distribution of risk is negotiated. In the modern world, impersonal bureaucracies and the market often form the only connections between capitalists, workers, and consumers. Questions about what constitutes an acceptable level of risk and who should bear the costs of accidents are routinely handled through regulation and litigation rather than face-to-face informal negotiation. Insurance and liability law have helped to create a market value for accidents and risk management. Evolving cultural ideas about masculinity and femininity, race, and ethnicity have also influenced who takes risks and who is at risk.

What risks are worth taking? Who gets to decide? How can risk be controlled? Who or what should be responsible when things go wrong? This is a book about the many ways Americans have answered these questions. It is also about how and why their responses have evolved over the past three centuries. Change took place in three stages. In early America, fire, disease, weather, and animals presented the most important dangers to property and human life. Experience and common sense overwhelmingly dominated the management of risk. New ideas, not new types of hazards, led to the first significant transformations in how people attempted to reduce their exposure to harm. In the eighteenth century, a few individuals—primarily adherents to the Enlightenment's faith that systematic analysis and experimentation could improve the human condition—introduced the first seeds of a more modern approach to risk. They promoted the use of statistics, new technologies, and new forms of social organization to manage familiar dangers. Despite their efforts, the vernacular continued to dominate most people's approach to risk, hanging on to old practices, such as using animals, even when translated into new contexts created by industrialization and urbanization.

In the following century, exposure to the dangers of factories and railroads convinced a growing number of Americans that piecemeal improvisations based on vernacular practices were hopelessly inadequate. The overwhelming violence of machines and the complexity of systems required a different approach. Reformers also questioned whether capitalists could be trusted to safeguard the lives of their employees and the broader public. One result of these critiques was increased intervention by the state to protect both individuals and the common good. New laws mandated not only safer practices in industrial workplaces but also the adoption of mechanical safety devices that controlled or bypassed human judgment. By the second half of the nineteenth century, state governments had begun to hire factory inspectors and other pioneering professional safety experts to offer advice and enforce regulation.

Not satisfied to restrict their efforts to industrial settings, experts and their allies preached the gospel of safety to a rapidly expanding audience after the turn of the century. A conceptual leap gave momentum to their proselytizing. The analysis and management of physical risk could be treated as a special kind of knowledge, separable from not only lived experience but also specific kinds of technological expertise. These men and women became key actors in constructing a society-wide safety movement.

The ambitions of safety experts, the growth of the regulatory state, a transformation in liability law, and the emergence of a twentieth-century consumer culture coincided to give shape to a third era. New techniques of design and production democratized exposure to an array of potentially dangerous consumer technologies. Mitigating the risks associated with twentieth-century consumption posed very different challenges than the natural risks of early America or the ongoing risks of industrial employment. Public officials and safety experts looked on with alarm as consumers quickly modified the vernacular culture of risk to encompass streetcars, automobiles, and handguns. They struggled against the sentiments, widespread among the public, that accidents happened to other people, that aggressive preventative measures were only necessary for the careless, inept, or criminal. Education and voluntary cooperation with expert prescriptions offered only a partial and uncertain solution to managing risk. Outside the structured, hierarchical environment of industrial workplaces, disciplining potentially dangerous behaviors required new strategies.

In the absence of other options, regulation, promulgated and enforced by local, state, and federal governments, became increasingly important. Before midcentury, most measures focused on controlling users of dangerous technologies. In the 1950s, safety advocates turned their gaze back toward manufacturers. They argued that the state should be able to force producers to anticipate and guard against uses and misuses of consumer goods that led to accidents. This mandate was reinforced by changes in liability laws. Over the course of the century, personal injury lawsuits took on an important role in both redistributing the costs of accidents and providing an incentive to both manufacturers and insurance companies for preventing accidents.

If Americans became more risk conscious in the twentieth century, they did not necessarily become more risk adverse. Risk taking had always been an attractive form of entertainment for some people. New innovations meant that the masses no longer needed to choose between safety and thrills. Consumers enthusiastically embraced the commercialization of risk taking as entertainment. They willingly paid to experience the excitement of riding roller coasters, watching daredevils, or being drawn into simulations of danger, first on the stage and later in the movies.

As Americans struggled to address familiar hazards from nature and then the less familiar risks of urbanization, industrialization, and new consumer technologies, resistance, conflict, and negotiation marked the way. Stakeholders debated whether lowering accident rates justified cost and inconvenience. Experts disagreed with each other about the best means to mitigate risk. Disputes also hinged on a larger set of social and political questions. Whose life or well-being was worth protecting and whose was not? What kinds of risks threatened the common good and were therefore an appropriate target of collective action, and which were private matters? How safe was safe enough? Consequently, the laws, devices, and practices that resulted almost always embodied compromise.

Disagreements also stemmed from the uncertainty inherent in trying to predict and control the future. Risk, by definition, describes a potentiality, a future event, something that might or might not happen. However, words like "safe" and "dangerous" signal a sense of certainty about relative degrees of risk. They can also indicate risk acceptance. More often than not, what we mean by "safe" is "safe enough." Who we are and how we have been taught to think about risk shape these judgments. As psychologists tell us, individual beliefs about the likelihood of particular kinds of accidents often diverge from estimates made by experts, resulting in defiance of expert prescriptions. Both contemporary psychological research and historical evidence also suggest that risk perception varies from individual to individual and between different social groups. These differences reflect personal experience, but also cultural values held in common and reinforced by others.[6]

Overview of the Book

The chapters that follow explain how early Americans managed risk through vernacular means; how and why an expert-dominated, modern risk culture developed and gained influence over everyday activities; how this interaction has shaped the risk perception and behavior of nonexperts (and vice versa); and why most of us do not always do what experts tell us to do. "Risk" is a capacious term, used to describe the uncertainties (usually negative) that arise from all kinds of activities, including the financial, political, and social.[7] This book focuses on the more specific threats of physical harm: threats to the body and, to a lesser extent, property. Accidents, rather than disasters, typically result from failures of prevention. Until problematized by particularly dramatic accidents that galvanized public outrage or, as in the case of the bridge painters, by a shift in the overall risk culture, these kinds of risks were viewed as normal, if not inevitable, and tolerated within limits (automobile accidents being our most familiar present-day example). Yet even such a restricted definition of risk still leaves open the possibility of seemingly endless examples. To make such an exploration manageable, most of the chapters that follow are organized around specific kinds of risks Americans found especially controversial or compelling at particular moments in time—risks that were "good to think with," to borrow anthropologist Claude Lévi-Strauss's famous phrase.[8] Part I, the section on risks from nature, includes fire, disease, weather, and animals. Part II deals mostly with railroads and factories, while part III considers automobiles, firearms, amusement parks, and household consumer products.

Organizing this book around a series of specific examples serves to underline several key points of my overall argument. The material characteristics of particular risks influenced how they might be managed. A horse could not be reengineered for safety in the same way as an automobile. Equally specific cultural meanings and social practices help to explain why, for example, the licensing of drivers and firearms involved very different criteria. Moreover, while people drew on many of the same tools in seeking to evaluate and mitigate specific kinds of risk, the outcome of their efforts did not necessarily turn out the same way.

It is also my contention that technological change did not, by itself, determine how Americans understood and managed risk. Instead, the physical characteristics and cultural meanings of specific kinds of human-made hazards intertwined in a process of "mutual shaping."[9] Living with railroads, for example, altered the way nineteenth-century Americans thought about risk. At the same time, Americans literally built their social values into those machines and the systems in which they operated, often emphasizing cheapness and expediency over durability and safety.[10] Taken separately, neither the characteristics of technology nor the ideas and actions of individuals and organizations can adequately explain what caused Americans to change how they negotiated risk. The full nature of the process becomes clear only by examining both technology and culture.

A Historian Writes about Risk

Type the keyword *risk* into a library catalog search engine, and a door opens to reveal an enormous number of books and articles. For more than three decades, sociologists, anthropologists, and psychologists have been researching and writing about the social dimensions of risk.[11] In recent years, they have also been joined by a small number of historians, although *safety* is still the preferred way of framing this topic.[12] A few more key strokes reveal the importance of the term *risk society*.[13] This concept, identified with the German sociologist and social theorist Ulrich Beck, has become the best-known and most influential aspect of this enormous and diverse literature. It is therefore worth saying something about its relationship to this book. Beck's fame stems from a series of books and articles, beginning with the 1986 volume translated from the German as *Risk Society*. He and other risk society theorists have argued that one of the most important characteristics of modern societies is that they invest huge quantities of resources, human and economic, in managing risk.[14]

This book owes an important conceptual debt to Beck for giving a name to this phenomenon. It also owes a debt to risk society theorists for framing their work in terms of a "before" and "after," thereby implying that there is a history to be told about risk societies. However, for most of these sociologists, history functions as little more than a rhetorical device. For those of us for whom the past is something more, such assertions cry out for empirical research: What was the "before" like? What were the historical processes through which it was transformed? What relationship does the past bear on the present? Beck has suggested, for instance, that "pre-industrial hazards, no matter how large and devastating, were 'strokes of fate' raining down on humankind from 'outside' and attributable to an 'other'—gods, demons or Nature"—and that premodern people were essentially fatalistic, reduced to cursing the gods over their uncontrollable suffering.[15] My research reveals a more complex picture. While many expressions of fatalism can be found in the historical record (as well as in the present), premodern people also recognized the importance of human agency. They might not have been able to do anything about preventing natural calamities, but they routinely assigned human responsibility for limiting the negative consequences of these events. We would not have survived as a species if this were not the case.

Much of the risk society literature also implies that historical change can be described in terms of structures and stages, and that the model is applicable across the industrialized world. While there is great value in recognizing metapatterns, I have chosen to make the case for a distinctive American risk society. The process could have turned out differently, and in fact has, in other places in the industrial world.[16] The establishment of risk-managing structures and the distribution of resources required individuals and groups to make decisions. Those decisions were informed by their personalities, life experiences, and historical contexts. In turn, those decisions created specific trajectories.

Beck has also claimed that we now live in a condition he describes as "reflexive modernity," distinct from both the industrial era that preceded it and some more distant preindustrial past. The difference lies in the kinds of risks we are exposed to, namely, the man-made, global threats such as environmental degradation or nuclear war that individuals can do nothing about.[17] However, the historical evidence laid out in my final chapter suggests that the global threats distinctive to the postwar world had little or no impact on either individual behavior or public policy regarding the kinds of everyday risks that are the subject of this book.

Join me then, for an exploration of the development of one familiar risk society, the United States. Roller coasters and lightning rods, fast horses and faster trains, puritan divines with a taste for statistics, and factory inspectors obsessed with technological fixes: these are some of our stories, the first of which is about fire. So, unbuckle your seat belt, take off your helmet, and venture into a world that preceded our own.

PART I

RISKS FROM NATURE

1

Fire Is Everybody's Problem

Fire existed long before human beings. Our species turned this chemical process into the most useful, ubiquitous, and dangerous technology of preindustrial societies. Domesticating fire exacerbated its risks. Brought inside to provide heat and light, uncontrolled fires could indiscriminately destroy flesh, bone, paper, and wood. The practice of constructing dwellings and whole cities of combustible materials further amplified risk.[1] A tiny careless or malicious act on the part of an individual could set a city ablaze, resulting in terrible suffering and loss for many people.

To use this essential technology without succumbing to it, early Americans relied on a combination of private and public practices, many of which had deep roots in European history. They viewed prevention of bodily injuries and fatalities from burns or asphyxiation as an individual responsibility, governed by vernacular precepts of carefulness and common sense. Across the colonies, community leaders intervened in circumstances where individual behaviors threatened the common good.[2] They prohibited storing gunpowder and other dangerous behaviors in dense urban areas, compelled chimney cleaning, and established a system of night watches.

With so much at stake, it is not surprising that the management of fire is one of the first places where more modern ways of understanding and controlling risk were implemented.[3] In the early decades of the eighteenth century, a small group of enlightened thinkers and entrepreneurs began looking for better ways of managing the problem of urban fires. They promoted a series of interconnected innovations: firefighting companies, fire engines, and fire insurance.

None of these practices were absolutely new.[4] But their specific characteristics and popularity heralded the spread of more subtle aspects of modernity: scientific ideas translated into widely used technology, social organizations specializing in the management of particular kinds of risks, and growing public faith in the peculiar abstraction of a risk pool.

Philadelphia, one of the largest cities in early America, offers an especially revealing window onto how early Americans dealt with the problem of fire. The city was home not only to Benjamin Franklin, inveterate improver and leading light of the American Enlightenment, but also to merchants, craftsmen, and toiling masses. In an increasingly crowded warren of streets and houses (and in the writings of other people who observed them) it was easy to glimpse the persistence of common practices and the halting evolution of the new.

Private Sorrows, Public Concerns, and the Prevention of Accidents

Elizabeth Drinker worried and occasionally obsessed about risk. Marriage to one of Philadelphia's wealthiest merchants insulated her from many of life's vicissitudes. But personal experience, observation, and the occasional newspaper account of a "melancholy accident" still provided plenty of material for her diary, often accompanied by snippets of moral judgments about the carelessness of others.[5] Poisonings, drownings, and runaway horses all fed her pen, but the dangers of fire appeared most often. Drinker's dual status as both a woman responsible for overseeing a household and a well-informed member of the city's educated elite gave her insight into both vernacular practices and innovations sponsored by her family, friends, and acquaintances.

In Drinker's world, open flames blazed everywhere. Household fireplaces provided heat throughout chilly Philadelphia winters. Servants cooked meals on a kitchen hearth. Wealthy enough to light her room with candles, Elizabeth could linger over her diary in the evening. Her poorer contemporaries mostly relied on rush lights: reeds dipped in animal fat to provide smoky, uncertain illumination. The view from the Drinkers' Front Street house often included plumes of smoke wafting upward from a variety of workshops: bread bakeries and ship's biscuit manufactories, soap boilers and blacksmiths.[6] Wooden ships tied up along the wharfs outside their front door hosted cooking fires and light for illumination, even at sea where an out-of-control fire imposed a certain death sentence. In the countryside, fire provided an essential tool for clearing and maintaining the land—a sight the Drinkers might have witnessed as they went to visit Henry Drinker's iron forge in rural New Jersey. The forge itself consumed huge quantities of charcoal manufactured in nearby forests and nearly burned to the ground twice in the time Drinker owned it.[7]

Although blacksmiths, iron workers, soap boilers, and other male workers certainly sustained burns as part of their work, women and children most often suffered fire-related injuries. The broad, open hearths of many colonial houses held a variety of dangers. Most women spent their days stooping or sitting next to the burning logs and glowing embers that heated their cooking pots. A moment of inattention was all it took to set the hem of a skirt or the sleeve of a blouse on fire. Those with enough presence of mind dropped to the ground to suffocate the flames or sought water to douse it. Others panicked and suffered the consequences. The *Pennsylvania Gazette* described the fate of one such young woman whose calico gown caught on fire while she was taking a teakettle off the flame. She ran to the neighbors for help, collapsing on the road because her frantic response had fanned the flames. Benjamin Franklin, the likely author of the article, somewhat callously editorialized that it was too bad that she had not thought to throw herself in the creek that ran next door to her house.[8]

Sitting close to the fire—a necessity in the winter—also had its dangers. Although beginning to go out of fashion, deep, open hearths contiguous with surrounding rooms remained common in eighteenth-century America. Within the hearth, pots hung on hooks suspended from a heavy pole of green wood called a "lug-pole" or "back-bar" running the width of the hearth. If allowed to dry out, these bars tended to break, splashing hot liquids and worse. Over time, householders replaced them with iron cranes, which were less likely to suddenly give way, but large pots of boiling liquid remained a danger.[9] Surviving images of early American hearths and inventories of accessories associated with fireplaces tell a similar story of the ubiquitous danger of fire in the home. Some households employed devices called "fenders" to prevent hot coals from rolling out onto wooden floorboards, but there is no evidence of any efforts to use mechanical or architectural devices to keep people *out* of fireplaces.[10] Instead, habitual carefulness offered the primary form of protection.

In this world constantly ablaze, fire posed a particular risk to those too young to grasp the threat lurking in flickering hearths and lamps. Small children risked terrible injury if a caretaker's attention wandered. Drinker noted "a very melancholy accident" in December of 1794. Jeremiah Warder lost a little daughter "between 2 and 3 years of age" when "she was left alone for a short time, and fell into the fier." Drinker's response epitomized a preindustrial risk culture that viewed the risk of bodily injury from fire as a private matter to be handled through carefulness and discipline. Responsibility for such accidents lay with mothers and other caregivers. "Young children should never be left alone," she opined, blaming a servant's heedlessness.[11] Harried mothers also tied toddlers to chairs or otherwise confined them to prevent injury.

Since fire was often the first source of potentially lethal accidents encountered by children, instruction about fire began young. Minor accidents offered an opportunity for impromptu lessons reinforced by physical punishment and verbal reminders. Some of these heuristics remain in the language in an idiomatic form, fire having become a metaphor for other dangers: "play with fire and you'll get burned," "a burnt child dreads the fire," and "stamp that fire out before it spreads."[12]

Over the next century, Philadelphians, along with other Americans, gradually exchanged their open hearths for cast iron stoves and candles for kerosene lamps, but little about the way they thought about or managed the risk of being burned changed. Domestic fire management remained a private concern.[13] In contrast, structural fires were too important to be left solely up to individuals, even in an era when the vernacular held sway.

"Several people have fire in their chimneys," Drinker noted during an unseasonably chilly June. For her, the sight of smoke provoked anxiety. Normally, the warmer days of spring and summer eased the threat of house fires (except from lightning, which we will address later). Winter was fire season and midnight its most dangerous hour as fires smoldered while householders slept. Structures caught fire for a variety of reasons. "Last Thursday night about 12 o'clock the house of Mr. Joseph Hyde of Lebanon was destroyed by fire." "Painful to relate," she wrote, the victims included "Mrs. Sarah Hyde and her daughter about fourteen years of age," as well as "Mr. Natl. Backhouse an old infirm man and an elderly Negro woman." Apparently, a log rolled out of the hearth in the middle of the night, igniting the floorboards. Drinker again added an admonishment, "People who are apt to leave their fire in a careless situation on going to bed, it is hoped, will be warned by this sad tale."[14]

The fire that destroyed the Hydes' residence apparently did not spread, but in the dense confines of early American cities and towns, a different outcome was also likely. Wind-whipped flames easily leaped between tightly packed structures. Wood, though more flammable than brick or stone, was attractive to builders because of its cheapness. Builders also saved money by not creating party walls: thick, external walls extending above the roofline between townhouses. As a result, roof fires roared through attic spaces, quickly engulfing whole city blocks.

Like many of her contemporaries, Drinker believed that blazing chimneys

Although no flames are visible, John Lewis Krimmel's study of a child kneeling on a hearth illustrates how easily clothing could catch on fire as women and girls worked. *John Lewis Krimmel, "Studies of cat, asparagus, various household utensils and cooking implements."* Sketchbooks, 1809–1921. *July 7, 1819. Courtesy, The Winterthur Library: Joseph Downs Collection of Manuscripts and Printed Ephemera.*

were the single most important cause of avoidable structural fires.[15] Through daily use, flammable tar and creosote built up as soot encrusted the flues. Eventually these deposits burst into flame or, less dramatically, smoldered, generating heat until roof shingles ignited. Changes in architecture exacerbated the problem of chimney fires. During Drinker's lifetime, a growing wood shortage, more elaborate house design, and changes in stove and fireplace design helped transform how people made fires in their houses. The Drinker townhouse on Philadelphia's Front Street was one of the grandest in the city. An imposing three-story brick structure, it contained multiple rooms with fireplaces that fed into a smaller number of chimneys.[16] By the mid-eighteenth century, better-off people like the Drinkers abandoned the broad hearths of an earlier era for new designs that brought with them a series of trade-offs. Narrower chimneys conserved heat rather than letting it escape, but they collected soot more quickly. Stoves, recommended by Benjamin Franklin for their efficiency, produced copious quantities of flammable creosote.[17]

"Our kitchen chimney took fire Yesterday before dinner but by the help of Isah and his Men, it was soon put out," Drinker wrote in 1767.[18] The distinctive crackling sound of deposits burning inside the flue and the color and quantity of smoke coming out of the chimney probably alerted the household. Wet blankets or other materials stuffed into openings could deprive the fire of oxygen, if someone discovered the problem quickly enough.[19] Otherwise, householders had to clamber up onto rooftops with water buckets and swabs to drown the flames before they spread. Drinker also kept an eye on other people's chimneys. "A very foul chimney in Water Street opposite the Bank Meeting House—took fire this evening—and occasion'd a great Hubbub," she noted in 1780.[20] A few years earlier, during the dark days of the Revolution, chimney sweeps proved hard to come by. Rather than risk a fire, the Drinkers stopped using the chimney in their main parlor, retreating to a back room to ward off the late November chill. Their neighbors exercised less caution. "A chimney in the alley on fire this afternoon—many more 'tis fear'd will be ere long, tis so difficult to get sweeps," Elizabeth wrote. Her prediction soon came true: "a chimney fire this afternoon, next Door, but one, where Mall liv'd—and one over the way yesterday," she noted on December 5.[21]

Drinker understood that the regular cleaning prevented chimney fires. Consequently, chimney sweeping constituted an important form of vernacular risk control. Some people, particularly the poor and those with the older style of broad hearths, cleaned their own chimneys, sometimes using a small tree pushed through the opening. The better off hired someone to do the job for them. During the colder months, sweeps walked the streets singing or shouting "Sweep O, Sweep O." If someone needed a chimney cleaned, they responded by hailing the sweep and striking a bargain for the work.[22] Sweeps delivered an essential service involving dirty, dangerous labor at low wages. They provide the first of many examples in this book of working-class people endangering their bodies in the service of protecting the lives and property of the more prosperous. In Philadelphia, as in the rest of English-speaking America, African Americans—in the South, often enslaved people; in the North, either slaves or later free blacks—did the job.[23] Richard Allen, a freedman who founded the Bethel African Methodist Episcopal Church in Philadelphia, made part of his living as a master sweep. If Allen cleaned a chimney himself, he likely used a brush and a spade-like device to loosen deposits in the flue while standing on the rooftop or on the hearth. He also sent apprentices up into the flues.[24]

Like other forms of prevention, cleaning chimneys involved uncertainties and trade-offs. How much soot built up depended on the design of the chimney, the type of fuel being burned, and a host of other even less measurable factors. The prudent might clean their chimneys (or have them cleaned) as often as four times a year, but this kind of effort involved time, inconvenience, and, if someone else did the job, money.[25] Those without resources or with the inclination to gamble or to harbor a willful obliviousness to risk might wait a good deal longer. Inevitably, stories circulated about sweeps who, intentionally or unintentionally, failed to clear chimneys of soot adequately. Elizabeth Drinker confided such a tale to her diary, annoyed by a fire in her back parlor chimney she had cleaned only the month before.[26] To add insult to injury, householders could be fined for chimney fires in newly (but presumably inadequately) cleaned chimneys, whether or not they did the cleaning themselves. Because a dirty or poorly cleaned chimney could set the whole town on fire, community leaders turned to the law to compel adequate prevention.

Over the course of the eighteenth century, the Pennsylvania Assembly, as well as local governing bodies, pushed through a series of ordinances designed to prevent fires by disciplining the behavior of residents and visitors. These statutes reflected commonly held beliefs about the causes of fires, balanced by political strategizing about which risks could or should be controlled through legislation. Laws not only required chimney cleaning but specified appropriate and inappropriate ways of doing so. The 1701 "Act for Preventing Accidents That Happen by Fire" forbade residents from deliberately setting fire to their chimneys to avoid the cost of a chimney sweep or the time and labor of doing the dirty job themselves. This practice, called "firing," remained commonplace in rural parts of the British Isles but was dangerous in the close quarters of cities.[27] People who neglected the task entirely could also be fined if two witnesses testified before a justice of the peace that their chimney was "so foul as to take fire and blaze out at the top."[28]

Legislators used their powers to try to ensure that chimney sweeps did a thorough job. In 1720, Philadelphia officials appointed an official chimney sweep, in imitation of Boston.[29] His monopoly was short-lived, however, because there was no easy and effective way to prevent his competitors from singing their wares and undercutting his prices.[30] In 1735, Benjamin Franklin, already interested in the problem of fire, published an article in the *Pennsylvania Gazette* suggesting that the mayor license sweeps.[31] It took nearly a half century for the General Assembly to respond. In 1779, it announced details of a new law in local newspapers. Master sweeps had to register with the clerk of the Philadelphia Contributionship, the fire insurance company founded by Franklin and discussed below. The clerk issued a certificate of registry and a tin or copper number to be affixed to the front of their caps. The point was to make sweeps accountable. A prudent householder would then write down the sweep's number so that if a chimney caught fire later or valuables disappeared from a house during the sweep's visit, he could theoretically be tracked down by authorities. As was typical for eighteenth-century laws, enforcement depended on a system of fines. Unregistered sweeps could be fined. Within the city limits, sweeps could not charge more than nine pence for cleaning a chimney with a single flue or fifteen pence for a two-flue chimney—a pittance for customers of the middling sort.[32] Sweeps could also be fined if a chimney caught fire within a month of being cleaned.[33]

Legislators also turned their pens to mandating other forms of prevention. Certain kinds of industries garnered a widespread reputation as the source for destructive fires (a reputation borne out in newspaper accounts). If Drinker kept a close eye on the baker and the soap boiler, so did the authorities.[34] The waterfront attracted particularly intense scrutiny. A 1721 statute focused on the wooden ships docked up and down the Delaware River waterfront. It forbade breaming (cleaning the bottoms of ships) with fire on the waterfront or docks, heating ships' supplies such as pitch and tar with a "blazing fire," and keeping lit fires on board after 8:00 in the evening when they were less likely to be closely supervised.[35] Experience and common sense suggested that cities built out of wood burned more readily than those constructed of stone and brick, but forcing someone to pull down a house and rebuild it in stone was a different matter than compelling a householder to clean a chimney. Laws requiring residents to build in brick or stone were widespread in the colonies but spottily enforced.

Conspicuously absent from the law books were laws regulating practices so widespread as to be ungovernable. The Great Boston Fire of 1711 began with a spark in a pile of oakum. Cities could not realistically ban storage of ubiquitous and essential materials such as oakum, straw, and hay. On the other hand, keeping large quantities of gunpowder in any one dwelling house—a cause of far fewer fires—attracted vigorous regulation because the public found the specter of an explosion within a city alarming and most people did not need a large amount close to hand.[36] Fire control could also be used as a pretext for regulating other kinds of undesirable public behaviors. Philadelphia, like New York, forbade residents from smoking in public. The 1721 law prohibited "the shooting of guns, throwing, casting and firing of squibs, serpents, rockets and other fireworks, within the city of Philadelphia."[37]

The law also took special aim at those who intentionally set fires. The specter of arson terrified early Americans. They were acutely aware that fire could be an intentional weapon, particularly attractive to the least powerful members of society, notably slaves. Along with everyone else, enslaved people used fire every day. Masters could restrict access to guns and other weapons, but this was entirely impractical with fire. In the eighteenth century, rumors of slave conspiracies often followed urban conflagrations. Most notoriously, New York magistrates arrested more than 150 slaves for conspiring to burn the city and murder white residents in the aftermath of a series of fires in the summer of 1741. Thirteen were eventually burned at the stake and another seventeen hanged for their purported crimes.[38] Articles about fires started by both slave arsonists and "careless negroes" regularly appeared in early American newspapers. So too did accounts of mysterious strangers, mad women, and criminals bent on covering their crimes—all blamed for unexplained fires.[39] Two hundred years later, it is impossible to separate out scapegoating from real causation in these accounts.

In early America, making and promulgating laws was cheap. Unpaid legislators gathered together to draft statutes, subsequently published in local newspapers to inform the population. These statutes show that lawmakers had a concrete, practical understanding about some of the common causes of fires and how to prevent them. A continual succession of new and revised statutes added over the course of the century suggests that people with influence believed them necessary and felt comforted by their existence. However, ineffective enforcement weakened the efficacy of these measures. In theory, citizens would discipline themselves once informed about what they were expected to do. In practice, only a literate few like Elizabeth Drinker might do so or even complain to authorities about lax neighbors, but for the majority, struggling each day to make ends meet, the importance of cleaning one's chimney probably barely registered until it was too late.

Firefighting and the Vernacular

"We were Allarm'd by the Cry of Fire at 11:00 at Night," Drinker wrote in 1759.[40] Crying fire constituted another communal responsibility. As with other aspects of vernacular fire control, anyone could participate. In practice, however, the night watch provided the first public line of defense if prevention failed. This ancient tradition followed Europeans across the Atlantic, taking on a characteristic voluntary form in the New World. In Philadelphia, the constable deputized a handful of unpaid householders to patrol the city streets looking for signs of trouble and admonishing citizens to behave in a lawful and orderly manner. When not dozing in the warm confines of a public house, the watch prowled the streets, sniffed the air, and peered into yards, basements, and windows. They acted as the eyes and ears and occasionally voice of collective efforts to protect the common good.[41]

Ad hoc firefighting constituted the universal practice in colonial America into the first decades of the eighteenth century; recruiting firefighters was one of the night watch's primary jobs. To rouse the slumbering, the watch reinforced their voices with the distinctive sound of a wooden ratchet or rattle. This sound must have provoked a special kind of fear in those burned out of their homes on earlier occasions.[42] The appropriate response to such an alarm, especially for men, was to hurry to the site of the fire and aid in extinguishing it. Householders grabbed fire buckets, ladders, and salvage bags. Although specialized over time, firefighting tools were, in their basic forms, common household objects. In a pinch, they could be improvised from materials on hand.

Once on site, volunteers drew on a body of common knowledge. They assumed that everyone knew what to do or could learn by observing. This approach provides a vivid example of the public side of vernacular risk culture. Individuals formed a bucket brigade, passing water hand to hand from the nearest well or body of water to the burning building. To put out a fire in a potter's workshop in nearby Elfrith's Alley, Drinker's neighbors broke down her fence to use the pump in her yard.[43] Others scaled ladders to reach smoldering roofs, which they swabbed with wet rags on the end of a stick. Some would-be fire firefighters showed up with axes and saws. They cut away parts of roofs to deprive the fire of fuel. Hook men pulled down burning walls. Others helped carry precious household goods to safety. Owners and tenants of nearby buildings wet down their own roofs and calculated whether to remove their possessions. If a fire began to spread, public officials gathered to make a dramatic decision. Law and custom empowered them to pull down buildings in the path of a fire or to use gunpowder to create a firebreak—a remarkable example of the way individuals' property (and therefore well-being) could be sacrificed for the greater good.[44]

The drawbacks of improvised firefighting became more and more apparent as cities grew larger. Responding to the alarm was viewed as a civil obligation for the physically able, but it was entirely voluntary. In smaller communities, self-interest as well as fellow feeling propelled neighbors to help neighbors. In the future, they might need the favor to be returned. In the greater anonymity of urban areas, reciprocity became less certain. Neighbors and bystanders could not necessarily be counted on to respond to the alarm. Firefighting was uncomfortable, dangerous work. Since fires often happened on cold winter nights, there was an added incentive to stay in bed and let someone else take care of the problem. If a well-equipped crowd showed up at a fire, they still had to organize themselves.

Fire scenes were often chaotic. Some direction came from constables, members of the watch, or specially appointed fire wardens, but this oversight was of limited effectiveness since the crowd of volunteers had never drilled together. The general confusion offered opportunities to those with less noble intentions. Salvaging household possessions was one of the important activities at fire scenes, but a fine line separated salvaging from looting. "Lost in the Hurry of Removing Goods at the late Fire, a large striking Watch, with a brass inside Case and a silver outside Case," Richard Smith advertised optimistically in 1734. The watch turned up later when a woman tried to pawn it. Most thieves escaped with their loot.[45] Sometimes crowds gathered not to help, but in the hope of free entertainment. Drinker's servant Polly Nugent went out on a December afternoon, telling her employer that she was going to chapel. Instead, she went to see a fire burning in an old wooden stable on Market Street. Her deception was revealed when she was run over by a man on horseback and badly injured.[46]

City governments responded in piecemeal fashion to the shortcomings of voluntary, vernacular firefighting with a hodgepodge of statutes. Boston initially provided a model for the rest of the colonies. A 1653 law required householders to have on hand a ladder and a twelve-foot pole with a swab on the end.[47] Many cities also required residents to possess two fire buckets. As these were to be made out of leather so that they would not catch on fire, they were far more expensive than the universal wooden bucket. This requirement imposed a significant financial burden for the many families that lived in two tiny rooms and counted a bed frame as their most valuable possession. Cities supplemented stocks of equipment that could be employed by volunteers, using fines collected during chimney inspections. Some legislators also tried using the law to compel participation in firefighting. In the aftermath of the Cornhill Fire of 1711, city fathers ordered Boston residents to "employe their utmost diligence and application" to extinguish fires. Fire wards were authorized to inform a justice of the peace about "disobedience, neglect, or refusal" on the part of individuals, who would be fined forty schillings (if they had the money) or imprisoned for ten days (if they did not).[48] It is unlikely that such a measure could be effectively enforced, however. Politicians' efforts to remedy the shortcomings of existing firefighting strategies are but one piece of evidence that few people were satisfied with the status quo. But it was not readily apparent what else could be done.

Improving Firefighting in Philadelphia

Perhaps not surprisingly, Benjamin Franklin fixed himself firmly at the center of public conversations about fire risk. From his perch as publisher of the *Pennsylvania Gazette* a few blocks west of the fire-prone Philadelphia waterfront, he regularly commented on the foibles of his fellow citizens and recommended measures to protect the common good. In the company of Philadelphia's leading citizens, including Elizabeth Drinker's husband, Henry, he facilitated the importation of English firefighting techniques and organized one of America's first fire insurance companies.

Franklin, Drinker, and their compatriots were motivated not just by the spirit of improvement and progress that informed the American Enlightenment, but also by their personal experiences with fire. When Franklin tackled Philadelphia's fire problem, he was a young, athletic man who had likely already hoisted his bucket at fires in Boston and Philadelphia. He was a person of his times—someone completely attuned to the meaning of the watch's fire rattle, ready with his buckets and salvage bag, looking (at least at first) only to make his culture's practices a little more reliable and effective.

In the aftermath of a serious fire that spread from one of Philadelphia's wharfs to the surrounding neighborhood in 1732, the young newspaper editor began reporting about other fires in order to rally his fellow citizens to join the bucket brigade. Initially, he identified inadequate public spiritedness as the core of the problem. Other young men and public officials just needed encouragement to do the right thing. After a fire in a wheelwright's quarters was put out before reaching the city prison, he lavished the volunteers with praise, repeating a formula that he used over and over. "By the extraordinary Diligence and Activity of the People, the Fire was at Length suppressed with only the entire Loss of those four shops."[49]

Franklin's campaign to spur volunteerism culminated with a secular sermon in the *Pennsylvania Gazette*, celebrating the heroic men who roll out of bed in the cold and dark to help their neighbors while chastising those who refused to contribute. Fires, he pointed out, afforded a public opportunity to display manly courage and virtue: "See there a gallant Man who has rescu'd Children from the Flames!—Another receives in his Arms a poor scorch'd Creature escaping out a Window!—Another is loaded with Papers and the best Furniture and secures for the Owner.—What daring Souls are cutting away the flaming Roof to stop the Fire's Progress to others!" Why do they make such an effort? "They do it not for the sake of Reward or Money or Fame: There is no Provision of either made for them. But they have a Reward in themselves, and they love one another."[50] Harsh characterizations awaited the less virtuous: men who lay "snoring in their beds after a debauch" or "prefer their own Ease at Home to the Safety of other People's Fortunes or Lives." Just as bad were other "wicked People" who make "haste to pilfer" or stand by as "idle Spectators." If pitching in was its own reward, in Philadelphia, the punishment for declining to lend a hand was hardly more substantial. One such bystander "receiv'ed a Bucket of Water on his impudent face" for refusing to join the effort. Franklin thought that the "drone" deserved "a Punishment much greater and more exemplary."[51]

A year later, Franklin had given the matter more thought. The first step was a volunteer fire company probably modeled after the Boston fire society first organized in 1717.[52] An unnamed "City in a Neighboring Province" had a "Club or Society of active Men belonging to each Fire Engine," he observed. Club members attended fires with the engine and kept it in good order. For this service, they received tax relief. Could Philadelphia do the same? Philadelphia (or at least its governing authorities), it turned out, would not. A more limited scheme would have to suffice.

Soon thereafter, Franklin and a group of friends organized the Union Fire Company for the purpose of "better preserving our Goods and Effects from Fire." They imagined the Fire Company as a society for mutual aid among the members, a better-functioning microcosm of common practice, and a supplement for an ineffectual city government.[53] Membership was limited to thirty individuals chosen from among men already prominent and politically active in Philadelphia. Members agreed that should one of their number die, they would act in his stead so long as the widow kept "Buckets & Bags in repair & causing same to be sent to every Fire." Their efforts helped ensure that others would help if her house caught fire, because they count on reciprocity from the fire company.[54] If peer pressure failed to compel participation, the offender also faced fines. The Fire Company's articles—part manifesto, part constitution—also include a sketch of vernacular practices: what kinds of buckets and salvage bags to have (leather and linen); how to get goods out of houses without being injured or killed (work in teams, candles in every room); and how to prevent thieves from entering houses. They did not include any information about how actually to extinguish fires, thus suggesting that the signatories thought it unnecessary to spell out what everybody already knew.

In 1734, the members agreed to a new provision signaling a shift in thinking about their mission. "As this Association is intended for a general Benefit," Franklin stated, "we do further agree, that when a Fire breaks out in any part of this City, though none of our Houses, Goods or Effects may be in any apparent Danger, we will nevertheless repair thither with our Buckets, Bags and Fire Hooks." Anyone, including total strangers, might benefit from their services, whether or not they could or would reciprocate. The members vowed to "give our utmost Assistance to such of our Fellow Citizens as may stand in need of it, in the same Manner as if they belonged to this Company."[55] This broader commitment generated logistical problems. Buckets and bags continued to hang in members' front hallways, but they realized that other pieces of equipment could not easily be carried long distances. Consequently, they began stashing ladders around the city: four at John Jones's house on Chestnut Street, two in an alley off of Second, one on the wall of the Work House, and one under the eaves of the shed covering the meat market, which members hurried to move when British troops occupied the city.[56]

In addition to mustering in response to the cry of fire, the Society met monthly to socialize and discuss firefighting techniques. During one of those meetings, Union members decided to raise the money for a fire engine by subscription. In July of 1749, Jacob Shoemaker Jr. and Thomas Say began collecting the funds. A month later, they had squeezed enough cash out of their friends and acquaintances to place an order for an engine to be built by Elias Bland in London. It arrived by ship the following spring—a barrel-sized pump mounted on a four-wheeled cart and a collection of leather fire hoses.[57]

Fire engines were the most significant technological innovation inserted into existing vernacular firefighting practices. These devices embodied in a modest form some of the ambitions of the scientific and mercantile revolutions. Engines provided a solution for two different problems: moving large quantities of water from wells or other water sources and allowing firefighters to dump water onto a fire without having to be in close proximity. In theory, using a mechanical pump made a great deal of sense; however, getting pumps to work more reliably and efficiently than buckets and swabs proved challenging.[58] The new acquisition performed beyond expectations. "The engine played to admiration, far exceeding all others there," the Company's minute book reported, with a hint of pride.[59]

As with many new technologies, the adoption of fire engines resulted in unintended consequences for the social organization of firefighting. The new technology demanded yet another layer of commitment from members of the fire company. Someone had to paint, oil, and repair the device's wooden body, metal and leather valves, and leather hoses. Deployment required a designated member to bring keys to unlock the storage shed. The cart was too heavy for one person to push or drag through the streets, so more members had to be present to move it. Attaching the hoses to the engine and to a water source and then operating the pumping mechanism required practice and teamwork. Although no one at the time planned it, this combination of new technology and the creation of organizations focused on the specific problem of extinguishing fires provided the basis for the next century's transformation of the profession. Firefighting, once everybody's responsibility, became the domain of specialists: first, specially trained members of firefighting clubs, and eventually, professional firefighters.

Almost from the beginning, Franklin's idea of a firefighting club proved very popular with the kind of virtuous (or at least ambitious and self-regarding) men he had imagined climbing out of their warm beds to help the community. In a short period of time, a number of other fire companies formed, including the Heart-in-Hand, the Britannia, and the Fellowship.[60] In October 1760, Elizabeth Drinker wrote that her husband, Henry, was going to a meeting of the fire company.[61] Like other members of the eighteenth-century elite, he believed that community leadership included standing in the front of the bucket brigade. As a property owner, he also had a lot to lose. In the meantime, Franklin and his friends were developing another plan to allow Philadelphians with adequate means to buy security of a different sort.

Insurance

Volunteer fire companies did not solve the fire problem any more than buying engines or forcing people to clean their chimneys. Instead, these organizations limited damage. In the 1740s, Franklin began pursuing another scheme that would allow property owners to recoup their losses when prevention, firefighting, and salvage failed. He and his fellow fire company members drew up the papers for a mutual fire insurance company. Realizing that their original plan to include only members of the fire company did not create a big enough risk pool, they opened up participation to anyone who could pay the premium and whose property was deemed free from unreasonable hazards.[62]

The story of eighteenth-century fire insurance is not, as one might expect, about the spread of statistical thinking, at least in the narrow sense. Rates were set not by actuaries, but by craftsmen who used their experience to determine how a building might burn and how much it would cost to rebuild it.[63] In effect, they applied a form of vernacular knowledge to the assessment of risk. Instead, enthusiasm for fire insurance evidences a subtle and complicated shift in mentalities on the part of those who both marketed and purchased it. The economic alchemy of insurance provided an innovative mechanism for separating the value of property (or a person, in the case of life insurance) from the integrity of the actual physical object. In an age when most wealth was invested and stored in buildings and other physical structures such as ships, a single fire could ruin a family's fortunes. If insured, value could still be recovered and then used to reconstruct the object or as capital for some other project. To make this strategy work, large numbers of investors had to trust that investment in insurance was worthwhile enough that they would regularly pay into a risk pool with the understanding that they might never get their money back.[64] Though typically portrayed as an act of prudence, buying fire insurance can also be seen as a respectable form of playing the odds, its growth in the eighteenth century not unrelated to that century's mania for games of chance, tontines, and lotteries.[65]

The new company was called the Philadelphia Contributionship. It was, by most accounts, the first successful fire insurance company in America. Like many of Franklin's other projects for improving Philadelphians' safety and security, the idea for a fire insurance company was not original. He borrowed most of the details, including the company's insignia, from a British model, the Contributionship for Insuring Houses from Loss of Fire, more commonly known as the Hand-in-Hand.[66] Plaques bearing this symbol are still displayed on some of Philadelphia's colonial-era houses. Company agents instructed homeowners to display fire marks prominently purportedly because fire companies would try harder to extinguish a blaze if they thought a reward might be forthcoming from an insurance company. More likely, their primary function was as a form of advertising—a prominently displayed symbol of public trust and a potent form of marketing for an otherwise immaterial product.[67]

The Contributionship began insuring houses in 1752 following a set of criteria set out in the company's deed of settlement. Homeowners who wanted to insure their property applied to the board. This submission prompted a visit by two board members to assess both the property's replacement value and how likely it was to burn or, worse yet, to burn uncontrollably, leading to a conflagration. In the early years of the Contributionship, master carpenters Joseph Fox and Samuel Rhoads did the assessments.[68] Significantly, this evaluation involved creating a taxonomy of architectural features that might limit the risk of a fire spreading or help firefighters control a small blaze such as a chimney fire. Contributionship directors based this taxonomy on experience and educated guesswork—effectively a codification of common sense. Initially, the company had a policy of insuring wooden houses only if they were built before 1752. By 1769, the company refused to insure any wooden buildings. Assessors also made notes about a more subtle set of architectural characteristics. How thick were the walls? They assessed houses with nine-inch party walls at half the rate of those with four-inch walls. Was there plaster and paint on the walls to act as a fire retardant? What was the condition of the shingling on the roof?[69] Inspectors also looked for the presence of a trap door onto the roof. This safety device made it easier to quickly extinguish chimney and roof fires. They were persnickety about its specific characteristics. Joseph Scull was refused insurance until the "stepladder and trapdoor [were] put in good order and iron rails made sufficient."[70]

A second insurance company, the Mutual Assurance Company or "Green Tree," was established in 1784 in competition with the Contributionship. Its founding resulted from a debate between prominent Philadelphians over what constituted an uninsurable risk. By the 1780s, Franklin had moved on and the board of the Contributionship decided not to insure properties fronted by large trees. They reasoned that foliage would impede firefighting efforts and that the trees themselves could be a fire hazard. The 1792 decision of the Insurance Company of North America to sell fire insurance to those who wished to keep their trees gave Philadelphians three companies to choose from.[71]

Several historians have pointed out that architectural criteria for insurability and for setting rates functioned as an "informal building code" that was probably more effective than statutes in making Philadelphia the most fireproof city in early America.[72] In Philadelphia, insurance proved remarkably popular. A half century after the Contributionship's founding, as many as one-third of all buildings in some neighborhoods were insured. Philadelphia also had more brick and stone buildings than other American cities of the same size.[73] Insurance policies influenced not only the way individual houses were built but also the character of neighborhoods. People were more likely to be turned down or charged higher rates if they lived next door to wooden buildings or to others engaging in dangerous trades.[74] Warnings that claims would not be paid on fires caused by neighbors engaging in illegal activities such as storing gunpowder or boiling tar encouraged citizens to press for the enforcement of regulations. Gradually, such activities shifted to the borders of the city.[75]

Although the availability of fire insurance in eighteenth-century Philadelphia mostly functioned as a positive incentive for policy owners to take preventative measures, fire insurance companies were not immune from moral hazard and fraud—recurrent themes as insurance became more common. Contemporary economists use the term *moral hazard* to describe the mostly unconscious tendency of people who are insured or otherwise protected from risk to exercise less care, knowing they can recover their losses. "Fraud" is defined as knowingly lying to obtain compensation.[76] Historically, the distinction was much more ambiguous because *moral hazard* also meant the temptation to commit immoral acts in order to obtain compensation.[77] In practice, sins of omission and commission could be difficult to distinguish from each other.

In drawing up the Deed of Settlement, Franklin created a long list of activities that precluded reimbursement in the event of a fire. The deed included a special provision that houses under construction could not be insured for their full value and policies would be voided if an unfinished house caught fire from the inside. This measure was intended to prevent unscrupulous property speculators or house builders from getting their money back if the property market seemed less than promising.[78] Other companies explicitly spelled out limitations of policies should fraud be discovered. The Massachusetts Mutual Fire Company, for example, reserved the right not to pay if "it should be proved that the insured purposely burnt the property, or neglected to extinguish the fire, when burning."[79] The availability of fire insurance provided an incentive to arson that had not existed before. The existence of fraud also provides a striking indicator of the way some people could quickly wrap their minds around the implications of new ways of managing risk: making decisions to engage in what previously would have been an economically irrational act of destroying their own property, finding ways to disguise their motives, and maneuvering through the loopholes in poorly thought-out contracts.[80]

At the end of the eighteenth century, significant changes in the management of fire could be recognized in the acceptance of insurance, the presence of fire engines and fire companies, and the design and materials of houses themselves. But in the streets of Philadelphia, it was also easy to see how old and new ways of thinking about and dealing with fire continued to coexist.

The Fragmentary Nature of Change

"Setting in [the] back parlor reading near 8'oclock, we were alarm'd by the noise of a fire Engine, and ringing of Bells, but had not heard the cry of fier, our house being deep and Entry long," Elizabeth Drinker wrote the day after Christmas, 1794. Looking out the back door, she saw fire in the distance, lighting up the night sky. Her husband set out to investigate, returning a while later to report that the still-unfinished steeple of the German Lutheran church was ablaze and the windows had blown out, showering a gathering crowd with glass. "I never saw so much of a fier before, never having been out of the house to see one, and so favour'd as never to be very near a house on fier," she explained, reflexively asserting her prudence. Instead, the Drinker household watched the church burn from the third floor of their house.[81] A huge crowd from the neighborhood, both men and women, turned out to help. No one had a ladder tall enough to reach the roof. The blazing steeple provided the only other access. Apparently, the engine also failed to throw water high enough to reach the flames. A bucket brigade formed: men passing the full buckets to the front and women passing the empties back. The crowd set about trying to dismantle and carry out the pipe organ as fire fell through the ceiling. Others bundled away parts of the library and the church's records and deeds. The more faint of heart busied themselves wetting down the roofs of nearby buildings and hoping for the best. In the morning, only a burnt-out shell of the once-magnificent church remained, though through the crowd's efforts, no other buildings burned. Investigations revealed that the Sexton, in a hurry to clear out the vestry fireplace after the Christmas service, had stored the ashes in a closet under the stairs from whence the fire originated. The church had no insurance.[82]

Benjamin Franklin did not live long enough to see or hear the excitement on Fourth Street. He had died in 1790. However, he would have recognized many elements of the scene and likely had something to say both celebrating the heroism of his fellow citizens and admonishing them for not more wholeheartedly embracing the innovations he promoted. The growth of cities in the nineteenth century set off an avalanche of changes based on the groundwork of fire companies, fire insurance, and building regulation that he helped introduce. New technologies such as fire hoses and eventually motorized fire engines transformed the nature of firefighting. Insurance companies grew increasingly sophisticated in predicting risk and compiling actuarial tables. Something also shifted in people's understanding of how they should participate in protecting themselves and their neighbors. None of Henry Drinker's or Benjamin Franklin's descen-

Many Philadelphians turned out to help fight the 1794 Lutheran Church Fire. *"Prospect of the new Lutheran church in Philadelphia," Courtesy, The Historical Society of Pennsylvania.*

dants would feel either social or moral pressure to rouse themselves out of bed at the sound of a fire engine. As the nineteenth century wore on, public officials increasingly discouraged bystanders from rushing in to salvage a neighbor's possessions or scuttling up a ladder to douse a chimney fire. On-the-spot volunteers were replaced first by the working-class men who dominated the hose companies and then, by the end of the century, by professional firefighters.[83]

Elizabeth Drinker hung on much longer than Franklin, living until 1817. As a house-bound invalid, she resorted to newspapers to provide "melancholy accidents" that could be reported to posterity, mostly women and children scalded or burned in their beds. These stories bore witness to not only what had changed but also what had not. Because most Americans continued to rely on open flames to provide heat and cook food, the private world of fire risk changed slowly. Behind closed doors, common sense remained the most common method for managing this everyday risk.

CONTROLLING FIRE IS AN AGE-OLD PROBLEM that spawned one of the most complex manifestations of premodern, vernacular risk management. American colonists brought with them from Europe not only a set of private habits and expectations about handling fire in homes and workshops but also a long tradition of community leaders using the law and social pressure to discipline risky behavior and draw reluctant citizens into collective efforts to prevent and fight the kinds of fires that threatened the common good. If one of the characteristics of a risk society is the use of collective organization to identify and control negative events that might happen in the future, the example of early American responses to fire illustrates the presence of that mentality long before the advent of other forms of modernity. By the same token, those practices grew out of a common, largely unexamined consensus—the vernacular—that assumed that everyone knew about cleaning chimneys and forming a bucket brigade. With so much at stake, it is not surprising that attention to the precepts of duty and carefulness was codified in the law and enforced by the state.

In the eighteenth century, American colonists, following the lead of their European cousins, began to rethink the management of fire. Although new technologies, notably the fire engine, most conspicuously embodied innovation, the real novelty lay in the insight that both firefighters and those at risk could be organized to anticipate the probability of fire. Property owners who bought fire insurance took a particularly momentous leap, demonstrating their faith in the viability of a risk pool with their hard-earned money. This rethinking mixed old and new in a slow evolution toward an early twentieth-century moment when new innovations in firefighting and fireproof buildings would coincide with the disappearance of open flames from everyday life.

Across the colonies, increasing numbers of people were also beginning to take another leap of faith, offering themselves and their families for inoculation against smallpox. In seeking to control the future possibility of infections, they abandoned not only common practice but also a kind of common sense in favor of the new authority of science and mathematics.

2

The Uncertainties of Disease

Nearly everyone who reached middle age in colonial America had survived multiple, potentially fatal infections. Rich and poor alike bore the scars of smallpox, measles, and any number of other afflictions. All had witnessed the sufferings of siblings, parents, children, and friends racked by fever, tortured by nameless intestinal maladies, or decimated by tuberculosis, known as "consumption." Death was also a familiar visitor. In Philadelphia, historians have estimated that one in four children died before the age of three. Malaria and yellow fever led to almost catastrophic mortality rates among both black and white residents of the Chesapeake. Overall, New Englanders were healthier, but eighteenth-century residents of towns like Ipswich and Salem still had a life expectancy of less than sixty if they survived to adulthood.[1]

In some ways, the story of how early Americans managed the risks of disease parallels their experience with fire. Like fire, disease was largely a private matter. Treating the sick mostly fell to women as part of their household or community duties. Formally trained physicians were rare, especially before the second half of the eighteenth century. Significantly, for our story, clergymen also stepped in to provide medical advice.[2] Only during epidemics was the state or some other organized body likely to take an interest in who was sick, how they managed their illnesses, and the number who had died. As with fire, these efforts to protect the common good were an acknowledgement that individual actions such as throwing garbage in the street or allowing a person with signs of smallpox to socialize could, if left ungoverned, intentionally or unintentionally sicken others.

Early Americans' understanding and response to disease, however, differed in significant ways from their reactions to fire. Paranoid New Yorkers might spin convoluted theories about *who* set a series of catastrophic fires and whether it was an accident, but the actual causal mechanisms of creating and extinguishing fires were well understood and could be reproduced at will: setting a glowing ember in a pile of straw would make a fire; putting water or dirt on the fire would extinguish it. Disease was a different matter.

Although scores of early Americans were convinced they understood both what caused disease and how to cure it, no absolute consensus existed. Still, theories, formal and informal, abounded. Following the ideas of the ancient world, many believed that disease was caused by imbalances in the body's humors. Some championed the relatively new "Miasmic theory"—an environmental explanation that certain climactic conditions and foul odors led to illness. A tiny minority suspected the detrimental presence of tiny "animicules" in the body. This idea, however, was still a long way away from germ theory as it emerged in the late nineteenth century.[3] Folk culture and Judeo-Christian religion added other, more supernatural explanations, including the malign influence of spirits, enchantments, alignment of the stars, and, of course, the mysterious workings of God's will.[4]

In a sense, all of these theories could be supported by observation and inference (as they still are in the twenty-first century when individuals claim from their own experience that a wet head can cause a cold or prayer can heal disease). Physicians and others could also demonstrate the ability to cure because common treatments such as bleeding and purging alleviated what we would call "symptoms"—the externally observable part of a disease. However, none of these treatments worked consistently, so people tended to keep experimenting until a change occurred. Moreover, no one had directly observed what a miasm or an animalcule or an imbalance in the humors did *inside* the body. The physical mechanisms that caused babies to go into convulsions or plague victims to grow black lumps in their armpits before giving up the ghost remained invisible behind a curtain of flesh.

These uncertainties made choices about how to prevent and treat disease difficult and sometimes agonizing. In the earliest colonial settlements, those options were largely shaped by long-standing European practices—composed of both protoscientific theories and vernacular ideas, tinctured by borrowings from native people. "Sufferers and healers," historian of medicine Roy Porter tells us, "operated within at least greatly overlapping, if not identical cognitive worlds"; the two cultures would not begin to significantly diverge until the nineteenth century. Even the book learning of medical school–educated physicians was readily available and understandable to other educated people.[5] Skill and experience marked the difference between most medical experts and their patients. By the end of the seventeenth century, the spirit of the Enlightenment and methods of the Scientific Revolution had reached even this rough frontier, bringing new ideas and new choices. Rather than solve the problem of disease, these new ideas introduced additional uncertainties: Whose ideas and prescriptions should one trust? Who had the correct answer? How could one decide?

In the story that follows, the seeming certainty of numbers provided a partial answer. Numerical analysis offered the possibility of determining the effectiveness of prevention and treatment without the necessity of actually understanding or observing the mechanisms of disease. Instead, observers (and advocates) could count and categorize victims and survivors to determine, through inference, what worked. Numbers could also be used to create trust by providing a seemingly objective basis for decision making. Probabilistic arguments were first used prescriptively to convince early American urban dwellers to undergo inoculation for smallpox. Looking backward from our own time in which statistics carry enormous authority, the fact that individuals gambled their lives (and, in some cases, the lives of families, servants, and slaves) on the basis of these kinds of arguments seems extraordinary. In fact, numbers alone were not enough to convince most people to undergo the process or to impose it on their children and servants. In practice, numbers were only one element in a more complicated process of decision making that transformed inoculation and eventually vaccination into at least some people's version of the medical vernacular.

New ideas and the smallpox bacillus both entered the Boston Harbor by ship in the 1720s. The result was both an epidemic and a social conflict over how best to protect individuals and the common good.

Smallpox and Inoculation

In early America, only family and friends bore witness to most people's struggles with infectious diseases. But when individual cases reached what we would now call "epidemic" proportions, they took on the character of a disease event—akin to disasters or large-scale accidents such as a later generation's train wrecks.[6] Like fires, diseases were most likely to become public disasters where large numbers of people lived in close quarters. Mayors, governors, selectmen, and justices of the peace used their legal power to enact quarantines, as well as to force citizens to clean streets, burn infected possessions, and otherwise take action to prevent the spread of disease. In communities where such diseases were not yet a permanent presence, they also acted to bar diseased people from the community or to isolate those that had already entered.[7]

In 1721, Boston was struck by a smallpox epidemic. For residents, it was a familiar event, imposing terrifying uncertainty, religious soul-searching, and hard choices about quarantine, treatment, and whether to flee the city. However, this epidemic was also different because at least some residents had an additional option for avoiding potential death or disfigurement—inoculation or "variolation." The availability of this new means of dealing with smallpox created a crisis of its own, pitting residents of the city against each other. The resulting debate helped introduce new ways of thinking about and managing risk into the colonies.

This is not, however, a straightforward story about the triumph of modernity. Instead, the story of how that crisis played itself out through the summer of 1721 and beyond illustrates the complexity of one part of early American risk culture as it first began to shift under the impact of the Enlightenment. Common practice mixed with new ideas, religion with science, local politics with an emerging global culture of both knowledge and disease; life and death decisions were made on the basis of statistical argument, but also observation, experience, and the recommendations and examples set by friends and people with standing in the community.

Inoculation was a folk practice largely unknown in Europe before the late seventeenth century but widely practiced in the Balkans, Central Asia, and Africa.[8] It involved inserting a small amount of pus from an infected person just under the skin to induce a mild case of the disease and subsequent immunity. However, information about the practice came to the American colonies by means of a small elite who had read about it through publications and correspondence of the nascent scientific revolution and through their questioning of African slaves. Their efforts to introduce the risk-managing technique of another culture ran hard up against the common practices of their own society, quarantine and isolation, because it involved purposefully infecting healthy people who not only would then become sick but also could pass on the infection to others.

The logic of inoculation—artificially inducing a mild case of smallpox to protect against a potentially more lethal "naturally acquired" case of the disease—defied the common sense of the time. Since epidemics were considered to be a threat to the common good, purposefully spreading a disease also had legal ramifications. Opponents to inoculation frequently used words such as "unlawful" and "felonious" because this was one of the few areas of everyday risk in which the state could and did intervene forcefully. Local governments had the power not only to confine infected individuals to their ships and houses but also to search for those individuals, door-to-door, if necessary.

Initially, experiments with this practice, often carried out in the midst of the panicky climate of epidemics, also offended the sensibilities of some of the new breed of experimental philosophers. They argued that there was not enough proof that inoculation, associated in their rhetoric with "old Greek women," was either safe or effective. Out of the struggle between inoculationists and anti-inoculationists came the first large-scale efforts to collect and analyze statistics measuring the efficacy of a risk management technique.

Cotton Mather as an Agent of Modernity

"About the Month of May, 1721," Cotton Mather later remembered, smallpox was "admitted into the City of Boston."[9] "Admitted" was a carefully chosen word. The disease had made its appearance in the form of two pox-spotted sailors aboard the British ship the *Sea Horse*, sailing from the West Indies. This ship had somehow eluded authorities who were authorized to prevent anyone from coming ashore from vessels carrying sufferers of "plague, small pox, and pestilential fever."[10] A desperate effort was made to contain the spread of the disease. Justices of the peace confined patients in their houses; the selectmen pressed twenty-six free black men into service to clean the streets in case the filth deposited there somehow harbored the disease.[11]

Despite these efforts, telltale signs began to appear among the population. Flu-like symptoms signaled the disease's onset: a splitting headache, fever, and aches and pains, especially in the back and midsection of the body. After four or five days, these symptoms abated, allowing patients the fleeting hope that they were suffering from some other malady. Then the fever returned and with it the pustules that gave the disease its name. The eruptions appeared first in the throat and the mouth and spread outward across the face, eventually reaching the torso and the limbs. Particularly dense pustules on the palms and the soles of the feet sometimes caused sheets of skin to simply peel away, leaving the victim unable to walk. The pain caused by inflammation was excruciating, akin to a bad burn. In fact, some people described feeling as if they had been set on fire.[12]

Compared with most other diseases, Europeans and Americans knew a good deal about smallpox—in part because its deadliness drew their attention. More importantly for our story, however, smallpox was an easily recognized disease without the aid of modern medical theories or technologies. Observation revealed that it could be passed from person to person by close personal contact or through the pus deposited on blankets, clothing, and other personal items. From this empirical evidence, observers concluded that the afflicted were most infectious when the symptoms (pustules) were at their worst (which is not true of all infectious diseases). They also knew that once stricken, individuals would not again endure the disease.

The experienced became deft readers of the disease's symptoms. They could usually visually distinguish smallpox from all the other poxes, particularly chicken pox, which also blistered the skin. They knew how to interpret the density of pustules to divine the infection's seriousness. They understood that almost no one survived what is now called hemorrhagic smallpox, characterized by horribly inflamed skin and an unbearably foul odor. If subcutaneous and internal bleeding did not kill the victim, secondary infection would.

A small number of pustules meant a better chance of survival and less long-lasting damage. Victims and relatives waited anxiously to see if pustules in the eyes would lead to blindness. Young people wondered if they would be so disfigured as to be unmarriageable. Some were lucky, some were not. George Washington caught the disease in Barbados in 1751. He recovered with no more damage than a few pock marks around his eyes.[13] Extensive facial pitting marking many survivors was sometimes noted in colonial newspapers by advertisers seeking runaway servants. For example, John Nutty, a twenty-four-year-old butcher by trade, was described by his master as "much disfigured with the Small Pox."[14] Modern epidemiologists have estimated that the disease leads to facial scarring in 75 percent of those infected, ranging from Washington's few pock marks to twisted mouths and lashless eyelids adhered together by scar tissue.[15] Much of the smooth, unblemished skin in early modern portrait paintings is a kind, artistic fiction.

By 1721, the "destroying angel," as Cotton Mather famously called it, was a familiar visitor to New England. In 1633, an epidemic struck Plymouth Colony, killing both Native Americans and European settlers. Boston suffered epidemics in 1636, 1659, 1666, 1677–78, 1689–90, and 1697–98.[16] As an adolescent, Mather described the scene during the 1678 epidemic. "Boston burying-places never filled so fast." Church bells tolled steadily for the dead: "6, 7, 8 or 9 in a day," he estimated.[17] Even for people accustomed to death, the impact on a city of less than seven thousand souls must have been terrifying. In places like London where the disease had become endemic, it was considered a childhood affliction like measles because most adults were survivors. In New England, it cut down whole families: mothers and babies, sturdy young men and aging matriarchs. Initially, no one made an accurate count of the infected and the dead, but eighteenth-century natural philosophers estimated and twentieth-century scientists concur that *variola major*, the virus that caused these epidemics, generally kills between 10 and 30 percent of those infected (although mortality ran as high as 90% among native Americans in this period).[18]

As an adult, Mather continued to think about smallpox in two seemingly contradictory ways. Like other Puritan clergymen of his time, he saw the working of divine will everywhere, including so-called plagues. Not surprisingly, this tough-minded theologian, notorious for writing in support of the Salem witch trials, explained disease as an expression of divine displeasure. In *The Angel of Bethesda*, his magnum opus on medicine, he wrote that smallpox was a "Scourge" that a "Holy and Righteous God has inflicted on a Sinful World."[19] To respond appropriately, he told his readers, they should "live unto God, and get into such a *State of Safety for Eternity*" as to "be *Ready* for whatever *Event* this *Distemper* may have upon them."[20]

Mather, however, was no simple-minded predestinarian counseling surrender to the inevitable. Like other (but not all) Puritan theologians, he also believed that because God had given mankind free will, people were obliged to use their knowledge, wits, and strength to preserve themselves.[21] Dangerous situations tested the pious. God required them to do more than get down on their knees and pray. Mather's theology also encompassed the idea that, most of the time, God worked through nature without violating nature's laws (which He had created in the first place). In other words, if God wanted to chastise people by dropping a rock on their heads, the rock would fall in accordance with the law of gravity and the accident as a whole would be explainable in mechanistic terms. Consequently, studying the laws of nature was a way of gaining a greater understanding of God—an obligation that Mather took up with enthusiasm.

His interest had been sparked at Harvard College, where he read "physik" and natural philosophy, as well as theology, with the idea that he might become a doctor if his stuttering rendered him unfit for the pulpit. By the early eighteenth century, Mather regularly corresponded with members of the British Royal Society and was an avid reader of the society's journal, the *Philosophical Transactions*. Through these actions, he inserted himself into the nexus of eighteenth-century natural philosophy. The Society and its journal constituted the most important "trading zone" (to use historian Peter Galison's term) for information about natural phenomena and scientific experiments in this period.[22] Members and would-be members sent in reports from the far corners of the Europeanized world on everything from sightings of mermaids to the first microscopic analysis of muscle tissue. Mather's contributions included conjectures about types of wood used to build Noah's ark, observations on medicinal herbs employed by Native Americans, and a description of the natural history of rattlesnakes.[23]

In 1714, the Royal Society published a brief report in which John Woodward, an English MD, described the contents of a letter from Dr. Emanuel Timoni on smallpox inoculation in Constantinople. The process, Timoni wrote, had come from Circassians, Georgians, and other "Asiatics" and had been practiced in Turkey for about forty years. He thought it a "happy Success." "The Operation has been perform'd on Persons of All Ages, Sexes, and different Temperaments . . . yet none have been found to die of the *Small-Pox*," which was, at the same time, "very mortal when it seized the Patient in the common way."[24]

This constituted new knowledge for most members of the Royal Society. However, Timoni's account confirmed information Mather had already acquired in talking to his slave Onesimus and others of African birth. Mather had received a confusing answer when he asked whether Onesimus had ever had smallpox. Further questioning revealed that, following a practice common in the region, he had been inoculated in Africa—a story that other Africans in Boston confirmed.[25]

Before considering what Mather did with this information, it is worth contemplating what the prevalence of this practice says about how premodern people understood and managed risk. Inoculation involved a purposeful gamble: inviting a bout of the disease that was likely (based on experience) to be relatively mild rather than possibly contracting the natural form of the disease with its more debilitating and potentially fatal symptoms. This choice was not taken lightly. Inoculated patients could and did die of the artificially induced disease, especially if it was aggravated by an underlying condition. Even if they did not succumb, inoculants could count on suffering symptoms that would make them feel very sick for up to a month. People with inoculated smallpox could also infect the unexposed with a more virulent natural version of the disease. For these reasons, inoculation seems to have been done mostly during epidemics when the risk of contracting natural smallpox was highest.

After reading Timoni's communication, Mather wrote to Woodward, who was one of his regular correspondents, inquiring why the practice was not more widely used in England. He told the doctor, "If I should live to see the *Small-Pox* again enter our City, I would immediately procure a Consult of our Physicians to Introduce a Practice, which may be of so very happy a Tendency."[26] As the 1721 epidemic worsened, Mather got his chance. He sent a letter, as well as abstracts of accounts by Royal Society correspondents, recommending the procedure to a number of Boston physicians. All except Dr. Zabdiel Boylston ignored him.[27] Boylston later explained that he remembered how narrowly he had escaped with his life during the last epidemic. "Now with my Wife and many others were gone out of Town to avoid the Distemper, and all Hope given up of preventing spreading of it," he resolved to make an "Experiment." Since he was immune as a consequence of already having survived the disease, Boylston "chose to make it (for Example sake) upon my own dear Child, and two of my servants."[28]

The procedure worked, and as the epidemic worsened, more and more people turned up on Boylston's doorstep inquiring about the operation. They came with children and "servants" (mostly African American and Native American slaves) in tow. Most were driven there by intense fear of the disease. Mr. Nathaniel Loring brought his only son "on Account of the Small-Pox having been very fatal to their Family."[29] According to Boylston, the child recovered quickly, rewarding his father's willingness to take a chance on the procedure. A Mrs. Dixwell was not so lucky. The "fat Gentlewoman" came to Boylston, not only because she was frightened by the number of sick people living around her, but more specifically because she had seen the corpse of someone who had died of hemorrhagic smallpox: "the Stench of whereof greatly offended and surprised her with Fear of Being infected." She died twenty-six days after being inoculated despite Boylston's desperate attempts to save her—the first of eight inoculants who did not survive the trial.[30]

Wracked with indecision, Mather held off having his two unexposed children, Samuel and Elizabeth, inoculated for almost two months. In his diary, he debated the pros and cons. Mather knew that ten people in his neighborhood had successfully undergone the procedure, but fear of public opinion held him back. "Our People, who have Satan remarkably filling their Hearts and their Tongues, will go on with Infinite Prejudices against me and my Ministry, if I suffer this Operation upon the Child," he rationalized.[31] Eventually, fear for his children, especially his son, won out (concerns for daughter Elizabeth seem to have been an afterthought). By mid-August, the disease had spread to Harvard College, where his son was enrolled. Sammy's roommate and "dearest Companion" had just died "of Small-pox taken in the common Way." The venerable Increase Mather also added his "urgent calls" that his grandson be inoculated.[32] Mather relented and then began to regret his decision as Samuel hung between life and death for nearly a week before finally beginning to recover.[33]

Eventually, Boylston inoculated 247 people. That left the other 12,117 residents of Boston, many of whom were not happy about the practice of deliberately giving their neighbors an infectious disease.[34] Public officials, selectmen, and justices of the peace reprimanded Boylston for spreading smallpox. Mather and Boylston's actions also infuriated Dr. William Douglass. Douglass had arrived from Europe a few years previously with an education acquired at some of the most important centers of learning, including Edinburgh and Leyden, as well as a medical degree from Utrecht granted in 1712.[35] Douglass rightly pointed out that little evidence existed beyond Timoni's observations in Turkey that the procedure actually worked. Prior to 1721, elite European physicians generally believed that it was unacceptably dangerous. But there was also a distinctly personal quality to Douglass's opposition. Mather had failed to follow medical etiquette by formally consulting with elite Boston physicians (although it could be argued that they had shared their opinion by ignoring his letter). Douglass probably viewed the two men as hopelessly and dangerously parochial.

In July, Douglass published a letter condemning the two men's actions in the *Boston News-Letter* under the pseudonym Philanthropos. The letter dripped with condescension, portraying inoculation as a method practiced by "Old Greek Women on Turks and others" but rejected by the "Learned" in England as "Wicked and Felonius." Boylston was a "Cutter for the Stone"—a barber surgeon who was "illiterate" and therefore "not capable of understanding the Writings of those Foreign Gentlemen" who had described the operation of inoculation (and, presumably, its shortcomings). According to Douglass, the process itself had nearly killed Boylston's son and threatened other members of the community who might contract the disease from the inoculated. Douglass laid most of the blame on the doctor, whom he portrayed as an ignorant and self-serving quack concerned only with money. Mather was more dangerous to have as an enemy. But without naming names, Douglass added a telling detail about him, suggesting who was an insider and who was an outsider in the elite circle of European natural philosophers. A "certain Learned Gentleman" had learned about the process because he had borrowed Douglass's copy of the *Transactions* (not having access to one of his own).[36]

Over the next few months, a war of words raged in the three Boston newspapers. While some of the most serious volleys appeared in the *News-Letter* and *Gazette*, Samuel Franklin's newly founded paper, the *New England Courant*, gave the most column space to the controversy.[37] Being a Franklin, Samuel could not resist having a little fun with the subject in the form of a fake letter to the editor—a rhetorical form that his little brother later perfected. "Horat" offered the horrific suggestion that rather than send troops to the frontier, a party of inoculators should be sent to confront hostile tribes armed with "Incision-Lancet, Pandora's Box, Nut Shell and Fillet." Their "ammunition" would be a "combination of Negro-Yaws and confluent Small-Pox"—an eerie preface to English employment of smallpox-laden blankets in the French and Indian wars.[38]

Meanwhile, the mob, that preindustrial force of moral order, also had something to say. Mather confided to his diary that he had become "an Object of their

Fury, their furious Obliquies and Invectives."[39] He already had scores of enemies for other reasons, but the inoculation controversy added to their ire. Deep in a mid-November night, one of them threw a grenade through Mather's window into a room where one of his relatives was staying while recovering from being inoculated. Either because it was badly made or, as Mather preferred to believe, because an "Angel of God" had intervened, the fuse fell out and the bomb failed to explode.[40] Franklin reported that a note had been tied to the fuse by a thread. It read "Cotton Mather, I was once one of your Meeting: But the cursed lie you told of you know who, made me leave you, you Dog. And Damn you, I'll inoculate you with this, with a Pox to you."[41]

Meanwhile, the epidemic continued through the autumn months. When it finally burned itself out, the selectmen made a final count of the casualties. Their report stated that 5,980 Boston residents had contracted the disease since April, when the epidemic started (a startling 51% of the town's population at the time); 842, or approximately 14 percent, had died. Another 850 fled the city (including members of the General Court).[42] These startlingly precise numbers, which later came to play an important role in statistical arguments about inoculation, had been compiled by clerks of the militia units authorized to conduct monthly door-to-door searches recording the sick and the dead—a novel exercise in an era in which governments counted very little of anything that did not directly involve trade, taxes, and dirty chimneys.[43]

Despite the fact that only 3 percent of Boylston's inoculation patients had died, local consensus held that inoculation was unlawful and endangered the public good. The next spring, the Massachusetts House of Representatives passed a "Bill to prevent the Spreading of the Infection of the Small-pox by the Practice of Inoculation." Although the governor and council turned it down, public officials in Boston and elsewhere in the colonies continued to take a dim view of the practice and to set up legal barriers wherever they could.[44] Meanwhile, the international community of natural philosophers had taken a different message from the Boston experiment. In the cosmopolitan courts and experimental halls of London, Boylston found a more receptive audience for a new kind of knowledge and the incremental beginnings of a culture of statistically justified expertise.

Communicating about Risk in the Atlantic World

Epidemics were not just local events. News traveled almost as fast as infection between London and the outposts of the Atlantic World. By June of 1721, reports of the Boston outbreak reached London. Returning ships brought another smallpox story, written for the illumination of the *Boston News-Letter*'s readers. Several physicians had made a presentation to King George I on inoculation. Recognizing that numbers and their own testimony would not be enough to convince the skeptical of the procedure's safety and efficacy, the doctors proposed experimenting on several condemned prisoners from Newgate prison. These hapless men "offered themselves to undergo the Experiment upon Condition of receiving His Majesty's most gracious Pardon." According to the article, the King was positively inclined but felt compelled to refer the matter to his solicitor general "to determine whether he can do it by Law."[45]

Inoculation was the talk of London that spring and summer. In April, Lady Montagu, wife of the former ambassador of Turkey, had her children inoculated to demonstrate publicly the efficacy and safety of the procedure. In August, the solicitor general gave his approval and the Newgate experiments began. When this procedure began to produce promising results, the Queen authorized an additional trial using orphans, before finally agreeing to have the Royal children inoculated.[46] The London papers also carried accounts of the Boston story. By and large, they painted a negative picture.[47] Enthusiasm ran higher among Royal Society members. As the year closed, correspondents reported on their experiments. Among them was Mather, who had begun writing up his version of events in August. The account was published as an anonymous pamphlet, which he sent off to London. Out of these exchanges, Boylston emerged as a hero.

In December 1724, Bolyston sailed for London, bearing a letter of introduction from Mather to James Jurin, secretary of the Royal Society and an MD.[48] Jurin was an enthusiastic and skillful quantifier, part of a small circle of Royal Society members fascinated by the power of numbers to provide an objective accounting of otherwise chaotic-seeming natural and human phenomena. Jurin had sworn to produce a yearly report compiling all available information on inoculation until enough was known to make an "impartial" assessment of the efficacy and safety of the procedure. The figures collected by the Boston selectmen, reports from across the British Isles, and a half century of information from London's weekly *Bills of Mortality* became grist for Jurin's ongoing project.

Jurin's most ambitious effort at statistical analysis had already appeared in print in 1724 when Boylston set sail for London. Like others at the same time, he compared the risks of dying of inoculation with mortality patterns from naturally acquired smallpox in order to prove impartially which was more deadly. But Jurin's efforts at harnessing the predictive power of statistics went beyond trying to settle the inoculation controversy. Like others, including Boylston, he suspected that underlying medical conditions and other factors made inoculation much more dangerous for some patients (such as the unfortunate Mrs. Dixwell) than others. Thus, he made a numbered list of cases illustrating "hazards" (such as a weak constitution or exposure to the natural form of the disease just prior) that might increase the chances of dying of inoculation. He then constructed a table focused on the numerical probability of dying of inoculation according to the number of risk factors involved.[49] Jurin's piece de resistance was an argument based on political economy; smallpox cost the society and the Crown money and manpower. He argued that the bills of mortality for the previous decade showed that an average of 2,287 Londoners died each year of natural smallpox. "It follows that if we substitute Inoculation for the natural Way, the Number of the Dead would be reduc'd seven Parts in eight," he argued. "Consequently 2000 persons that are yearly cut off . . . might be preserv'd to their King and Country."[50]

Boylston lacked Jurin's mathematical training and ambition, but they were men with like minds and, by at least some accounts, became good friends. Jurin also introduced Boylston to Sir Hans Sloane, one of the court physicians and probably the most influential advocate of inoculation in Britain.[51] Sloane and others encouraged Boylston to write up a "report" on the Boston inoculations. Unlike Jurin, Boylston had actually treated the people enumerated and described in the report. He was, first and foremost, a physician who had made an unpopular decision to provide a treatment he trusted despite intense pressure to do otherwise. Jurin's calculations about the risks of inoculations (which suggested a lower death rate) must also have made him aware that in the frenzy of the epidemic he had inoculated people who were not good candidates for the treatment. Compared with Jurin's pamphlets, Bolyston's report was therefore much more of a manifesto and justification of prior actions.

In his preface, Boylston stated that he hopes his account will "influence some few at least to save their Lives this Way, seeing from Time to Time, in the Weekly Bills, the Numbers who died of the Small Pox." "Weekly Bills" were bills of mortality. His imagined audience was numerate persons who anxiously scanned the weekly lists looking for a spike in the number of deaths from London's endemic smallpox—a warning that, prior to inoculation, they could act on only by leaving the city. Beginning in 1704, a version of these bills would have been familiar to Boylston from yearly lists published in the *Boston News-Letter*.[52]

The battle over inoculation would drag on in both places for decades more. But among those who were won over was William Douglass. In 1730, he published his own dissertation on the subject. "We find by some years Experience," he admitted, that smallpox "received by Inoculations is not so Fatal and the Symptoms frequently more mild, than in the accidental Contagion." In other ways, while recognizing the effect of inoculation, he still stressed the important role of elite physicians. As the process was still dangerous, it "therefore requires Discretion in applying it to proper Subjects, and Judgment in managing the Distemper so receiv'd."[53] In other words, inoculation should be done only by select doctors, preferably graduates of medical schools, not of their fathers' practices.

Trust and Popular Acceptance of Inoculation and Vaccination

"The Small Pox in town & proves very mortal," Elizabeth Drinker noted. By 1762, when she made the entry, news about smallpox deaths had already become an anxious refrain in her diary.[54] Between 1756 and the beginning of the Revolution five major epidemics swept through Philadelphia. British troops and refugees might have marched the contagion into the Quaker city when it became a major transit point during the Seven Years War. Or, the disease might have arrived on board any one of many ships arriving each season.[55] By the 1750s, the size and connectedness of the port city made it impossible to identify a single ship or soldier as the source of contagion, the way Bostonians had done a generation earlier. By the same token, isolation and quarantine seemed less and less viable as strategies for containing epidemics. By some calculations, smallpox killed more Philadelphians than any other single cause during this period, accounting for as many as 25 percent of burials.[56]

Some city residents experimented with inoculation in the 1730s, but the practice initially failed to take hold.[57] Fear generated by this new wave of epidemics created renewed interest among the city's elite, including Drinker. In the late 1750s, she began gathering information about the process. There is no evidence that Drinker underwent inoculation, suggesting she may have been a smallpox survivor. Instead, she seems to have been deciding what she would do once she began to have children. Her approach and eventual decisions offer a rare window into the way one (admittedly atypical) layperson approached the difficult question of whether it was better to actively choose to acquire smallpox or risk that good habits and community controls would provide better odds.

Drinker paid attention to tables of mortality and other statistical arguments presented in the press and in the popular medical literature, but she also wanted other kinds of information and reassurance.[58] Her own observations played an important role. She repeatedly visited friends whose children either had been inoculated or had caught smallpox the natural way, noting differences in symptoms. She also attended at least one inoculation. In April of 1760, she watched as prominent Philadelphia physicians Cadwalader Evans and John Redman inoculated two daughters of her friends Joshua and Katty Howell.[59] Her diary offers no details of the occasion, but a characteristically age-of-Enlightenment moment can be imagined: quietly well-dressed people gathered in a neighbor's home to witness a scientific experiment.[60] The attending physicians might have narrated the procedure, perhaps drawing out the simple action of making an incision and slipping a bit of infectious material under the young patient's skin. For people accustomed to bloodletting as a routine medical procedure, the minimal violence of the insertion must have been reassuring, along with the presence of other people known for their prudence and good judgment: the children's parents and the doctors, one of whom would become Drinker's personal physician and close friend. Then, as now, people were more willing to take risks if they were surrounded by their friends, or had received their friends' approval.[61]

Once convinced of the efficacy of the process, Drinker had another decision to make: whom should she trust to inoculate her children? She and her contemporaries believed that selection of the proper technique and skill in its application might decide whether a patient lived or died. By the time Drinker began thinking about the issue, physicians had turned inoculation into a far more complicated process than just inserting a bit of infected material under the skin. They also recognized the need to convince medical consumers, such as Drinker, of the superiority of their particular approach.

Inoculation provided an importance source of status and income for Philadelphia physicians of Drinker's acquaintance. Benjamin Rush, a close friend of Drinker's husband, Henry, credited his growing reputation as an expert in the Suttonian method as an important attractant of patients in the early years of his practice. As Rush described it in his memoir, the "mode of infecting the arm by a small puncture, instead of a long incision, was a very popular one, and brought me many patients, some of whom continued to employ me in other diseases."[62] Other doctors distinguished themselves from the competition, while also justifying their continued involvement (and subsequent bills for services), by creating an elaborate set of treatments to prepare patients for inoculation and care for them during the ensuing illness. These typically involved prescribing medicines ("physick") and bleeding to support the patient's humoral balance. A physician publicizing his method in the *Pennsylvania Gazette* prescribed a dose of "refrigerant" powder to cool the blood, as well as frequent purging. He recommended bleeding for adults who were very "plethorick" (hot blooded).[63]

When Drinker's oldest daughter, Sally, passed her first birthday, Dr. Redman came to the Drinker house to do the inoculation. Drinker prepared the child's body by giving her "physick"—coaxing her to take nine pills in the weeks before the procedure, some of which purposefully induced vomiting. The decision purposefully to cause her own daughter suffering must have weighed on Drinker's mind. She consoled herself by keeping track of Sally's symptoms. Some observations fit the generally qualitative tone of eighteenth-century diaries. But in an effort to be scientific, Drinker also counted all the pustules on her daughter's body: "about 30 in the whole," she noted.[64] She would follow the same pattern for her next two children.[65] In due course, Sally recovered. "Daniel Drinker buryed his little Daughter Sally this afternoon. She died of Small Pox," Elizabeth Drinker noted two years later.[66] The implications of the sad death of a cousin who shared her daughter's name but whose parents had made a different choice were probably not lost on her.

Although inoculation spread first through the upper classes, within a decade Philadelphia physicians also carried out the procedure in the households of bookbinders, barbers, sailmakers, and other craftsmen.[67] Expense as well as ignorance remained a barrier, however. The Enlightenment types responded. There were scattered efforts to engage in mass inoculations, including Benjamin Franklin's "Society for Inoculating the Poor Gratis," established in 1774, and George Washington's well-known efforts to inoculate his soldiers during the American Revolution, but these reached only part of the population.[68] Quarantine remained an important tool for managing the disease.

How widespread inoculation became and why it spread remain matters of conjecture and controversy among historians.[69] Most significantly for our story, historian of medicine Sarah Gronim has been able to document how a broader population recast inoculation as a form of what she calls "local knowledge." Gronim argues that "New Yorkers integrated it imaginatively into common ideas about the body and disease, reconceptualized its theological meaning, and incorporated it into familiar social relations of healing."[70] Once decided upon, inoculation became the norm in Drinker's family for several generations—a rite of passage for her children and their children. Experience and habit replaced careful study in settling the matter. The expense of the doctor's visits, the discomfort involved in preparatory medicines, the insertion of infected material, and the pain of the disease itself were accepted as necessary and safe enough.[71]

Although inoculation had made the transition from unacceptable to acceptable risk, introducing live *variola* virus into healthy individuals continued to be a gamble for both patients and the communities within which they lived. At the very end of the eighteenth century, Edward Jenner discovered that inoculation with pus from people suffering from cowpox (a related but much milder animal disease) offered an alternative for such high-stakes wagering. Vaccination conferred immunity to smallpox without the accompanying risk of communicating the disease or dying of complications. Appropriately, perhaps, Jenner belonged to one of the last generations of country doctors who were also natural philosophers comfortable with translating one community's common knowledge into a science-based medical practice.

Jenner's ideas came to the United States through his treatise *Inquiry into the Causes and Effects of the Variolae Vaccine*. After reading Jenner's article, Dr. Benjamin Waterhouse, a Harvard professor, procured cowpox matter from England. In a nice piece of historical symmetry, the first American vaccinations took place in Cambridge and Boston, where Zabdiel Boylston had slipped a bit of infected matter under the skin of a servant eighty years prior. Like the Philadelphia inoculationists Redman and Rush, Waterhouse had more than philanthropy in mind in promoting vaccination. To the outrage of some of his contemporaries, he imagined building a fortune from selling cowpox to other practitioners and setting up a vaccination hospital. Yet another Enlightenment figure, Thomas Jefferson, then president, intervened. He convinced Waterhouse to share the material, to be propagated in the southern United States.[72]

The greater safety and effectiveness of vaccination suggested a new possibility: the state might require vaccination as a means to protect the common good. After an investigation of the safety and efficacy of the process, the British Parliament passed a law mandating infant vaccination in 1853.[73] The United States experimented with a less sweeping series of initiatives by state and local authorities with an 1809 Massachusetts law mandating infant vaccination. Other states put laws on the books requiring vaccinations for public school children or during epidemics. For a time, immigration authorities also required a certificate of vaccination for incoming travelers.[74]

Such measures met vociferous, organized opposition. Anti-vaccinationists, as they styled themselves, argued that smallpox vaccinations were ineffective and dangerous. They carried out their campaigns in public, marshalling their own scientific and statistical arguments—debating modernity on its own terms. They also brought public officials into court on the grounds that mandatory vaccinations violated the individual's right to make decisions about trading one bodily risk for another. Some judges thought they had a point. After all, vaccinations stood alone among public health measures in requiring individuals to submit to a potentially risky medical procedure for the purposes of prevention. More typical public health regulation required people to abstain from doing something that threatened the common good (such as making social contacts while sick).

Although the leaders of the anti-vaccination movement typically adhered to what one historian of the movement has called "irregular" medical theories such as water cures and homeopathy, they counted among their ranks victims of "accidents of vaccination"—typically parents of children who had died from secondary infections they believed had been transmitted during the vaccination process.[75] In the second half of the nineteenth century, a number of states repealed laws or modified them in response to legal pressures.[76] Even the worldwide elimination of smallpox as a functioning disease has not put an end to the bigger argument about risks and rights and who gets to choose.

LIKE EVERYTHING ELSE ABOUT THIS RISK STORY, the details reveal something more complex than the simple triumph of rationality over fatalism, modern mentalities over premodern ones. Even the vast majority of people whose worldview was still firmly rooted in the premodern did not necessarily just surrender themselves to disease, no matter how little they understood about their affliction or how strongly they felt their suffering somehow represented a divine chastisement. They strived to maintain their health, prayed for relief, and sought out treatment when sick even when it was painful or made them feel worse.

Enlightenment thinkers, dedicated to the improvement of everyday life, predictably turned their attention to disease as a cause of human suffering. But their interventions added to, rather than displaced, older ways of understanding and managing disease risk. Statistical comparisons of therapies, the most novel and modern of their interventions, became one more tool that people like Elizabeth Drinker could use to make decisions. Numbers alone could not, however, convince most people to take a chance on inoculation or other risky medical procedures. As risk psychologists point out and ensuing chapters will document, most people resist seeing themselves statistically when it comes to risk.[77] Moreover, medical procedures rarely have a zero probability of negative consequences. With inoculation and then vaccination individuals are asked to accept a small risk in exchange for a larger probability that they will be protected from greater harm and that their choice will contribute to the common good. The nineteenth-century backlash against vaccination (and its continuing echoes in the twentieth and twenty-first centuries), not to mention the continued popularity of alternative therapies, also evidences the particular vulnerability of medical descriptions and prescriptions about risk to vernacular reinterpretation and challenge from both medical practitioners and patients.[78]

Although originating in nature, both fire and smallpox were, in a sense, human-made threats. Like the ferocious animals to which it was sometimes compared, fire was most dangerous when it slipped out of the grasp of human keepers who fed and controlled it. The *variola* virus could not exist without a human host and could not propagate without the human habits of living in tight quarters and traveling over large distances. In a sense, practices such as setting out on the ocean in small ships and choosing to live in places with extremes of temperature and precipitation exacerbated the risks of weather. But in other ways, meteorological phenomena seemed more inscrutable and more distant from human agency. This did not inhibit either vernacular thinkers or natural philosophers from trying to predict and prevent its worst dangers. Nor did it inhibit the entrepreneurial from recognizing that money was to be made from uncertainty.

3

Doing Something about the Weather

The news from Philadelphia for November 23, 1732, was brief: "It has been very cold this Week past, that our River is full of frining Ice, and no vessell can go up or down, a Thing rarely happening so early in the year. Many People are ill with violent Colds, and Wood is risen to an excessive price."[1] By midsummer, snow-covered streets existed only in overheated imaginations. Benjamin Franklin described July of 1734 as the hottest in memory. "The Weather has been so excessive hot here for a Week past, that a great Number of People have fainted and fallen into Convulsions, and several have died in a few Hours after they were taken." Despite the heat, the harvest had to be gathered in. Farmers and their workers mostly walked to work and carried out their tasks with hand tools—the scythe and reaper—that required great physical effort to use. "From the Country round about we hear that a great many of the Harvest People faint in the fields," Franklin told his readers. The unnamed "Negro Woman" who died over her wash tub in Newcastle, Delaware, probably numbered among the many people with no alternative to working in the oppressive heat.[2]

Of all the risks confronting early Americans, the vagaries of weather were the most constant, most pervasive, and least susceptible to human manipulation. Until smoke from factories and automobiles began to raise the temperature of the atmosphere, human activities had little effect on creating weather events. Even the most careless or malicious person could not make a blizzard or a flood in the same way someone might set a fire. No one could prevent a harbor full of ice or sweltering weather at harvest time. Instead, managing risk depended on anticipating the specific weather conditions and then protecting one's self by finding shelter, reefing one's sails, or cutting extra wood. Vernacular knowledge based on careful observation and experience guided day-to-day decision making. To plan for longer-term ventures such as planting crops or setting out on a journey, early Americans depended on collective memory and other forms of common sense. As we will see, they also sometimes turned to protoscientific predictions based on astronomy for guidance and reassurance.

The problems of prediction and protection also attracted the attention of natural philosophers. Their experiments initiated a slow transformation in the way both scientists and the public understood the weather, laying the groundwork for the creation of scientific forecasting in the nineteenth century.[3] Franklin's investigation into the meteorological phenomenon of lightning had a more immediate impact. His experiments led to the first widely used, science-based risk-managing technology: the lightning rod. Popular acceptance of Franklin's theory that lightning was a form of electricity paved the way for widespread use of this device. The commercialization of lightning rods in the nineteenth century also provides an early example of another characteristic of modern risk societies: the creation and adoption of technologies that had no other purpose than protection from potential dangers.

Prediction

"Rain all night and all this day—some say, that they never knew such a long spell of wett weather in this month, but I rather think that their memories fail them, and that they may have said the same before," Elizabeth Drinker wrote at the end of a damp May, judgmental as usual.[4] Remembering the weather had a more significant function than just making conversation. Older people offered an important resource within their communities for understanding what constituted normal weather. Experience informed decisions about when it was safe to plant crops or butcher a hog or set out on a journey. Lack of collective knowledge about local weather patterns was one reason why colonization could be so dangerous. Individuals had no objective standards against which they could measure their subjective experience. Therefore, collective memory—the hottest summer, the wettest May, the hardest winter—became the de facto yardstick.[5]

Drinker's contemporaries, as well as generations of their descendants, supplemented the weather wisdom of memory with other practices. "Red sky at morning, sailor take warning," counseled a familiar adage.[6] This weather sign, indicating an approaching storm, was a familiar part of vernacular weather prediction. Weather signs included not only the short-term indicators of fair and foul weather, but also longer-term evidence signaling the character and timing of coming seasons. Farmers, sailors, travelers, and others paid attention to the color of the sky, the wooliness of caterpillars, or the diligence of squirrels, because their safety and livelihood depended on knowing what would happen next. Though her livelihood did not directly depend on her prognosticating abilities, Drinker sometimes tried her hand at reading the signs. "The moon was red last night, which is no sign of rain," she wrote during a hot, August morning. It rained anyway, prompting a bit of self-criticism: "I often, or frequently, find myself mistaken in my weather-wisdom."[7]

For those with a more abstract turn of mind, almanacs offered an alternative form of forecasting. These compendia of useful information typically contained tables called "ephemerides" (or ephemeris) that used astrological calculations to predict the weather a year in advance. Readers could look up a specific date. Presented alongside numbers and symbols describing the position of heavenly bodies would typically be a brief forecast. Benjamin Banneker, an African American almanac maker from Maryland, predicted that the first of June, 1792, would be sultry and warm under the influence of Mars and Venus. Seven days later, the growing influence of Mercury brought a change in the weather in the form of thunder and rain.[8]

Ephemerides were based on the ancient and widely accepted theory that heavenly bodies—the moon, planets, meteors, and stars—powerfully influenced the weather.[9] The premise was less illogical than it seems to our modern sensibilities. From ancient times, astronomers had been able to consistently predict the movement of stars and planets, vesting their discipline with particular power and prestige. The leap came in positing that celestial bodies governed terrestrial events. In practice, both almanac makers and users put particular emphasis on the predictive power of special dates, usually linked to events in the Christian calendar. A vestigial version of this survives in the American custom of predicting the duration of spring on Groundhog Day (the same date as Candlemass), although the ritual of looking for an animal's shadow is more closely linked to reading of weather signs than astrology.[10]

The creation of ephemerides remained serious scholarly business through the seventeenth century. Harvard students dominated the creation and printing of American almanacs. Future ministers with a taste for mathematics ("philomaths") toiled over the complicated calculations. They presented their work with little explanation to an audience that already knew what the tables meant. By the dawn of the eighteenth century, astrological prediction gradually fell out of favor among natural philosophers. Not surprisingly, the predictions could not withstand comparison with actual weather records gradually compiled and analyzed by a new generation of natural philosophers searching for better predictive techniques.[11]

Changes in how a few elite thinkers understood the causes of the weather had little direct effect on the great majority of people's taste for astrological prognostication. Almanacs grew increasingly popular as the century wore on, served by the growth of print culture and the ease with which even the educated accommodated contradictory ideas about causality. The most famous of eighteenth-century American almanac makers, Benjamin Franklin, was not above handily turning this intellectual inconsistency to profit. *Poor Richard's Almanack* gave readers conventional tables of ephemerides, but Franklin added sly witticisms making fun of their unpredictability. For instance, in the introduction to the 1737 *Almanack*, Franklin's alter ego, Richard Saunders, asks the reader for a little leeway in his forecasts: "a day or two before and a day or two after the precise day against which the Weather is set; and if it does not come to pass accordingly," Saunders suggested that the reader blame the printer (Franklin, of course), who "may have transpos'd or misplac'd it."[12]

Into the mid-nineteenth century, readers seem to have considered weather forecasts an essential part of almanacs—they shied away from buying versions that did not contain them.[13] But whether or not they based their decision making on those predictions is less clear. Except among a few true believers, tables of ephemerides probably held little authority in most people's decision making, compared to weather signs and craft and collective knowledge. They were relied on only in the absence of any other way of knowing what might happen.

Understanding and predicting the weather also became an obsession among many of the natural philosophers we have already met, including Franklin, Mather, and James Jurin. The most zealous of their number believed that if they collected and analyzed enough information, the laws of nature regarding weather would be revealed and accurate prediction would become possible.[14] Systematic observation constituted the centerpiece of their method, just as it did for farmers and fisherman. But they tackled the problem with an additional tool: quantification. Numbers allowed comparison over time and between places. Eventually such comparisons made it possible to predict storms and other meteorological events.

Central to their investigations were two measuring devices, the barometer and the thermometer, which could be used to assign numbers to what previously could only be described subjectively. "I am sorry it is not in my power to begin immediately the course of observations you proposed in your last

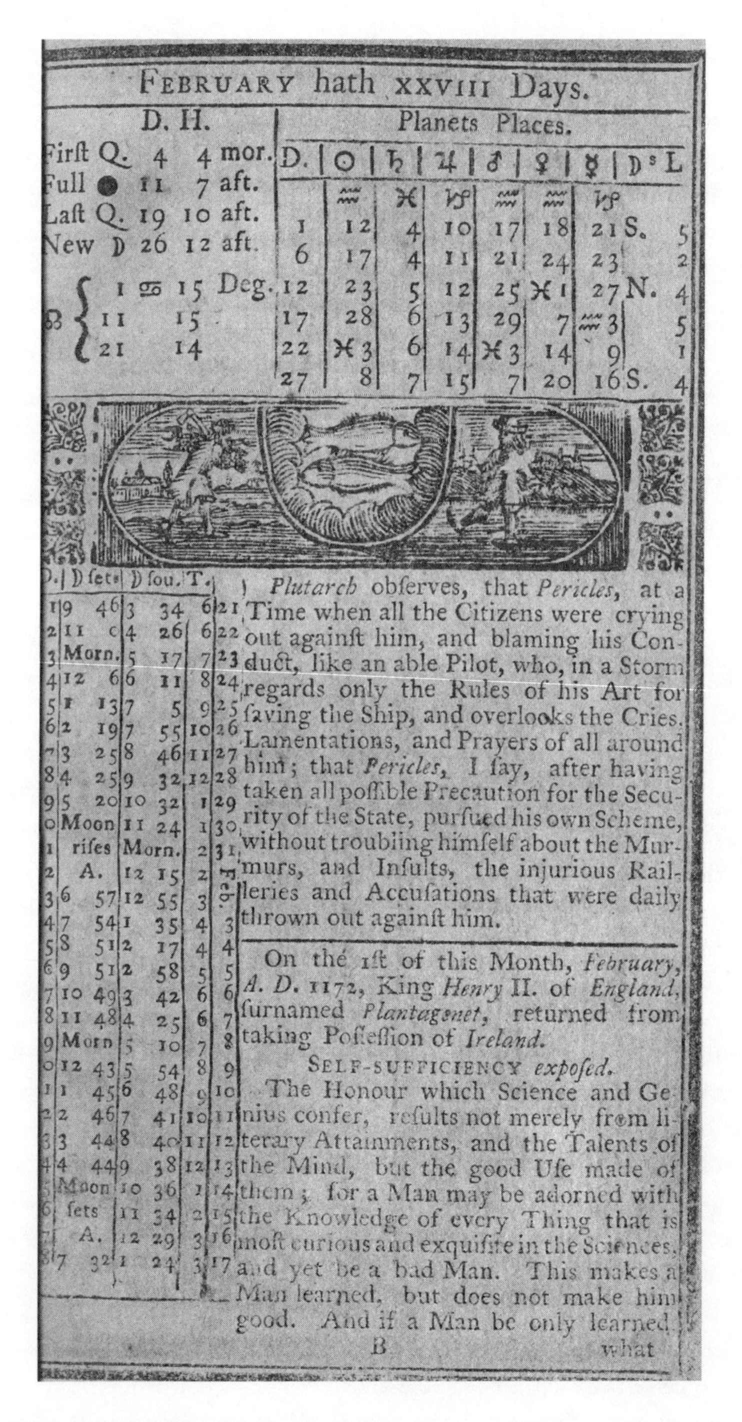

FEBRUARY hath XXVIII Days.

	D.	H.	
First Q.	4	4	mor.
Full ●	11	7	aft.
Last Q.	19	10	aft.
New ☽	26	12	aft.
☊ 1	♋ 15	Deg.	
☊ 11	15		
☊ 21	14		

Planets Places.

D.	☉	♄	♃	♂	♀	☿	☽s L
	♒	♓	♑	♒	♒	♑	
1	12	4	10	17	18	21	S. 5
6	17	4	11	21	24	23	2
12	23	5	12	25	♓ 1	27	N. 4
17	28	6	13	29	7	♒ 3	5
22	♓ 3	6	14	♓ 3	14	9	1
27	8	7	15	7	20	16	S. 4

D.	☽ sets		☽ sou.		T.	
1	9	46	3	34	6	21
2	11	0	4	26	6	22
3	Morn.		5	17	7	23
4	12	6	6	11	8	24
5	1	13	7	5	9	25
6	2	19	7	55	10	26
7	3	25	8	46	11	27
8	4	25	9	32	12	28
9	5	20	10	32	1	29
0	Moon		11	24	1	30
1	rises		Morn.		2	31
2	A.		12	15	2	Feb.
3	6	57	12	55	3	
4	7	54	1	35	4	3
5	8	51	2	17	4	4
6	9	51	2	58	5	5
7	10	49	3	42	6	6
8	11	48	4	25	6	7
9	Morn		5	10	7	8
0	12	43	5	54	8	9
1	1	45	6	48	9	10
2	2	46	7	41	10	11
3	3	44	8	40	11	12
4	4	44	9	38	12	13
5	Moon		10	36	1	14
6	sets		11	34	2	15
7	A.		12	29	3	16
8	7	32	1	24	3	17

Plutarch obſerves, that *Pericles*, at a Time when all the Citizens were crying out againſt him, and blaming his Conduct, like an able Pilot, who, in a Storm regards only the Rules of his Art for ſaving the Ship, and overlooks the Cries, Lamentations, and Prayers of all around him; that *Pericles*, I ſay, after having taken all poſſible Precaution for the Security of the State, purſued his own Scheme, without troubling himſelf about the Murmurs, and Inſults, the injurious Railleries and Accuſations that were daily thrown out againſt him.

On the 1ſt of this Month, *February*, *A. D.* 1172, King *Henry* II. of *England*, ſurnamed *Plantagenet*, returned from taking Poſſeſſion of *Ireland*.

SELF-SUFFICIENCY *expoſed*.

The Honour which Science and Genius confer, reſults not merely from literary Attainments, and the Talents of the Mind, but the good Uſe made of them; for a Man may be adorned with the Knowledge of every Thing that is moſt curious and exquiſite in the Sciences, and yet be a bad Man. This makes a Man learned, but does not make him good. And if a Man be only learned

B what

This table of ephemerides predicted the weather for February 1758. *Benjamin Franklin,* Poor Richard improved: being an almanack and ephemeris for the year 1758, *The Library Company of Philadelphia.*

letter," Thomas Mann Randolph wrote to Thomas Jefferson in 1790. The reason was a lack of measuring devices. "I have not a thermometer even, at present, but shall provide myself with one, and as soon as possible, with a Barometer."[15] These bits of "philosophical furniture" first became available in the colonies in the 1720s but spread slowly, probably because they were expensive.[16] Jefferson bought himself a barometer from Sparhawk's in Philadelphia on July 5, 1776, perhaps as reward to himself for his labors in declaring American independence.[17]

Jefferson and his fellow data collectors were on the right track to better predictions, but they lacked both adequate amounts of data and the means of sharing it in a timely way. In the nineteenth century, weather prediction improved because meteorologists created the capability of modeling weather systems using large amounts of data sent over the telegraph. Quantification gradually gave observers a more effective means of describing and predicting the behavior of nature. Like epidemiological statistics, numerical data about weather eventually allowed scientists to "see" patterns that were not visible to observers until the much-later introduction of satellites.

In the shorter term, Franklin's experiments aimed at understanding the physical characteristics of electricity led to the Enlightenment's first truly revolutionary contribution to the understanding and management of weather risks. As with fire, the fact that people might be willing to pay for better forms of protection was not lost on those with entrepreneurial ambitions.

Lightning Protection

Mark Twain is often credited with noting that "everybody talks about the weather, but nobody does anything about it."[18] This observation may be true, but that has not precluded seemingly endless experimentation to solve this fundamental problem. Most experiments yielded little of practical value. Lightning rods—yet another product of Franklin's fertile mind—are a notable exception. Poised along the rooflines of houses, barns, forts, and churches, these curious devices, born of Enlightenment investigations into natural philosophy, were arguably the first science-based safety device. They gave a physical form to scientific theories, reshaping how ordinary people understood and dealt with the natural environment. They were also part of an emerging market revolution in which everything, including the fear of lightning-created fires and property destruction, was subject to commodification.

In 1754, Charles Woodmason, a South Carolina plantation owner, sent Benjamin Franklin a fan letter written in a distinctively eighteenth-century form that Woodmason called a "Poetical Epistle":

> No fire I fear my dwelling shou'd invade,
> No bolt transfix me, in the dreadful shade;
> No falling steeple trembles from on high,
> No shivering organs now in fragments fly,
> Nature disarms, and teaches storms to spare. . . .[19]

The unnamed agent of these miraculous acts turns out not to be the Almighty, as the language might imply, but rather a very simple device and Franklin, its creator, at least as far as most people were concerned. Franklin's insight was that lightning was an electrical discharge, similar in nature to the sparks he and other experimenters generated using frictional generating devices and Leyden jars (a kind of electrical condenser). He initially imagined the lightning rod as part of a scientific experiment to test whether thunderclouds contained electricity. Although the physics of the experiment is actually more complicated, his contemporaries thought of the rod as drawing an electrical charge out of the clouds. Like rainwater in a downspout, the "electrical fluid" could then be carried to further apparatuses or harmlessly into the ground by means of a metal cable or chain.[20] Franklin installed a lightning rod on the chimney of his Philadelphia townhouse in 1752. A lightning strike rang a bell inside the house and left a charge in a Leyden jar in the cellar.[21]

Franklin quickly realized, however, that the device might also be used to protect buildings from the damage caused by lightning—a fact he announced in the 1753 edition of *Poor Richard's Almanack*. In the region of Philadelphia, lightning strikes were (and still are) relatively common during summer thunderstorms. The result, when they struck a building, was often fire—a problem close to Franklin's heart. Chimneys and church steeples—often the tallest features in the local landscape—suffered the most damage. Lightning superheated the mortar in such structures, causing it to blow apart. Civic and military leaders also worried about the terrifying possibility that lightning might ignite an explosion in the huge quantities of gunpowder stockpiled to fight the century's many wars. The 1769 detonation of one hundred tons of powder stored in the Church of San Nazaro in Brescia, Italy, resulting in more than three thousand casualties, attracted international attention.[22]

The first lightning rods began appearing atop buildings almost immediately after Franklin publicly announced his ideas. Writing to Swiss philosophe Horace-Benedict de Saussure in 1772, Franklin pictured rods as used on "private houses in every street of the principal Towns . . . on Churches, Public Buildings, Magazines of Powder, and Gentlemen's Seats in the Country."[23] However, the implied ubiquity of his invention was something of an exaggeration. In Franklin's time, lightning rods were a phenomenon mostly of cities and a few plantations in the South.[24] William Shippen III, one of the founders of Philadelphia's College of Physicians and chief surgeon for the Continental Army, typifies the kind of person who invested in lightning protection in its earliest years. A 1785 engraving of his house at the corner of Fourth and Spruce Streets, a few blocks from Franklin's dwelling, clearly shows an air terminal projecting from one of the chimneys from which descends a lead, described as "fixt to the wall by iron staples."[25]

Not surprisingly, the elite were first to learn of Franklin's ideas. Until the 1820s, theories about electricity and information about the new technology spread primarily through private correspondence among gentlemen interested in natural philosophy, as well as through a few publications such as the *Gentlemen's Magazine* and the presentations of itinerant lecture demonstrators. These individuals traveled a circuit of the great coastal cities. Ebenezer Kinnersley, the most famous of these early popularizers, regularly lectured on electricity in Philadelphia, Boston, and New York between the 1750s and 1770s. Published as a broadside, his syllabus promised instructions on how to "secure Houses, Ships, &c." through the use of lightning rods.[26] Kinnersley and other speakers undoubtedly addressed a select audience—people such as Shippen who had the time, literacy, money, and proximity to attend.

Franklin and his correspondents viewed the lightning rod as a kind of quintessential Enlightenment technology—a boon to mankind born of the marriage of scientific theory and experimentation embodied in a relatively simple technology. However, that boon was only apparent to those who shared an Enlightenment mind-set. These individuals were convinced that science not only described the physical world but offered a means to control and direct it. Even among those acquainted with the existence of the device, not everyone thought it was a good idea. It was not outside the realm of reason to worry that intentionally attracting a large electric charge and drawing it down the side of a building and into the ground might have unforeseen consequences. Like purposefully infecting someone with smallpox, "drawing down the electric fire" seemed to defy common sense. Moreover, even advocates realized that only trial and error could establish exactly how to construct an effective system. Shippen's house was struck by lightning despite the presence of lightning protection. A July 1781 lightning strike traveled through the house, following the wires composing the servant bell system, in the process blasting plaster off the walls and ceiling.[27]

William Shippen cannot have been the only homeowner to experience the consequences of doing it wrong.

Some historians have also pointed to theologically based objections to lightning rods as evidence of the "long war between science and religion."[28] But this interpretation is likely an imposition of a conflict that did not really start until after Darwin. Cotton Mather was not alone among eighteenth-century preachers in being able to entertain religious and scientific ideas of causality simultaneously and hence prevention. In 1755, one of Franklin's correspondents, John Winthrop, became embroiled in a war of words over the efficacy of lightning rods and the question of whether installing one was an ill-advised end run around God's plan. Winthrop taught natural philosophy and mathematics at Harvard and was the descendant and namesake of the original governor of Massachusetts. He became an outspoken advocate of electric experimentation and of Franklin's devices. His sparring partner was the Reverend Thomas Prince, who had published a short pamphlet in which he argued that lightning rods could potentially cause earthquakes by drawing electricity into the ground. Prince, a preacher as well as a student of geology, could not resist suggesting that employing lightning protection might be not only materially but also spiritually dangerous—a challenge to divine providence. The *Boston Gazette* provided a battleground for their intellectual jousting, just as it had for inoculation thirty years earlier.

As the debate unfolded, it became apparent that Prince was an enthusiastic but not very competent natural philosopher. Winthrop showed with devastating effect that his opponent did not know much about electricity, as leading investigators understood it at that moment. Eventually Prince changed his story. "I was never against erecting them," he claimed, as long as it was done "with a due Submission to the sovereign Will and Power and Government of GOD in Nature in humble Hopes of greater Safety, and with a becoming Trust in Him, and not in *them*."[29]

The exchange must have left Winthrop on the defensive. A decade later, he wrote Franklin that he had "read in the *Philosophical Transactions* the account of the effects of lightning on St. Bride's steeple." Apparently, the church was considering rebuilding the steeple without lightning protection. Winthrop told Franklin, "Tis amazing to me, that after the full demonstration you had given, of the identity of lightning and of electricity, and the power of metalline conductors, they should ever think of repairing that steeple without such conductors. How astonishing is the force of prejudice even in an age of so much knowledge and free inquiry!" Franklin responded, "It is perhaps not so extraordinary that unlearned men, such as commonly compose our church vestries, should not yet be acquainted with, and sensible of the benefit of metal conductors."[30]

The lightning rods and accompanying grounding apparatus installed by Franklin and his admirers were essentially homemade, one-of-a-kind systems. Blacksmiths, mechanics, and handy homeowners built them using available forms of iron and wood, much as a carpenter shapes door latches out of generic forms of material. In the 1753 version of *Poor Richard's Almanack*, Franklin recommended constructing the rod out of the "rod-iron used by nailers."[31] By the early decades of the nineteenth century, some makers employed factory-made wire or purchased glass insulators, but these systems remained essentially homemade.[32] Lightning rods also continued to be rare, especially in rural areas. Even if they knew about the existence of these devices, most farmers and householders had neither the cash for the raw materials to make a system nor access to technical explanations of how to do so.

By the mid-nineteenth century, dozens of factories produced lightning rod systems.[33] Their glass globes and elaborate points and projections gradually replaced Franklin's plain, utilitarian designs. Growing popularity of lightning rod systems coincided with what historians have called the "market revolution"—a period in which increasing numbers of Americans (including, significantly, small-town and rural northerners) regularly participated in the cash economy and imbibed the values of a market society.[34] It was also propelled by what historian David Jaffee calls the "village enlightenment": an antebellum process of the "democratization of knowledge" through the distribution of printed materials and the spread of educational institutions across the northeastern United States.[35] Popularization of scientific ideas about electricity reframed the perception of lightning risks for a new audience, simultaneously intensifying fears of lightning strikes and proposing a solution. Given the possible profits, increasing numbers of inventors, manufacturers, and salesmen (some more honest than others) committed themselves to creating and selling lightning protection.

Commodifying Safety

In the late summer of 1853, Herman Melville spent a few months living just outside the village of Pittsfield in the Berkshire Mountains of Massachusetts. Chronically short of money, he was gathering inspiration for short stories that could be sold to popular magazines. The following spring, drawing on his experiences in Pittsfield, he wrote "The Lightning-Rod Man," which appeared in the

August 1854 issue of *Putnam's Monthly Magazine*.[36] Told from the perspective of a nameless householder, Melville's brief narrative recounts the visit of a lightning rod salesman during a summer thunderstorm. Refusing his host's offer of a place near the fire, the visitor plants himself in the center of the room, holding tightly to "a polished copper rod, four feet long, lengthwise attached to a neat wooden staff, by the insertion into two balls of greenish glass, ringed with copper bands [insulators]." "The metal rod," Melville informs the reader, "terminated at the top tripodwise, in three keen tines, brightly gilt."[37] "Jupiter Tonans," as the narrator calls him, then attempts to frighten his host into buying his wares. "I warn you, sir, quit the hearth. . . . Are you so horribly ignorant then as to not know, that by far the most dangerous part of a house during such a terrific tempest as this, is the fire-place?"[38] Both men then fall into an argument about the efficacy of rods and the relative danger of lightning. Finally, the narrator grows so frustrated by the salesman's evasive patter that he breaks the rod and kicks his visitor out into the storm, berating him with a speech about the hubris of testing God's will by employing technology.[39]

Lightning rod salesmen were a relatively new phenomenon in 1854. Unlike most of their traveling brethren (and, more rarely, sisters), they faced a distinctive challenge in peddling their wares. Lightning rods (or, more accurately, lightning protection systems, of which the rod is the most visible and symbolic part) are a safety technology (like seat belts and fire alarms and radon detectors), falling into a category of devices primarily useful for lessening risk. As commodities, safety technologies can be quite difficult to sell. People are hesitant to invest their hard-earned money in a form of insurance they may never need. Unlike fire or life insurance (at least in their twentieth-century forms), safety technologies can also fail or, if faulty or badly designed, actually make an accident worse. Nineteenth-century lightning rod systems seem to have been particularly prone to failure. Melville's narrator, for example, brings up the fact that lightning had struck a local church steeple armed with a rod only the week before—a rod the lightning rod man had in fact installed.[40] Because of the problems of uncertainty and failure, people who sell safety technologies tend to rely on two strategies to bolster their message: creating a heightened perception of risk so that potential customers can imagine themselves as potential victims (a strategy with great potential for abuse), or combining safety with some other value—aesthetic beauty, social status, and others.

The success of various kinds of risk-mitigating devices also depends on whether they can catch the imagination of an era. Most pre-twentieth-century safety devices (guards around fireplaces, safety harness on horses, etc.) were mechanical—it was obvious from looking at them how they work. Lightning rods were anomalous because their design was based on a new set of scientific theories. The scientifically untrained could not immediately see how they work. An individual required some understanding of, or at least belief in, electricity to perceive that a lightning rod was more than a metal spike decorating a building. Electricity, as a natural phenomenon and a subject of scientific discovery, attracted intense popular interest in the eighteenth and nineteenth centuries. Not coincidentally, the growth of popular interest in lightning rods coincided with the first period of widespread exposure to scientific ideas.[41] As technological fixes, lightning rods also nicely suited a world in which everyone was more exposed and therefore more aware of the risks associated with weather.[42]

Beginning after 1815 and with increasing frequency in the 1830s, these ideas spread to small-town and rural folk, especially in the north, setting the stage for Melville. Journalists, editors, and letter writers began using the pages of popular journals, ranging from the *Genesee Farmer* and *Gardner's Journal* to the *New York Religious Chronicle*, to discuss new theories of electricity. These writers assumed readers' familiarity with electrical apparatuses such as the Leyden jar. They also engaged in arguments about why readers should install lightning rods, gave directions for erecting rods, and provided explanations for failures.[43] This enthusiasm for and access to technical and scientific knowledge are apparent in Melville's story. Both protagonists couch their argument in highly technical terms—discussing, for example, the relative conductivity of various substances (wood, metal, cloth) and the difference between the upstroke and downstroke of lightning.[44] Melville himself was both a product of and a participant in the village enlightenment process. So too were the Berkshire farmers and townspeople from Pittsfield with whom he mingled during that thundery summer of 1853. They enjoyed ready access to a lending library, the Pittsfield Library Society (founded 1801), and flocked to the Lyceum to hear lectures from such luminaries as Ralph Waldo Emerson, Henry Ward Beecher, and Horace Greeley.[45]

Before the 1850s, farmers and householders likely made and installed their own lightning rods (if they had them at all) or commissioned a blacksmith or local mechanic to do it under their supervision. Knowledge about how to do this spread informally between neighbors or through the pages of popular publications. A reader who wrote to the *Genesee Farmer* in 1832 asking for directions for choosing and installing rods on his barn received an outpouring of advice.[46] Several readers responded by not only prescribing the height of the rod, how to fasten it to the building, and how to ground it, but also providing the scientific rationale for each specific detail.[47] Strikingly, these articles and editorials often added keeping up with scientific and technological advances to the moral obligations of prudent men. "The farmer who neglects this mode of preserving his house or his barn can have little claim on the charity or commiseration of the public," the editor opined.[48] He urged farmers to employ "the aids philosophy and science have provided."[49] Similar advice and exhortation appear in a wide variety of farmers' journals, religious papers, and *Scientific American*.[50] In this instance, the conversation about self-help through the acquisition of useful knowledge proved short-lived. It evaporated after 1850, as these devices became commodified, that is, patented, manufactured in factories, and sold by specialized companies.[51]

Rural participants in both the village enlightenment and the market revolution provided a tempting target for a new group of opportunists: inventors with dreams of exploiting patents, manufacturers, and eager salesmen. Unlike their urban counterparts, rural people were cautious entrants into consumer cultures, worried about the opinions of their neighbors, as well as their limited supply of money.[52] Lightning rods, with their preexisting meanings of science, rationality, and prudence and their conspicuous position on the tops of houses and barns, nicely bridged older values of conserving property with the emerging market culture of status through consumption and display. As with inoculation and vaccination, community acceptance, particularly by elites, subtly shifted collective risk perception and paved the way for more widespread adoption.

The devices these entrepreneurs peddled often bore little resemblance to "the blunt, rusty iron rods which are in so general use."[53] They manifested whorls, tridents, side appendages, glass balls, twisted rods, hollow rods, square rods, round rods, oval rods—in short, every shape the human imagination could conjure up. The patent system's requirements provide a hidden reason for all this baroque ornamentation. The first "lightning conductor" or "lightning rod" patents began appearing in the 1840s. Since everyone (including patent examiners) knew that Franklin had invented the lightning rod nearly one hundred years prior, patentees typically claimed some form of "improvement" on the original device. Added elements or novel shapes allowed patentees and manufacturers to distinguish their products from those of their rivals and from previously patented devices. Most, of course, claimed that each added bend and whorl and bit of gilding was utilitarian, adding to the efficacy of the rod in guiding "electric fluid" safely into the ground, but of course, it did not hurt that they were decorative and looked expensive.

The elaborate designs of these rods also added value in a different way. Even if lightning never struck, they were still worth owning for the decorative and status value added to an individual's house and barn. To have a lightning rod at all

Lightning rods from the second half of the nineteenth century were often highly ornamented. *North American Lightning Rod Company catalog. Courtesy, Chicago Historical Society.*

made one kind of statement, but to have a rod with extra curlicues and gilding, harmonizing (if that is the right word) with increasingly elaborate Victorian architecture, was quite another.

Patents in and of themselves do not prove actual use by consumers since most patented devices never make it to the marketplace. But at least some of these early patentees rapidly moved to exploit their intellectual property. In 1851, G. W. Otis of Lynn, Massachusetts, patented a square rod with multiple points each tipped with gold. Soon thereafter, he licensed the rod to the Lyon Manufacturing Company of New York.[54] The owner, Lucius Lyon, mounted an interesting effort piggybacking on both the institutions of the Enlightenment and an emerging market culture. Not only did he exploit the authority of the patent system (it must be good if it is patented), but he also solicited testimonials from ministers, professors of science, and heads of various schools. He entered the rod into prize competitions at various fairs, including the New York Crystal Palace, where it reputedly won a prize. Lyon was already the author of a treatise on lightning protection, further adding to his authority.[55] Lyon's marketing strategy also played on the aesthetic dimensions of Otis's rods. He garnered a testimonial from a "Prof. J. Ennis" for the company's brochure, which stated that "as architectural taste is now a prominent feature in nearly all buildings, this new arrangement possesses a special recommendation. It adds decidedly to the beauty of the building, and thus serves the double purpose of protection and decoration."[56]

Undoubtedly, some farmers, mechanics, and clever people continued to make and erect their own lightning rods, and some lightning rod manufacturers sold their product through mail order or through the general store.[57] But most companies preferred the kind of salesman described in Melville's story (although presumably with a better ability to get along with customers). These "lightning rod men" not only peddled the system but also supervised its installation, which meant they needed a variety of skills and contacts.[58]

It is not entirely clear why lightning rod companies chose this form of marketing. In general, peddling was a very common form of sales in this period and offered a cheap way for companies to sell over a wide geographic area.[59] Part of the reason may also have to do with the nature of the product. Enlightened farmers could erect their own rods. Lightning rod men knocked on the doors of people who had not yet been convinced they needed lightning protection. Unless their houses or barns had already been hit, they probably would not have seen the urgency of running to obtaining a rod from the general store. And, if even their houses had been hit, they might be more likely to abide by the folk wisdom suggesting that lightning never strikes the same spot twice. Given the mixed press around lightning rods, sales pitches offered the opportunity to explain why the particular system in question would not fail.

To add value to their product and to slow competition from customers and local mechanics, lightning rod companies strove to create at least the impression of a monopoly on expertise. Writing in 1879, crusader John Phin railed against the lack of information available to those who wished to make and install their own rods: "It is true that we have a few pamphlets and one small book; but they can hardly be dignified with the name of treatises on the subject since they are all written in the interest of some particular patent, and were never intended to give such information as would enable an ordinary, intelligent mechanic to erect a lightning rod for himself."[60] Phin was also furious about the mystique surrounding patented devices. "Almost all the lightning-rods sold by itinerant vendors are patented; and may therefore be worth while to remind our readers that *all essential requisites for perfect protection may be embodied in a rod which does not infringe on any patent.*"[61]

Because lightning rod men rapidly gained a very bad reputation as swindlers, they could not necessarily embody the kind of authority that would sell their product.[62] Instead, they came armed with the kinds of pamphlets Phin derided, as well as warranties that gave the authority of the legal system to their sales pitches. They frequently included a lengthy technical discussion of the nature of lightning and lightning conduction, testimonials from famous or authoritative men—typically scientists, heads of schools, and government officials—and quotes from books and articles about lightning.[63] At least some salesmen also carried elaborately illustrated books detailing the decorative elements that could be added to a system, thereby reinforcing the decorative side of the sales pitch.[64]

How many people bought these systems? No one knows. Commercially made lightning protection systems were expensive, costing between $65 and $200—5%–10% of the cost of building a medium-priced house in the same period, which would have priced many cash-strapped homeowners out of the market.[65] On the other hand, for people acutely aware of the risks of lightning and very afraid of fire, this price may have seemed like a good bargain. Certainly, the surviving physical evidence of fancy rods and many small companies suggests there was plenty of business to go around.[66]

Lightning rods are still with us, but they no longer carry the powerful meanings that inspired eighteenth-century gentlemen to versify their virtues and nineteenth-century farmers to debate the best way to erect them. Modern rods are subtle—architects do their best to hide them, and they are nearly culturally invisible as well. Twenty-first-century people do not think a lot about electricity. We rarely open our newspapers to read about lightning strikes, although clearly they still occur. When we do think about lightning, it is often in terms of surge protectors, not lightning rods. Fear of electricity no longer speaks to our collective cultural imagination.

KNOWLEDGE SHAPES RISK PERCEPTION. It also shapes risk-taking and risk-managing behavior. In early America, most people knew about the weather through the evidence of their own senses. Observation and collective wisdom shaped their beliefs about what was likely to happen in the future and how they could best protect themselves. The efforts of natural philosophers to systematically observe and describe meteorological phenomena led to a different way of knowing about weather—one based on analysis of quantitative data compiled over time and across space. Popularization of new theories about electricity leading to the adoption of lightning rods as a risk-managing technology offers a particularly dramatic example of how new ideas can change the choices individuals make about managing risk.

In the modern world, scientific ways of knowing the weather have become naturalized and accepted into popular culture. Say "It was a hundred degrees today with 90 percent humidity," and most adult Americans will feel the heat—a cognitive leap that would have been completely alien even to educated eighteenth-century people like Elizabeth Drinker and Cotton Mather. More importantly, we routinely turn to meteorological experts to make both mundane (shall I bring an umbrella?) and life-and-death decisions about the weather (should I evacuate before the storm arrives?). Trust in scientific weather prediction is predicated on accepting that forces we cannot see—high and low pressure fronts, winds in the stratosphere, a few degrees of warming far out to sea—are the causes of what will happen today, tomorrow, and sometime next week. While science-based systems of protection remain important, forecasting has become the more important commodity.

Significantly, concern about weather risk did not generate the kind of regulatory law governing infectious diseases and fire. In this period, regulation was reserved for human-generated or human-propagated risks, particularly those that threatened the common good. In some vague sense, failure to erect lightning rods might qualify, but almost no one seems to have thought about it in this fashion. Moreover, most individual errors in managing weather risk—setting out on a snowy night, putting seeds in the ground before the last frost—redounded only to that individual or someone in his or her immediate circle. They were situated as private matters and, moreover, uncontrollable and unexplainable bad luck or divine punishment. No form of human-enforced discipline was likely to make any difference. In contrast, managing the risks of animals, the subject of the next chapter, revolved around the business of discipline, properly and improperly applied.

4

Animal Risk for a Modern Age

Animals think for themselves. Like us, they have their own individual emotions, drives, and interpretations of the world. Domestication captured this quality for human use.[1] In choosing to live and work side by side with animals, human beings not only turned animals into technology but also created an important source of risk in everyday life.[2] Although the seeming unpredictability of fire sometimes prompted early Americans to talk as if it were a living entity, most understood the fundamental differences between animate and inanimate risk. Animals are aware of and, if properly trained, cooperative with humans. But animals can also make choices that confound human expectations. Sometimes those choices benefit humans. Unlike a machine, an animal might choose to save a human life at its own peril. It might also panic, lashing out in an unexpected effort to defend itself, resulting in disaster for all involved.

Frequent encounters with domesticated animals were the norm in early America. People accepted that a certain amount of risk necessarily resulted from using animals for food and work. This acceptance began to wane in the middle decades of the nineteenth century. New laws banned garbage-eating pigs from city streets and prohibited drovers from moving their flocks on public roads. Cheaper forms of fencing made it economical to confine grazing animals. Railroads made it dangerous not to do so.[3] Gradually, most kinds of animals were segregated from people who were not their owners or handlers and banished from public spaces.

Horses proved the notable exception. The number of draft and riding animals on America's roads and especially in cities increased until the century's end.[4] The results were predictable: crashes, collisions, small children and grown men kicked and run over.[5] Because these animals played a key role in linking together a developing transportation system, restricting their movements to control risk was untenable. Instead, decision making was largely left up to the individual. While a few laws, some dating back to colonial times, restricted street racing and other dangerous practices, to a remarkable degree people could do whatever they liked—drive down the middle of the street, hand the reins to a child—without formal qualification or oversight.[6]

The ways Americans dealt with the dangers of horses demonstrate how vernacular practices around one particular kind of risk evolved in the context of industrialization and urbanization. At the center of this process stood a body of knowledge and skill about how to interpret and direct the behavior of another species. True to the innovative and entrepreneurial spirit of the nineteenth century, inventors and professional horse trainers touted a wide array of devices and techniques, which they promised would make training and using horses significantly easier and safer. In reality, very little of what they had to offer was truly new. All of the basic techniques for balancing the utility and dangerousness of horses, such as breeding, restraint, and training, predated industrialization. If the spread of Enlightenment and scientific ideas had an effect at all, it was in propagating a philosophy of kindness toward animals, which made it safer to be a horse and perhaps indirectly helped humans by lessening the population of abused horses that might behave in dangerous ways.[7]

Instead, changes in the way Americans managed horse-related risk grew out of the values and divisions of a capitalist society. The collective societal decision to allow virtually unrestrained use of horses in public spaces prioritized speed and convenience over safety and health. An enormous market for horse-related labor encouraged specialization of knowledge. Willingness to accept risk also became commodified. Working-class men filled the ranks of people who took on the dangerous job of training young horses. They were paid for their willingness to assume risk and, less importantly, for their skill. Tradesmen and members of the middle class aspired to be knowledgeable consumers and users of already-broke horses. They turned to the mechanisms of the law and the market to limit their physical and financial risk. In a society that admired skillful handling of potentially dangerous horses, wealthy (and not so wealthy) hobbyists gained social status while having fun driving fast horses on city streets.

The Characteristics of Horse-Related Risk

In a horse-dependent society, even non-riders and drivers had strong awareness of the risks and rewards of using these animals. Part of that knowledge came from sharing public and private spaces with animals that could weigh half a ton or more. Many people had struggled to control a horse or observed someone else doing so. They knew the damage that could be caused by a "runaway" that decided to take off down a crowded street with a wagon whipping around behind it.[8]

Humankind's long association with horses has made us somewhat immune to how remarkable it is that anyone would dare to sit on the back of or in a wheeled vehicle hitched to a thousand-pound mammal that can run thirty miles an hour and is armed with a set of rock-hard hooves suitable for cracking the skull of a pursuing predator.[9] Horses have a number of instinctive behaviors that are potentially dangerous to riders, drivers, and pedestrians. Because they are prey animals, their natural impulse is to flee from danger, sometimes at a dead run. Horses will "spook" or jump sideways or backward in reaction to something unexpected in their path. Many a distracted or uncertain rider has ended up on the ground because a deer or a rabbit or a person came out of the bushes at the side of the road. The introduction of railroads and then automobiles further contributed to the list of things that frighten horses. Some horses will also purposefully try to shed unwanted riders, usually by bucking or rearing—behaviors that evolved to dislodge predators from their backs.

At the same time, a horse with the right personality and training might be counted on to protect the well-being of its human partner, in a way that no machine could. Our ancestors placed extraordinarily physical and mental demands on horses, asking them to set aside instinct and trust their human partners in the face of obvious threats to their well-being. Successful cavalry mounts learned to ignore gunfire and proceed calmly past the bodies of dead horses. Teamsters' horses threaded their way through the tight confines of city streets and then stood patiently while goods were loaded and unloaded. Coach horses tiptoed across rickety bridges and onto crowded, unstable ferries.[10] Many people took horses for granted. Others, however, developed a deep admiration for the courage and intelligence of these animals. *Friendship* was a word commonly used to describe the bond between people and horses. Soldiers and coachmen and others who did dangerous jobs with equine partners developed a particular appreciation for the willingness of some horses to protect their riders and drivers or to put themselves in danger because they were asked to do so.

Until electricity and the internal combustion engine finally displaced animal energy, Americans depended on draft animals both for transportation and to power various kinds of mechanical devices, such as mills. Before Americans began to improve their roads and build a market economy, many pragmatists preferred not to deal with horses because, compared with other draft animals, they were relatively expensive, volatile, and prone to injury and sickness. Farmers often chose oxen to pull their plows and wagons. They weighed the ox's durability, ease of care, and handling against the horse's greater speed and opted to take their time.[11] Given the poor condition of the roads, many travelers walked or traveled by water. If they did use horses, they tended to ride rather than drive them. Instead of an invigorating canter across the countryside, most trips probably involved slow plodding over hill and dale. However, horses (and mules) had one important advantage over oxen or travel by foot—they were faster, which mattered because speed became an increasingly important value as industrialization and a networked market economy took hold.

No one knows for sure how many horses were in use in eighteenth-century America, but historians generally agree that, between 1790 and 1840, the number probably grew faster than the human population. By midcentury, one historian has estimated there was "one horse for every four or five people," and the number of horses substantially outnumbered oxen and other draft animals.[12] The ways horses were used also changed. Taking advantage of better roads and new designs that made wheeled vehicles speedier and more comfortable, more people drove than rode. Driving was also a practical choice because one horse could transport multiple people and their baggage. As the century wore on, horses became more and more important not only for transportation but also as a source of power for everything from mills to farm equipment.[13]

Wherever there were horses, there were accidents. Some of the most dramatic involved runaways. The driver of the Richmond stage handed over the reins to a passenger to hold while he replaced a sick horse. For unknown reasons, Mr. Stokes, the passenger, let go of the reins. The horses took off galloping at full speed with the stage full of frightened people behind them, running over the driver and crushing his legs. Most of the passengers jumped for their lives. The battered carriage and frightened horses were later found some ways down the road.[14] A half century later, Eadweard Muybridge, the photographer, joined the growing number of Americans injured in stagecoach accidents. His biographer speculates that the resulting head injury precipitated a personality change that eventually led him to murder his wife's lover.[15]

The pleasant Sunday drive could also result in tragedy. Looking back on his childhood, former slave Mazique Sanco remembered his first job: "When I was ten Dr. Flemming gave me to his crippled mother-in-law for a foot boy. She got crippled in a runaway accident when her husband was killed," he recounted. "He had two fine horses, firey and spirited as could be had. He called them Ash and Dash, and one day he and his wife were out driving and the horses ran the carriage into a big pine tree, and Mr. Dean was killed instantly, and Mrs. Dean couldn't ever help herself again."[16] Antebellum American newspapers echoed Sanco's narrative over and over again. "Horace Greeley, Editor of the Tribune, was thrown from his carriage," an abolitionist paper reported in 1849. The horse had become "unruly" and run away with the vehicle. Greeley, it was reported, "was severely bruised and cut on the head, but is now convalescent, and able to attend to business."[17]

The poor condition of many roads and especially bridges also led to accidents that were often no fault of the horses themselves. In fact, some people owed their lives to horses' sense of self-preservation. When the Lewis Town ferry sank in the middle of the Indian River with a load of people and animals, one female passenger had the presence of mind to grab hold of a horse's tail while he swam to shore.[18] In the wintertime, many people enjoyed sleigh riding on frozen lakes and rivers that held their own hazards. On a chilly Saturday afternoon in 1856, James Calendar drove a sleigh containing his wife and two young girls over an air hole. The ice collapsed beneath the weight of the sleigh. Members of a skating club used the cords and reels they had brought along as a safety measure to rescue the passengers, but the horsc and sleigh disappeared beneath the ice not to be seen again.[19] The introduction of the railroad in the 1830s added a particularly lethal element to driving. While many horses were frightened by trains, drivers were not cautious enough. Most notoriously, they raced locomotives to grade crossings, with predictably dire results.

Dramatic accidents involving well-to-do people attracted the most attention, but men who worked with horses for a living were more frequently victims of horse-related accidents. In the antebellum South, enslaved African Americans did much of the actual work of training and caring for animals. In the North, the job fell to farmers and wage laborers. By the late nineteenth century, half of Boston's teamsters were Irish. African Americans constituted the majority of Pittsburgh's drivers.[20] After the Civil War, these two social groups were also well represented in the ranks of western cowboys.[21] A nation run on horsepower required tens of thousands of workers to shoe horses, drive streetcars, clean stalls, and administer veterinary care. Working-class and enslaved men also played a pivotal role in the first and most dangerous steps in making a horse safe and useful.

Making and Buying Safe Horses

"A well broken horse will prove a kind servant; an unruly horse is a very dangerous companion. You are more safe with a steam engine," the *Massachusetts Ploughman* advised in 1844.[22] The comment was not meant as an endorsement of inanimate power, but rather of properly training young horses. Unlike a steam engine, a horse could not be reengineered to make it safer. The closest analogies were selective breeding for tractability and castration ("gelding") to control the hormones that could make stallions aggressive and unpredictable. Although nineteenth-century inventors turned their minds to patenting a variety of devices to control horses, most never caught on because mechanical restraints invariably inhibited horses' ability to do work.

Selection and training were the two most important means used to produce safer horses. Each had its risks. Mistakes in matching horses to particular occupations or to the skills and temperament of individual riders and drivers could endanger not only the horse and its master but also innocent bystanders. Training held even more risks for both animals and humans because it required the trainer to gradually replace horses' instinctive reactions with predictable responses to human directives.

"Breaking" was the term often used to describe the process through which a horse that had not previously been ridden or driven was introduced to his or her new occupation. "If you suffer them to gain the mastery you are at their mercy, to which you are not safe in trusting," the *Ploughman* explained, reflecting a common (but not entirely universal) assumption that breaking the horse's will to resist orders from humans was an essential first step in training. A horse that was obedient and knew its job was said to be "broke." That also meant that it was safe to use, at least by knowledgeable horsemen and women.[23]

Some horse owners regularly broke their own horses, but many others preferred to hand over the job to someone else—slaves, employees, or their own teenage sons—sometimes because they had more skill but often because they were willing to take on the physical risks. The ranks of those who worked with horses for money were dominated by the same kinds of working-class men who increasingly put their bodies at risk in other high-risk occupations. While some gravitated to this work because they loved horses and had real skill with them, others just needed the job. The latter group helped give horse breakers a reputation for impatience and brutality. The result was frightened or aggressive horses that were a danger to themselves and others long after they had been "sold down the road," as horse traders liked to say about getting rid of a problematic horse. Sometimes these characteristics were obvious to future users, especially if they knew the signs. Sometimes they lay hidden in a horse's psyche until triggered by a particular gesture or stimulus. In horse training and horse trading, financial pressures, ignorance, and masculine bravado undercut adherence to ideals of the vernacular culture of risk such as carefulness and self-discipline which made for safe horses.

Breaking horses was Leroy Daniels's first job. At sixteen, he ran away from his father's farm and had to find work. His first employer left him alone on a Montana ranch with seventy wild mustangs penned up in a box canyon. Daniels made progress with his charges until he accidentally drove a half-broke team into a temporary gate. "I went up into the air and came down on my back," he later told the cousin who helped him write his memoirs. Daniels lay where he had fallen until his employer returned from town eleven hours later. "I guess he thought I was dead because he stopped his rig and got out and came over and gave me a kick in the ribs. That brought me to."[24]

Daniels survived his fall and his boss, going on to a career as a horse dealer. In his youth, he had belonged to one of the last generations of itinerant horse breakers who traveled through some parts of the United States offering their services to local farmers, professional horse breeders, and urban horse dealers. Like many of his peers, he had learned most of his techniques on the family farm by working alongside his father and uncle and the rest through experimentation and experience. In his old age, he insisted that he "gentled" horses rather than breaking them, avoiding the methods of old-time "rough riders," including what another writer described as "riding in swamps and severe whipping."[25] He thought that horses that survived these techniques were unreliable or dangerous to handle. His description of his gentling techniques suggests that the difference may have been more about the degree of force rather than the absence of it. Like many of his peers, he was also under pressure to get the job done quickly.

If a horse was intended to work in harness or race, training typically began around the age of two or in the spring after the horse's second winter. General wisdom suggested that it was in the best interest of horses to postpone riding until age three or four when the animal was more physically and mentally mature. However, horse owners often took a chance and started sooner. In their calculations about the risks to horses of premature breaking versus cost, financial considerations often won out. They wanted to see young horses broken as soon as they were physically capable of working. Otherwise, the animals just stood around eating while requiring expensive attention from horse shoers and sometimes veterinarians. One correspondent to the *Southern Planter* calculated that it would cost farmers an extra $340 (a substantial sum in the 1840s) to wait until a horse fully matured at five rather than putting it to work at three years of age. The writer thought that waiting was a waste of money that would never be recouped later in a horse's life.[26]

The risk of hurrying the process was failure in "taming" the horse or permanent injury, which rendered it unusable. Owners also engaged in calculations about the horse's potential value versus the cost of feeding and training it. They waited longer and spent more time and therefore money on training particularly valuable animals.[27] Cheap horses usually got the minimum. The western practice of bronco busting was the ultimate expression of this calculation. Riding a wild horse to exhaustion did not always work. The practice also resulted in a significant number of horses and cowboys with broken necks. But a seemingly endless supply of mustangs and young men could be had for next to nothing, so the technique made economic sense.

Breaking was easiest if young horses had been gently handled since birth. But this was more the exception than the rule, except with the most valuable horses. Instead, trainers often started with animals that had previously experienced limited interaction with people. As a consequence, many horses initially saw no reason to obey what many of them probably viewed as a two-legged predator. When tied up, they pulled back; when harnessed, they attempted to kick the cart to pieces or lay down in the traces; when saddled and mounted, they bucked, reared, or bolted in an effort to shed their riders. Others did not wait until the work began, taking offensive action by trying to bite, kick, or stampede anyone trying to work around them on the ground.

Daniels, like many of his peers, started off by doing what animal trainers would now call desensitizing horses to stimuli that were likely to set off undesirable behaviors. Tying horses to a "snubbing post" with a stout rope was a favorite technique. The horse would pull and pull until it wore itself out. This step in the process was followed by "sacking." Daniels used an old canvas raincoat, which he would shake in the horse's face "until the horse gave up being afraid of the noise." Then he would saddle the horse, tie its head down on its chest (some people tied it to the tail) so it would fall down if it bucked; he then climbed on. The accident that ended Daniels's first job happened because he wore down driving horses by attaching them to a wagon with an extra long tongue (so they could not kick the wagon) and letting them run across the prairie.[28] Although Daniels did not say so, horses did fall down. They also regularly broke their necks pulling against the snubbing post.[29] In addition, horse breakers fell off and were injured or killed in untold numbers. These were the same classes of men who died in huge numbers in shipwrecks and mine cave-ins and, by mid-nineteenth century, fell off the tops of trains where they were working as brakemen. But because this was agricultural work, no one kept track of their numbers. As we will see in later chapters, quantification of workplace accidents emerged out of the organized labor movement, which had no relationship with horse breakers and farm workers.

A few horse tamers or horse educators reached a level of minor celebrity for their ability to reduce even confirmed "mankillers" to a state of subservience. They turned their abilities into careers conducting demonstrations and writing books. Many shared at least a partial reliance on mechanical devices that allowed them to "throw" difficult horses when they acted up. Unlike many of the techniques of horse breakers like Daniels, these devices were attractive because they promised to allow the trainer to assert dominance at a safe distance and without resorting to a whip. Typically, this technique involved using ropes attached to one or both of the horse's front feet and then run through a pulley attached to a surcingle that encircled the horse's body like a belt. When the horse reared or threatened to kick, the trainer yanked on the ropes pulling the horse's feet out from under him. For good measure, some trainers liked to sit on a prostrate horse to emphasize who was the boss. John S. Rarey, one of the most famous of these characters, described his approach as modern and scientific. In the 1850s, he compared it to an earlier generation of "horse whisperers" who had achieved their ends through mysterious means, including blowing in the horse's nose and whispering magic words into its ears.[30]

Throwing horses was particularly attractive to people trying to train horses quickly without getting hurt themselves. Others were not so sure these techniques were worth the potential risks to both horse and human. The editor of the *Ohio Farmer* declared that "one exhibition of modern horse education was sufficient for us. At it we saw more than enough to confirm us in our opinion that to throw a horse is not only physically dangerous, but morally degrading to him." In the moral economy of horse training, he thought that ignorant, unskilled, or cruel trainers should have to risk paying a price for their methods and lack of discipline. "If the old-fashioned 'rough rider' pushed severity too far, his limbs and even his life often paid the penalty of his want of feeling and judgment." In contrast, "the horse educator is a sly, cunning, cowardly wretch, who disables a

Some horse trainers used specialized devices to subdue particularly difficult animals. This series of images demonstrates John Rarey's method of "throwing" a horse. *John Rarey,* The Complete Horse Tamer and Farrier *(1857). University of Delaware Special Collections.*

noble animal whose chief fault is that he is too good for the use a cruel and exacting master endeavors to put him to."[31]

The author of one early twentieth-century horsemanship manual suggested that such "rapid" methods were characteristically American. He weighed the costs and benefits of them against "Old World" techniques of gradually introducing lessons and reinforcing them through repetition. In terms of safety, he thought European methods were more dangerous for trainers and "rapid" techniques resulted in more injuries to horses.[32] In a country where horseflesh was so cheap that the British army found it worthwhile to purchase remounts in Chicago, the economic logic makes some sense (although there was a subculture of American horsemen and women who did utilize these slower methods, whether or not they knew of their origins).

The goal of people such as Daniels was a horse that knew the basic signals and carried a rider or pulled a wagon without (too much) protest. Horses learned much of the rest of what they needed to know on the job (or were resold if they did not catch on fast enough). A responsive, "educated" mouth and elegant head carriage were valued in fancy carriage horses and pleasure horses ridden by the wealthy, but these qualities were unnecessary and largely unappreciated in most work horses. They needed to know that different pressures on the bit signaled whether to turn left or right, or to stop.

Many people, lacking either the time or the knowledge to make a safe and useful horse for themselves, opted to buy horses that were already "broke." In theory, this approach allowed knowledgeable consumers to buy their way out of some of the risks of horse ownership and use. However, this strategy involved its own uncertainties. Broke enough for one kind of job or one kind of user did not necessarily mean broke enough for other purposes. A poor fit between the temperament and abilities of horse and user also created its own risks. Horse dealers were particularly notorious for pawning off problem horses on unsuspecting buyers, but individual owners also had incentives to lie about or at least overlook an animal's shortcomings. "Caveat emptor"—buyer beware, the Supreme Court ruled in the 1839 landmark case *McFarland v. Newman*. Newman had bought a horse with a runny nose from McFarland. It turned out to be symptomatic of glanders, a highly infectious and lethal disease. Newman could not recover the cost of the horse because he had not negotiated a warranty. "He who is so simple as to contract without a specification of the terms, is not a fit subject of judicial guardianship," wrote Justice Gibson.[33]

Savvier buyers sometimes asked for and got short-term warranties for health and soundness.[34] Shopping for a horse or, worse yet, a team of horses was still filled with uncertainties that were far more complex and subjective than detecting a lethal disease. In the same decade as *McFarland*, one such unfortunate person detailed his efforts to buy a horse that was both safe and enjoyable to ride to humorous effect. He bought his first mount from a Quaker because he judged members of the Society of Friends to be "shrewd" in "their judgment of horseflesh." The animal in question was "a well-bred, gay little animal full of life and spirit." However, the buyer was, by his own estimate, more than thirty pounds heavier than the seller. The poor little horse kept falling down under the burden of the additional weight. Stumbling was a serious concern, frequently mentioned in horsemanship manuals, because it could result in a slow-motion fall that might injure horse or rider—a frightening prospect on a lonely road.

Our hero was relieved to sell the horse for what he had paid for him, resolving to buy a bigger, heavier animal. This new animal, however, responded to neither whip nor spurs. "Having the greatest aversion to a horse that 'won't go'; it is an eternal trial of one's temper," he explained. A trip to the veterinarian revealed that the horse wouldn't "go" because it had a serious respiratory disease. A third horse, bought from "a very respectable stablekeeper," would neither "pass nor be passed by stage, omnibus or hackney coach." A fourth horse was advertised as "so docile a lady might drive him with a pack-thread," but it would not go at all with a gentleman in the saddle. The fifth and final horse, described by the seller as appropriate for a dandy, bolted at the touch of the author's spurs and ran home to his stable. "I would as soon fondle a mad dog as take such another dance with a dandy," he decided, although still without a mount.[35]

Gentlemen purchasing a horse for pleasure could afford to buy and sell a few before finding the right one. They could also afford to pay a premium for a good mount or driving horse. A bad choice by a tradesman or farmer could be disastrous, physically and financially. Some relationships between horse and human improved with time as each learned to trust the other and found ways to work around fears and idiosyncrasies. Familiarity also allowed users to predict when their horses would spook or engage in other dangerous behavior. They could then take preventative measures (or at least get a tighter grip on the reins).

Familiarity was not a tool available to people who rented horses from livery stables. In nineteenth-century cities, these businesses let horses for a wide range of purposes, from going for a Sunday drive to transporting a sick or injured person to a hospital.[36] Many people must have experienced a moment of anxiety upon first stepping out into the street driving or riding an unknown animal. It is, however, difficult to find evidence of accidents involving misbehavior of livery horses. Livery stable owners had a legal obligation to provide their clients with horses that were, as one lawyer put it, "kind" and "suitable" for the purpose for which they were rented.[37] Moreover, most livery horses knew their jobs and did not exert any more effort than necessary. As a consequence, most resisted efforts to get them to do anything potentially dangerous, despite the efforts of riders and drivers hoping to show their prowess in handling a fiery steed. Surveying the parade of carriages passing through Central Park on the first hot Sunday of the year, a *New York Times* reporter remarked that earlier in the spring the park had been crowded by the carriages of the wealthy drawn by "high-spirited animals that keep the Park police continually on the alert." Summer brought out hired "hacks, cabs, and buggies" drawn by "machine-like moving animals of the livery stable type." He thought it significant that no accidents were reported to the park police that day.[38]

While most horse owners prioritized safety, soundness, and overall usefulness in buying horses or offering them as rentals to the public, a smaller subculture had other ideas. They preferred to sacrifice safety for speed. A few in their ranks were willing to spend enormous amounts of money to hurtle along roads and byways, dusting friends, competitors, and bystanders as they went. Their story introduces a new theme: the commodification of risk as entertainment.

Fast Horses and Powerful Men

Before the spread of automobiles, horses offered exhibitionists, thrill seekers, and the status-conscious an irresistible opportunity to publicly display both bravado and knowledgeable consumption. They composed the nexus of a series of interlocking cultures of recreational risk taking. All over America, men and boys were obsessed with fast horses. Farm boys raced their ponies across the prairie in the West, while plantation owners wagered their fortunes in the South. "A fast-rising cloud of dust far down a Virginia road probably alerted the common planter that he was about to encounter a social superior," the historian T. H. Breen noted. "A horse was an extension of its owner; indeed, a man was only as good as his horse."[39]

Although laws reaching back into the colonial era banned racing of saddle horses on Northern tracks and city streets, these statutes did not preclude a friendly competition between two gentlemen driving their roadsters on a Sunday afternoon—an activity that had its epicenter in the burgeoning metropolis of New York City. While it lasted, amateur road horse racing bridged the values of a slowly disappearing preindustrial risk culture in which participants took real physical risks to display their skill and mastery over nature and a newer culture in which risk taking was becoming a form of entertainment, formalized, commodified, and overseen by professionals.[40]

In his own time, Robert Bonner dominated the ranks of men who drove fast horses through the streets of Manhattan. Typical of trotting enthusiasts, he was an aggressive, competitive, self-made man. Bonner had emigrated from Londonderry in the North of Ireland as a child. Like Benjamin Franklin, he began his career as a teenage printer's devil (apprentice). While still in his twenties, he bought an obscure mercantile paper and turned it into an extremely successful family weekly, the *New York Ledger*, which featured stories and articles by the leading literary lights of the day. The pressures of his rapid success, Bonner later said, gave him headaches and dizzy spells. His doctor recommended horseback riding—the preferred recreation of an older generation of New York's upper classes. Bonner found it "entirely too violent." Because of his working-class background, he probably had not ridden previously. Instead, in 1856, he took up driving.[41]

Bonner resided on the east side of Manhattan, near the corner of Fifth Avenue and Fifty-Sixth Street, at the edge of a city that was rapidly expanding northward. Down the street, he built a posh stable to keep his horses.[42] Of a morning, he would walk down to the stable where a groom would hand him the reins of a single horse or pair, typically hitched to a light wagon. He would then drive toward the cluster of shanty towns that Olmstead flattened to make Central Park in 1857, and from there north into an increasingly rural landscape toward an old Dutch road, then called Harlem Lane, where men of similar interests gathered to exercise their horses and perhaps engage in "brushes" or informal races.[43] The horses driven by the enthusiasts of Harlem Lane were mostly what were known as "trotters" (or, more rarely, "pacers"). They were raced not at a gallop, like thoroughbreds, but at a slower two-beat gait. Trotting was the normal fast gait for all horses pulling wagons because it was more controllable than the faster gallop, but part of the fun for Bonner and his friends was seeing how fast they could go before their horses broke into a gallop.

Nineteenth-century writers liked to describe harness racing as "democratic." In comparison with turf racing, dominated by Southern plantation owners, who spent fabulous sums for thoroughbred horses trained and ridden by others, it was. The *Brooklyn Eagle* described the crowd at a similar gathering place as including not only "gentlemen of leisure" but also "men of limited means who spend all their spare cash and sometimes more upon horse flesh."[44] Other than trainers who were out exercising their employers' horses, everyone drove their own horses (rather than sitting next to or behind a professional driver as the carriage crowd often did). This required skill and nerve. The same energy that inspired road horses to speed made them high-strung and unpredictable. By the 1850s, the fastest horses in America had been clocked trotting at more than twenty-five miles per hour on a track.[45] Road horses could not go quite so fast, but they went fast enough. Pushing them to these speeds on an open road meant risking collisions or being run away with if the horse broke out of a trot and into a gallop.

If even the butcher and the milkman could participate, they could not really compete because it took a lot of money to buy faster horses and skillful employees to choose, train, and care for those horses. Bonner learned this lesson at the hands of Cornelius "Commodore" Vanderbilt. In their first brushes, Vanderbilt repeatedly left Bonner in his dust. Bonner responded by buying the best horses he could find, often at grossly inflated prices. Contemporaries estimated that, by the end of his life, he had purchased nearly a million dollars' worth of horseflesh, spending tens of thousands on individual animals with a particular reputation for speed.[46] Spending huge sums of money for living things that might choose not to perform or might die suddenly was only one of many ways in which harness racing was all about risk.

"The passion for betting on races pervades the whole community, high and low," the *New York Evangelist* told its readers, causing the "ruin of clerks, messengers, and mechanics."[47] Telegraphic reporting made it possible to bet without even going to the track, though huge crowds went anyway to watch match races between celebrated horses. Bonner was a rock-ribbed Presbyterian who shared the religious press's view that gambling was immoral. He stood out among his contemporaries for his steadfast refusal to put his horses in a situation that could be gambled upon (Vanderbilt, in contrast, was notorious in his taste for wagering). As Bonner explained to a reporter, after the purchase of his first fast (and expensive) horse, "I made up my mind on the spot to show that a gentleman could own fast horses without using them for gambling purposes."[48]

The primary attraction of fast horses for Bonner (as well as Vanderbilt, Russell Sage, Ulysses S. Grant, and others) was the excitement of wheeling along at high speed on the razor's edge of losing control of a powerful horse. The thrill was reinforced by informal competition, public display of mastery over a high-status animal, and knowledgeable (and conspicuous) consumption. If they needed any further ego gratification, all of this was widely reported on by the press and even memorialized in a series of Currier and Ives prints.

All of these characteristics were epitomized in Bonner's purchase and display of Dexter, one of the most famous trotting horses in nineteenth-century Amer-

Currier and Ives captured the excitement and danger present when New York horsemen came to show off their fast trotters on Harlem Lane. Commodore Vanderbilt is the gray-haired man in the top hat; Robert Bonner is the bearded man to the right of Vanderbilt. *Currier and Ives, "Fast Trotters on Harlem Lane N.Y. Commodore Vanderbilt with Myron Perry and Daisy Burns. Bonner with Dexter" (1870). Eno Collection, Miriam and Ira D. Wallach Division of Art, Prints and Photographs, The New York Public Library, Astor, Lenox and Tilden Foundations.*

ica. Bonner bought Dexter in 1867 after watching him set a new speed record at a track in Buffalo, New York. He offered $35,000, the highest price paid up to that time for a trotting horse.[49] Dexter was already a celebrity, but Bonner made him more famous. Dexter's distinctive profile and style of trotting became the model for the wind vanes that farmers bought by the thousands to decorate the roofs of their barns. A few years after his purchase, P. T. Barnum repeatedly wrote Bonner asking whether he would like the horse to be exhibited as part of his traveling museum and menagerie. Dexter's fame was such, Barnum wrote, that "scores of people would be glad to see him." Bonner declined.[50]

Unlike most road horses, Dexter was a stallion (and remained one after Bonner bought him). In exhibition races, a rider on a mare galloped in front of him to "keep up his emulation and determination to conquer."[51] Most people did not drive stallions, especially in crowded urban areas, because of the risk that the stallion would try to mount a mare or fight with other horses, so Bonner enjoyed additional admiration for successfully taking an intact male horse out on the road.

In the nineteenth-century equivalent of a photo opportunity, Currier and Ives issued another lithograph of Bonner, this time sitting beside president-elect Grant in a wagon behind Dexter. A reporter described the image for his readers: "'Taking the Reins' introduces us to a glimpse of Harlem Lane, spinning along which at a tremendous pace is the renowned horse Dexter. President Grant steers the horse, the reins having been handed over to him by Mr. Robert Bonner, who sits beside him." The reporter could not resist commenting that "the manner in which the President holds the reins marks the experienced driver, and he appears to be conversing easily with his companion as the headlong steed whirls them along over the dusty road."[52]

Public accounts typically focused on Bonner's mastery of all things equine. But, as in the rest of America, other men worked behind the scenes to make his triumphant outings possible. By 1875, Bonner owned nearly eighty horses—divided between his city stables and a farm in Tarrytown, where Edward Everett (named after the famous orator who was one of Bonner's authors) was available to breed to mares belonging to Bonner and others.[53] He employed a series of men to manage the Tarrytown farm and supervise the breaking and training of young horses. Other men and boys cleaned stalls and groomed and exercised horses. Special animals had their own grooms, akin to body servants. When Vanderbilt sold Bonner his famous mare Maud S., Bonner also hired Maud's trainer, William Bair, and an African American man (probably a former slave) who went by the single name of Grant. The papers reported that Grant had "been in charge of her [Maud] since she was 3 years old."[54] The departure of Edward Everett's groom prompted a hurried search for someone who could "get along" with the stallion.[55] Veterinarians and shoers also contributed their expertise (and submitted their bills).

In the early years of the nineteenth century, the driving habits of New Yorkers attracted little public comment. Cartmen and butchers, omnibus and stage drivers, cabmen and recreational drivers were largely left to sort out the uses of the road. By the time Bonner took up the reins of his first fast horses, increasing urban congestion and perhaps a shift in the way Americans thought about risk resulted in increasing concern about what was called (in the law and in public discourse) "reckless driving." Until city ordinances began to set speed limits in the late 1880s (typical of a period that was beginning to try to quantify risk, even if drivers themselves had no way to measure speed without a speedometer), reckless driving was a subjective term.[56] Police officers, judges, juries, bystanders, and, of course, drivers decided how fast was too fast, whether a driver was exercising adequate caution, and whether an accident was unavoidable or the result of carelessness or malfeasance.

Not surprisingly, the law was applied differently to different kinds of people. New York City authorities were generally extremely tolerant of the driving habits of elite men like Bonner (the same cannot be said for Washington, DC, where Ulysses S. Grant was arrested for driving too fast up Fourteenth Street while he was president).[57] Informal racing, particularly in the northern reaches of Manhattan, was treated by authorities as an exception to prohibitions against speeding, despite more and more incidents and accidents as the century wore on. In 1889, an irate reader wrote in to the *New York Times*, "Only last Sunday my wife had a very narrow escape from being run over by one of those reckless 'sports' at One Hundred and Thirtieth Street and Seventh Avenue." The writer was particularly put out by the behavior of a mounted policeman stationed at the corner who "seemed to be more taken up with self-admiration and his own importance than the duty to which he was detailed." "Persons of influence" who were charged with reckless driving also found that they need not deal with the same justice system as cartmen and stage drivers. Scandal ensued in 1880 when it was revealed by the *New York Times* that a Justice Wandell had opened up his court after hours in an open lot so that a prominent person arrested for reckless driving could be immediately arraigned and not have to spend the night in jail.[58]

Short of homicide, most of the consequences of fast driving could be handled quietly and often privately. Financial damages for wrecks were often paid in cash. Bonner compensated the owner of a cigar store forty dollars for damages to his wagon after a collision on Eighth Avenue.[59] Other horse owners ended up in civil court, particularly if there were injuries involved (and they had enough resources to be worth suing). In 1865, the New York Supreme Court ruled in favor of William Walters, who sued H. D. Hull for damages resulting from Hull's alleged reckless behavior during an informal "brush" on the road between Bayside and Flushing (Hull tried to squeeze his horse between Walters and a loaded wagon that Walters was trying to pass).[60]

This was, after all, the "gilded age," in which money translated into influence, legal or otherwise. But even Bonner and his fellow drivers found that their authority was no match for the streetcar companies and developers relentlessly pushing northward, creating congestion in the streets. Central Park offered a refuge for carriage drivers willing to proceed at a stately pace, but men who wanted to trot their horses at full speed were pushed farther and farther from their homes and stables. A prominent group of enthusiasts hatched a plan. In the early 1890s, they convinced the New York legislature to pass a law authorizing the creation of a "speedway" or linear trotting track conveniently located along the west side of Central Park. The Park Commission had all but agreed when information about the plan became public. With the help of the *New York Times*, petitions were circulated and outrage loudly voiced.

For a brief moment, the privileged use of public resources, long enjoyed by Bonner and his friends, was seemingly being called into question. Seven thousand people signed a petition. Their spokespeople argued that a "racetrack" in the park would benefit a small minority of trotting horse owners (they estimated 250 using Health Department records) at the expense of everyone else. The track would take play space away from children and would attract "toughs and rowdies." Both sides used the word "safety" to justify their positions.[61] Horse owners argued that they needed a space where they could safely "let out" their horses without having to worry about cross traffic or being "run over by bicycles."[62] Their opponents portrayed the track as a threat to both the morals and physical well-being of women and children.[63]

In the end, the track was not built in the park. Instead, by 1897, work had begun on the Harlem River Speedway, an enormous public works project that created a track winding along the Harlem River from 155th Street to Dyckman Avenue—a distance of two miles. Access to the speedway was limited to separate roadsters from slow-moving wagons and pedestrians.[64]

Though no one knew it, the day of the urban equine roadster was almost over. Within a generation, the children and grandchildren of Bonner and Vanderbilt would be roaring through the streets in motorcars—scaring horses and running over small children. Rich men like Bonner no longer had the incentive to learn to drive a fast horse or the impulse to spend money on such a hobby. Although speed and conspicuous consumption were common denominators between the two pastimes, automobiles required a different set of skills and elicited another kind of strategy when it came to balancing risk and utility.

THE EVOLUTION OF METHODS FOR MANAGING risks from domesticated animals in the eighteenth and nineteenth centuries offers an example of a vernacular risk culture that grew and changed with industrialization but did not become the subject of expert scientific or technical interventions. Safety around animals continued to depend first and foremost on the ability to judge and control behavior. Even today, despite enormous amounts of scientific research on animal behavior, the most influential "dog whisperers" and "horse whisperers" are not that much different from the Rareys of the nineteenth century—self-taught people with a special gift for communicating with and controlling animals. Moreover, the rest of us still learn (or fail to learn) how to deal with animals by trial and error, observation, and instruction from parents, teachers, and friends.

What has changed is the economic necessity of managing animals in public spaces. The widespread adoption of the automobile marked an important turning point in Americans' growing intolerance of being exposed to the risks of animals belonging to strangers. Since there is a limit to which animal risk can be made predictable, we have opted instead to confine domestic animals to private spaces or to require owners and handlers to exercise a higher and higher degree of control in public.

By the time Robert Bonner took to the road in 1856, public discussions about risk had already begun to find a new focus: the destructive power of machines. First the smoke-belching, mindless juggernaut of the locomotive rumbling across the landscape and then the more hidden risks of machine shops and steel mills attracted attention and debate. A striking amount of both language and habits of the vernacular risk culture was adopted into this new mechanical world. But it was gradually eclipsed by the emergence of a new kind of expert-driven approach to risk that remade some of the tools of the Enlightenment—quantification, redistribution of the costs of accident through insurance, and state-sponsored regulation—and created others almost from scratch, including tort law and the technological fix.

PART II

INDUSTRIALIZING RISK

5

Railroads, or Why Risk in a System Is Different

The scene was startlingly new: two locomotives moments after impact, still belching smoke; bodies tumbling down an embankment; distressed passengers frozen in panicked disbelief. Reputedly America's first head-on railroad collision, the violent meeting of two Portsmouth and Roanoke locomotives offered a slow-motion portent of lethal wrecks yet to come.[1] Soon after, investigators reconstructed the chain of events. Rounding a hill, a lumber train met an equally heavily loaded passenger train traveling on the same track. The collision barely damaged the locomotives, but one of the passenger carriages derailed, spilling people down the hill and crushing others. Three died immediately, and thirty others sustained serious injuries.[2]

If observers could agree about what happened, little consensus emerged about who or what was to blame. Did fault lie with the men driving the trains? If so, had they failed in their duty because of lax supervision or a lack of skill or some combination thereof? In theory, the lumber train might have waited on a siding, thus preventing the accident. Rumors spread that the driver was inexperienced—a barber's apprentice who only recently had learned to operate a locomotive. Others speculated about mechanical failure. Perhaps the lumber train's heavy load proved too much for its brakes. Possibly a rail sprang loose, catapulting the ill-fated carriage off the tracks. What about the passengers? Some jumped to safety, while others remained to be crushed in their seats. Were the ones who kept their places responsible for lacking the foresight to save their own lives? Double tracking would have also made such an accident impossible, if railroad investors had been willing to spend the money.[3]

This 1837 accident on the Portsmouth and Roanoke Railroad was reputed to be the first head-on railroad collision in American history. *S. A. Howland,* Steamboat Disasters and Railroad Accidents in the United States *(1840), Hagley Museum and Library.*

A public inquest was conducted. This court hearing, on the causes and circumstances of the passengers' deaths, produced an unsatisfying conclusion. Legally, no one could be held responsible. The editor of the *New York Spectator* responded with a burst of frustration: "Where then was the fault? Either there must have been mis-management by *somebody*, or travel on rail-roads is liable to dangers which no vigilance or caution can prevent."[4] For early railroad passengers, travel on this new conveyance offered a frightening prospect: the violence of machines, without the possibility of controlling that risk or even knowing who was ultimately responsible, let alone how to hold them accountable.

Railroads represent the first widespread example of what scholars would later call "complex socio-technological systems."[5] These ensembles of humans and machines are particularly characteristic of modern and postmodern risk societies. As the New York editor observed in 1837, accidents in systems frequently follow different patterns from accidents involving animals and tools.[6] Multiple causes, both human and technological, are often involved. Systems amplify individual misjudgments and mechanical failures, making apparent the negative consequences of relying too much on the carefulness, judgment, and experience of individuals. People who precipitate systemic accidents do not necessarily suffer the consequences, because they may be miles away.

Railroads helped to teach Americans how to manage risk in complex systems. Over time, it became apparent that predictability was the key. Replacing people with technology offered one effective means to achieve this end—if the technology could be rendered reliable. However, even if technological substitutions worked, the "human factor," as a later generation of safety engineers would call it, could not be eliminated because people could not be taken out of the system. Nineteenth-century railroads were labor-intensive enterprises that required skills, knowledge, and often backbreaking work to function. People also entered the system as passengers, bystanders, and trespassers, creating additional uncertainties. Ultimately, controlling railroad risk required both new technologies and new techniques for disciplining human actors to behave appropriately and predictably.

From the opening of the first American railroad in 1828, almost one hundred years of extensive debate and innovation proved necessary to bring railroad risk down to a generally accepted level, and far longer to reach the very low accident rate we enjoy today.[7] As late as 1912, railroad accidents remained the single most important cause of accidental death in America.[8] Lessons were learned slowly, impeded by resistance from a wide range of actors who viewed safety as a value to be weighed against cost, revenue, convenience, and personal autonomy.

For much of the nineteenth century, courts, legislatures, and the public tolerated a degree of risk that later generations would find unacceptable, fearing that putting too high a price on preventing or compensating accidents would hamper economic and industrial growth.[9] Deploying new safety technologies also proved extraordinarily complicated and plagued by uncertainties.[10]

At the same time, pressure for change mounted. Initially, it was fired by a sense of frustration and dread exemplified by the *New York Spectator* editorial, as well as by the public suspicion that railroad corporations prioritized profits over passenger's lives. By the 1880s, intense labor conflict and public concern about the social costs of industrial accidents refocused attention on railroad workers. Railroads were one of the first industries to attract extensive scrutiny and regulation by the state in the name of protecting workers and passengers. Regulators wielded a battery of modern tools, notably the collection of statistics and the use of technological fixes, each of which had first been experimented with in the eighteenth century in very different contexts. The threat of liability for personal injury also first became a significant factor in decision making about the management of risk in the context of passenger travel on railroads.

Long before anyone began to count accidents or provide specific instructions about safety, a puff of smoke and the rattle of metal on metal signaled the necessity of learning to live with a new kind of risk. In the absence of any alternative, the first generation of railroad workers, passengers, and bystanders viewed the railroads in the context of what they already knew. Experience with roads, horses, stagecoaches, steamboats, and other aspects of antebellum life shaped their behavior and their expectations. Their improvisations resulted in a kind of railroad vernacular that would linger in both the behavior of individuals and the technology itself long after it had lost its original utility.

Learning to Live with the Railroad

Abolitionist William Lloyd Garrison loved railroad travel. In 1849, he paused from condemning slavery to enthuse about the virtues of passenger trains. "Who can estimate the comfort, advantage, enjoyment of this mode of conveyance? What an incalculable amount of animal suffering is to be saved by it to the world!"[11] The railroad was clearly better than the other forms of transportation that carried him on his travels on behalf of the abolitionist movement. Over long distances, it moved significantly faster than carriages or saddle horses (not to mention the creeping pace of canal boats). Metal tracks and ballasted rail beds lifted trains above the mud and snow that often rendered roads impassable.

While railroad accidents did happen, their scale seemed minor compared to the shipwrecks and steamboat explosions. "Nothing has yet occurred on any of our railroads so fraught with horrors as the remembrance of the fate of the Lexington or the Atlantic or many a similar disaster on the western rivers," Garrison noted.[12]

This comparison with other forms of transportation fueled what later generations would call "railroad fever." In a society that viewed economic development as the key to independence and growth, the railroad looked like the path to the future. Railroad fever inspired passengers like Garrison to buy tickets. But more importantly, railroads could carry raw materials and manufactured goods across the nation's vast distances, not only creating new markets but also raising property values along the railroad routes. The promise of economic benefits opened investor's pockets and spurred the pens of state legislators to authorize the building of dozens of railroads in the 1830s and 1840s. By 1840, there were rail lines in twenty-two of the Union's twenty-six states, totaling nearly three thousand miles of track, most of it east of the Appalachians. In the next ten years, the total mileage tripled, extending outward toward the Mississippi and growing denser in New England and the Mid-Atlantic.[13] In the east, the fever began to break by the late 1840s, suppressed by familiarity, disappointment, and economic uncertainties. Farther west, it lasted late into the century, sustained by hopes of economic prosperity and a connection to the outside world.

All this enthusiasm attached to an ensemble of startlingly rudimentary and unreliable technologies. The first locomotives involved little more than boilers strapped down to a wooden platform on wheels. Builders mounted wagons or coach bodies on platforms, loosely linking them together with chains. Inevitably, cars crashed together when starting or stopping. Most early locomotives did not even have brakes. Technological improvements came gradually, sometimes with unintended consequences. When inertia proved inadequate to halt trains, railroads adopted the awkward expedient of brakemen manually applying brakes on individual cars in response to a signal from the engineer. They also replaced chains with the link and pin coupler that proved the nemesis of thousands of railway workers.[14]

By the late 1840s, when Garrison rode the railroad, it had taken on a more stable and familiar form. But mechanical failures remained routine. Boilers exploded or went dry, axles and driving gears broke, and smoke stacks spewed a steady rain of cinders and ash on passengers, cargo, and nearby fields. Wooden passenger cars rode up on each other or "telescoped" in collisions and then turned into blazing infernos lit by the spilled embers from makeshift wood stoves used to provide heat. Infrastructure that failed to keep pace with the growing size and speed of trains was also a persistent problem. As locomotives became more powerful, able to reach speeds of sixty miles per hour in the late 1840s, trains in regular service still crept along at less than twenty miles per hour, slowed by the threat of poorly constructed tracks—the cause of frequent slow-motion derailments. Short on capital, American railroads also resorted to building single-track lines to be shared by trains traveling in both directions. In doing so, they explicitly rejected the prudent British practice of double tracking to avoid head-on collisions. Instead, a complicated set of rules dictated when a train would pull over on a siding to let traffic coming in the other direction pass.[15]

Railroad employees filled the gaps. To a remarkable extent, the completion of any railroad journey depended on the skill, ingenuity, courage, and sheer hard physical labor of a variety of workers. Engine drivers nursed unpredictable equipment along unreliable roads. Firemen bent their backs to the task of shoveling tons of coal into the firebox, taking care to arrange it to burn properly. Brakemen waited for the signal to scramble up the cars to set the brakes. Sometime in the first decade, railroads also added the position of the conductor. On passenger trains, conductors collected tickets. But more importantly, the conductor was the master of the abstract tools of timetables and train rules that kept trains traveling on the same tracks from running into each other.

Little in antebellum life prepared workers for service on a train crew. Some railroad managers looked to steam packets and stagecoaches for potential employees, particularly conductors, reasoning that work on other kinds of public conveyances provided translatable skills. They also sought out mechanics with knowledge about stationary steam engines to drive trains. But this kind of prior experience proved only partly helpful. Moreover, railroad managers could not find enough out-of-work stagecoach drivers and mechanics to fill all the available positions—hence the Portsmouth and Roanoke's employment of a former barber's apprentice. Before the Civil War, most railroad workers were men in their twenties and thirties who had come of age on farms and in small towns.[16] While nominally high literacy rates prevailed in the United States, rote schooling methods did not prepare potential railroaders for the kinds of abstract reasoning that train rules and timetables required. Some men came with such physical impairments as color blindness or hearing loss that, left unrevealed or even unknown, could contribute to an accident.[17]

Into the late nineteenth century, most railroad companies also put little effort into formal training or testing. The earliest train operators were self-taught. For the rest of the century, most railroad employees continued to learn on the job, teaching themselves through experimentation or by watching one another. Out of these beginnings evolved a system of hiring, promotion, and informal apprenticeship that persisted even after railroads became huge complex bureaucracies running hundreds of trains a day. Many would-be engine drivers learned how to drive a train by taking a job as a fireman, watching the engineer at the controls, and eventually relieving him through longer and longer stretches of track.[18] As in other forms of informal apprenticeship, skills sometimes passed from father to son or through other family connections. In the antebellum South, where African American slaves often filled the backbreaking job of fireman, this informal learning process could backfire on white masters. In 1856, the *Memphis Evening News* reported that a "negro fireman employed on the Somerville Branch Railroad" took his first steps to freedom when he "stole the locomotive and taking on seven or eight other Negroes, ran away with it." The fugitives abandoned the locomotive outside the city and disappeared into the woods.[19]

Learning by doing meant that incompetence was often revealed only after accidents happened. Some learners did not survive the learning process. Many others bore the scars of a momentary miscalculation or lapse of attention for the rest of their lives. Railroad work also wore out bodies in more subtle ways. Even the prestigious work of driving a locomotive required physical strength, stamina, and the ability to endure long hours in an unheated open cab.

Early American railroads relied primarily on timetables to guide workers in making decisions about how to move trains through the system of tracks, switches, and sidings without colliding. These tools provided the first significant intervention in the improvised manner in which the earliest railroads were run. Timetables made the larger systemic operation of the railroad legible to individuals embedded within it. A conductor could "see" where other trains were supposed to be in the system. If he relied on his senses to convey this kind of information, it would usually be too late to avoid collisions.

Since mechanical failures often caused delays and the company reserved the right to add additional trains to serve demand, written or oral train orders and "train rules" supplemented these tools. Train rules dictated which trains should have priority and what crews should do when they were uncertain about where other traffic was in their part of the system. Ignorance or neglect of train rules likely contributed to the wreck described at the beginning of this chapter. Making this operational method work required not only discipline but also the ability to figure out what often-arcane rules actually meant. Engine drivers and conductors also quickly became acquainted with one of the most durable and essential ways that the vernacular was built into the seemingly mechanistic operations of railroads. They learned that some rules mattered more than others and that they often could not stay on schedule if they followed the rules exactly.

Reliance on individual judgment even more strongly marked antebellum railroads' approaches to the public's safety. Left to their own devices, passengers and bystanders fell back on prior experience, reasoning by analogy, and common sense. Not surprisingly, all these strategies proved unreliable.[20] Pedestrians' habit of using railroad tracks as pathways persisted into the end of the nineteenth century, resulting in hundreds of casualties each year. In an age when walking was the dominant mode of transportation, even respectable, middle-class people routinely traveled long distances on foot. More than one person met an untimely end because the tracks seemed like a better thoroughfare than muddy or snow-bound roads. Valentine Gay, "a respectable citizen of Lyman, Maine," died traveling along the path of the Lowell Railroad. Gay was walking "on the left hand track and seeing a merchandising train coming down stepped on the other track, not perceiving that a passenger train was immediately behind him."[21]

Henry David Thoreau, perhaps the most famous observer of early American railroads, often followed the railroad "causeway" when traveling between Walden Pond and Concord Center.[22] Thoreau knew the risks of pedestrians using such a thoroughfare. But he trusted in his own ability to sense impending threats. In *Walden*, the locomotive's warning to "get off the tracks" is repeatedly used as both a literal description and a metaphor to describe why individuals needed to pay attention to what was coming up behind them.[23] Railroad tracks also proved enticing to pedestrians because of the way they fit into the landscape. In many places, rail beds lay nearly flush with the surrounding fields and woods. The rail line Thoreau walked along still survives. It is startling to see how much it melts into the woods and how closely it skirts the pond itself (although a fence now discourages pedestrians from emulating his walk into Concord). Intersections with existing roads formed a new kind of crossroads ("grade crossing") often buried in the surrounding countryside.[24]

Most railway companies made little effort to keep pedestrians off the tracks because they and the courts considered this a matter of individual responsibility.[25] On his way to visit the Lowell textile mills, Charles Dickens rode the same line where poor Valentine Gay had been hit five years earlier. He expressed shock at the lack of protections for pedestrians and other kinds of traffic at crossroads compared to English railways. Where the train crossed the turnpike "there is no gate, no policeman, no signal," he told his readers, "nothing but a rough wooden arch, on which is painted 'WHEN THE BELL RINGS, LOOK OUT FOR THE LOCOMO-

TIVE.'"[26] This warning was more protection than could be found at most rural crossings and testament to Massachusetts legislators' efforts to emulate the British model. By the mid-1830s, a number of states required bells but did not mandate fences or signs. Railroads did install cowcatchers on their own initiative because cows (unlike human beings) were massive enough to derail early trains and because, for reasons detailed below, collisions with cows were more likely to produce lawsuits than collisions with human beings.

Previous experience with other forms of transportation and a belief in the power of individual judgment also shaped behavior among the railroads' first generation of passengers. Like packet boat or stagecoach travelers, they made little distinction between work space and passenger space when choosing a place to ride. Young men found it particularly exciting to perch on the locomotive's platform or to ride on the steps of the cars. Passengers swung on and off of moving trains. In the era before rail travel, passengers on all kinds of common carriers necessarily took an active role in ensuring that they would get where they were going. Stagecoach passengers expected to get out and walk and sometimes even push when the grade was too steep or the road too deep for the horses. They also sometimes jumped clear of the coaches to avoid disastrous crashes caused by runaways.

These behaviors carried over into rail travel. They made a certain amount of sense when trains were small and slow. Early railroad passengers, particularly able-bodied men, sometimes helped to put derailed trains back on the tracks. More anxious passengers kept an eye out for signs of trouble, preparing to jump for their lives. Into the 1850s the sensibleness of this course of action was reified by law. The courts still allowed passengers who were injured while jumping out of trains that they thought were in imminent danger of wrecking to sue the railroad for damages, even if the anticipated wreck did not actually occur. The law specified that it was enough to show that the act had been prompted by some kind of action (e.g., one of those slow-motion derailments) that the railroad could have prevented.[27]

However, it turned out that being a railroad passenger required a different combination of active and passive behavior than earlier modes of transportation. Then, as now, people probably learned how to behave on trains by watching other more experienced passengers or by soliciting advice. Snippets of evidence survive to suggest the content of those conversations. In mid-nineteenth-century America, more and more people relied on advice literature to know how to negotiate unfamiliar situations. Riding the train was no exception. Dionysius Lardner's *Railway Economy* included a section on "Plain Rules for Railway

Travelers." Lardner's first rule was "don't try to descend while the train is moving." Other admonitions included "never sit in an unusual place" and "beware of yielding to the sudden impulse to spring from the carriage to recover your hat which has blown off or a parcel dropped."[28]

Railroad employees also provided information to passengers, principally in the form of verbal warnings. The law and many companies' policies required conductors and station crews to inform passengers when they were endangering themselves by sticking arms and legs out of windows or lingering near the tracks in the presence of a moving train. In early personal injury cases, lawyers for injured passengers could respond to the railroads' claims that their clients had contributed to their own injuries by evoking the principle of "failure to warn." But there is plenty of evidence that many railroad conductors and station managers treated railroad risk as casually as some of their passengers. Railroad employees put an overflow of passengers in freight cars or the caboose or left them standing on the steps of cars. They asked male passengers to disembark from moving trains to spare the engineer from coming to a full stop. They gave the go-ahead even when the location of an oncoming train was uncertain. As a consequence, accidents happened.[29]

As long as they could, railroad managers held onto an approach emphasizing rules and discipline to control workers and personal intervention to fend off lawsuits when disaster did strike.[30] The approach was cheap to implement and comfortably suited the values of people who had come of age in a world without large machines moving through space faster than a horse could run. While these companies remained small, local, and with relatively light traffic, the strategy was tenable. In the 1840s and 1850s, public outrage over increasingly dramatic accidents led to a different approach.

The Collision Crisis, Regulation, and the Technological Fix

Hundreds of people were likely killed or mutilated in the first decades of American railroading. But they were struck down in ones and twos—fitting a preexisting pattern in which fire, runaway horses, and exposure took a steady but unremarkable toll on human life. While dramatic collisions like that of the Roanoke and Portsmouth attracted some concern, the overall response was resignation: accidents manifested an infant industry's growing pains. Existing institutions such as the courts and legislatures were assumed to be adequate to the task of holding railroads and their employees responsible and redressing victims' losses. Whatever risk attached to railroad travel was worthwhile given the dangers and discomforts of other forms of transportation.

This initial period of risk acceptance was gradually eclipsed by changes in the number and character of railroad accidents. In 1853, a series of highly publicized wrecks, remembered by railroad historians as "the collision crisis," brought a definitive shift in public opinion.[31] In April, a freight train running behind schedule slammed into a passenger train at Grand Crossing, Michigan, leaving twenty-one people dead. Two weeks later, an express train shot through an open drawbridge at Norwalk, Connecticut, killing forty-six passengers and injuring scores of others. August brought more tragedy. A heavily loaded excursion train traveling toward Providence, Rhode Island, raced down a single line of track in an effort to beat the regular train to a switch. The conductor thought he had a four-minute buffer. Instead, it turned out his watch was running slow and the regular train was four minutes ahead of schedule. The two trains collided, resulting in fourteen casualties.[32]

In each of these cases, railroad workers could, in some sense, have been held responsible. At Grand Crossing, the freight train's driver had misunderstood the rules about which train had precedence. In Norwalk, the driver ignored a signal showing that the bridge was open. The Norwalk conductor also failed to keep his watch set to the correct time. However, a wide range of commentators laid blame squarely at the feet of corporations that overworked their employees, required them to take chances, and resisted spending money on safety devices. As Americans became more familiar with the new technology, risk management based solely on managerial discipline and individual responsibility seemed to make less and less sense. The *American Railroad Journal* declared, "Give us a double track, an electric telegraph accessible at each station, signals that can be seen and understood, and we will venture the risk of accidents."[33] In a series of more inflammatory editorials, the *National Era* decried the "inhuman monopolies" that committed "railroad murders" through negligence and greed.[34]

It is likely that passenger train wrecks never killed or injured as many people as individual incidents involving workers crushed between cars, passengers jumping on and off trains, and trespassers struck at grade crossings and while walking on the tracks.[35] But in the aftermath of the 1853 collision crisis, passenger train wrecks captured the imagination of legislators, journalists, and the reading public (not to mention judges and juries) in a way that other accidents did not. Gory journalistic accounts fed a distinctive kind of fear related to putting an individual's personal safety in the hands of unseen and unknowable persons.

No amount of carefulness or attention to duty on the part of passengers would make any difference if a faceless worker fell asleep at a switch or failed to set his watch to the correct time. As people abandoned the notion that they could control the outcome of wrecks by anticipating them and jumping clear or exhorting the driver or picking a safer seat, as the earliest railroad passengers had done, fear of passenger train wrecks shifted into a mental and cultural category of involuntary risk, prompting a higher level of what risk psychologists call "dread."[36]

The collision crisis represented a decisive shift in public discourse and collective risk perception. A consensus formed: railroad travel was unnecessarily dangerous—not because of its ordinary day-to-day risks, but because of the possibility of catastrophic wrecks. On the other hand, if burgeoning ridership provided any indication, few people willingly relinquished the conveniences of train travel. Instead, journalists, legislators, and influential members of the middle class pushed for greater safety. As common carriers, railroads already had a legal obligation to protect passengers. However, this duty was enforced primarily by the opportunity for individuals to sue for compensation after accidents happened.[37] Growing public concern about wrecks and other kinds of accidents led to a gradual redefinition of railroad safety as part of the common good and therefore a matter for state intervention through regulation.[38] As we have already seen, such interventions have a long history, predating industrialization. The problem of railroad risk propagated a new model that would become increasingly important in the twentieth century: using the police power of the state exercised through a specialized agency. These "bureaus" or "departments" or "boards" served the important function of imposing the methods and mentalities of a modern risk culture, particularly the collection of data about accidents and mandates for new, risk-controlling technologies on private corporations.

Nineteenth-century understandings of federalism dictated that state and local governments (rather than federal agencies) should provide primary oversight, despite the increasingly networked, long-distance nature of the American railroad system. Although state railroad commissions existed before the Civil War, the establishment of the Massachusetts Railroad Commission in 1869 created a model for state-sponsored investigation and regulation that endured well into the twentieth century.[39] State legislatures moved much more slowly to protect those who received a pay packet from the railroads. Accepting the idea that workers knew the risks of their jobs and had willingly assumed them, legislators and railroad commissions dabbled in only a few experiments—the Massachusetts Commission, for example, briefly required wooden guards to warn brakemen on the tops of cars of upcoming low tunnels in the early 1870s.[40] Passenger safety received most of the attention until the mid-1880s, when a growing labor movement made its influence felt. The 1893 passage of a federal Safety Appliance Act signaled a decisive political shift in both worker protection and the regulatory role of the federal government (although it did not necessarily carry the same meaning in the rail yards as in the halls of Congress).[41]

Safety advocates recommended, and the laws sometimes required, a wide variety of measures, ranging from uniforms for passenger train crews (in order that they could be identified in an emergency) to better bridges. But in the four decades following the collision crisis, they put particular stock in the adoption of new technologies, most importantly train control systems that used the telegraph to prevent collisions; new car designs that lessened damage from impacts; automatic couplers that eliminated the need for workers to stand between cars when attaching them together; and the air brake, which allowed the engineer to stop an entire train with the turn of a lever, eliminating the need for brakemen riding the tops of cars.

Technological fixes made sense as a means of making railroads as systems more predictable. For legislators, they also offered a tangible type of reform, the implementation of which could easily be assessed. Technological remedies also suited a culture infatuated with invention as a panacea for all kinds of problems. Beginning in the late 1850s, the number of patents granted for safety devices increased exponentially, testifying to the hopes of inventors seeking to commodify public anxiety about railroad risk. By 1875, the federal government had registered more than nine hundred patents for a single device: the safety coupler—a device designed to prevent railroad workers from being crushed while assembling and disassembling trains. Twelve years later, that number increased to 6,500.[42] But the actual implementation of these ideas, particularly the systems technologies of air brakes and automatic couplers that needed to be installed on all cars to be effective, presented enormous technological and economic challenges.

A number of scholars have described the arduous process required to turn ideas into actual railroad appliances that would function consistently under the strain of use. They have also analyzed the reasons why railroad companies sometimes resisted efforts to compel adoption, especially with regard to freight trains.[43] Although individual railroads bore a strong resemblance to each other, they differed in many essential details such as the way cars were linked together. Car design and building were crafts in which individuals held strong ideas about the best ways to do things. In the post–Civil War era, lack of standardization posed a growing problem as once-isolated railroad companies became increasingly networked. This was particularly true in the freight industry, where a system called "car interchange" meant that trains could be composed of cars from many different companies. To be effective, systemic technologies such as air brakes and couplers needed to be installed on all the cars in a train—an enormous cost when multiplied out over an entire system. Licensing fees to inventors further added to costs.[44] On the other hand, the existence of these devices created pressure inside and outside of railroad companies for their adoption.[45]

The railroad managers who ushered in standardization and other technological changes increasingly shared with railroad commissioners and other reformers an outlook that favored science-based engineering and systematic testing and analysis. But many working inside the industry also straddled the world of vernacular practice and the increasingly abstract world of mechanical engineering. Master car builders, the men who decided which devices should be used on railroad cars, queried engine drivers and rail workers, hoping to learn from their experiences and ad hoc solutions to specific problems.[46] They also looked for numbers to measure performance, risk, and cost. Like others who in an earlier era might have been called craftsmen or mechanics, they sought to professionalize—gathering together in professional organizations and learning to speak the language of scientific engineering. At the same time, some of the most influential railroad reformers had no practical experience riding the rails except as passengers. This lack of hands-on knowledge and skill did not stop them from prescribing and sometimes mandating specific technologies. Charles Francis Adams, who played a crucial role in inventing railroad regulation through the influential Massachusetts Railroad Commission in the 1870s, was among the first of a new type. Like many of the high-profile safety reformers who would follow him in the early and mid-twentieth century, he trained as a lawyer. His knowledge of railroads and opinions about what should be done to improve them came from reading reports, gathering data, and talking to railroad managers.[47]

On the rails and in the yards, workers and managers viewed new safety technologies in light of a more complicated reality. Alabama trainman and local union official J. J. Thomas first encountered a freight train equipped with air brakes during the Brotherhood of Locomotive Engineers' 1887 Chicago convention. Delegates were invited to watch a demonstration staged by the Westinghouse Company. Rather than signal brakemen standing on the tops of individual cars to apply brakes, the engineer could control braking for an entire train from the cab. In a memoir published nearly twenty years later, Thomas questioned the device's necessity and real purpose. While advocates saw the air brake as a means to save the lives of thousands of men who otherwise would have been employed to ride the tops of cars, Thomas suspected that it represented one more means through which railroads could extract more work from their employees. As a good union man, he felt compelled to tell his readers that "since air has been introduced, the engine crew has been held responsible for everything pertaining to train service except registering and seeing that the markers are displayed, the stove kept hot, and the boxes all cushioned in the cabooses." Recalling the brake demonstration set him to romanticizing the railroads of his youth where crews "working in harmony" used hand brakes to stop at every little station on the Selma, Rome, and Dalton road where he began his career. Good judgment, he implied, was a better guarantee of safety than a mechanical device.[48]

Thomas's ambivalence about the new safety device was far from unique among railroad workers. The campaign for safety device laws did not arise from the grassroots or find strong support among the rank and file. While the various railroad brotherhoods spoke out in favor of safety devices, it took a great deal of politicking to elicit their endorsement of the 1893 Railroad Safety Appliance Act, through which the federal government finally stepped in to require air brakes on trains traveling between states.[49] Like Thomas, many workers feared that so-called safety devices would be used by management either to create more work or to eliminate jobs. Their fears were not entirely unfounded. Brakemen, in particular, came to regret the introduction of air brakes. Railroads that installed the new technology found that they could make do with far fewer brakemen on each train.[50]

As it turns out, the adoption of air brakes also did not initially lower the overall collision rate as expected. Looking at their own internal statistics, railroad insiders concluded that the number of collisions was, in fact, increasing in the 1890s, becoming the fastest-growing category of railroad accident.[51] Between 1896 and 1897, worker casualties from collisions doubled and the number of passengers killed tripled.[52] This upswing in accidents seems to have been caused by two phenomena that commonly undermined the safety gains of new risk-mitigating technologies. The widespread adoption of air brakes coincided with a significant increase in railroad traffic. As a consequence, railroads used the improved stopping distances of air brakes as a justification for running trains closer together in the system. In a classic risk trade-off, they used these devices to increase productivity rather than to make the system safer.[53] The sense of security provided by air brakes also seems to have encouraged engine drivers to run at higher speeds, further contributing to the rising accident rate.[54]

Safety advocates again looked for a technological solution. They hoped that "block signaling," the use of strategically positioned stoplights to keep trains from simultaneously entering the same stretch of track from opposite directions, would provide a technological means to eliminate human errors leading to collisions. In theory, the stop signal tripped by one train entering a section of track would prevent the next train from entering until a safe distance between them had been established. But many railroads put off implementing any form of block signaling because of the expense involved. Where the technique was adopted, observers found that engineers and signalmen sometimes treated the stop signal as a warning rather than a command—a practice that was tolerated because of the perceived need to move heavy traffic through the system efficiently.[55]

Confronted by the limits of technological fixes, railroad managers gave renewed attention to what was sometimes called "the human element." In the late 1860s, the railroad brotherhoods had embarked on a campaign of self-policing, ejecting known drunkards, thieves, and the grossly incompetent in order to earn the confidence of the railroad corporations. They promised to deliver "responsible, reliable men," and, to a certain extent, they did.[56] A few railroad companies also introduced piecemeal improvements. In the late 1880s, a growing number of railroad companies, confronted by rising accident rates, returned to this enduring theme with a sense of urgency. They argued that inadequate *discipline* lurked at the center of the accident problem.

The Culture of Trainmen and the Problem of Discipline

Finding more effective ways to control the human element in accidents was no simple matter. It required not only identifying and codifying a set of behaviors that would improve safety but also renegotiating a long-standing bargain between employees and employers and challenging essential elements of trainmen's work culture. Built on a complicated mix of expediency, economics, and nineteenth-century ideas about learning, masculinity, and personal responsibility, this arrangement proved extremely difficult to remake. The exact nature of the bargain depended on the job, but it was epitomized and ultimately problematized in the role of the engineer. Engineers habitually used their own judgment in walking a line between following the rules and getting trains through the system on time. Supervisors also stood ready to forgive minor mistakes if the engineer generally succeeded and had a good reputation and a long work history. They also let engineers prepare their replacements in whatever way they saw fit, perpetuating a vernacular culture with no formal qualifications, screening, or system of training. It was taken for granted that being a white male was a necessary prequalification.[57]

"Duty has been my watchword," early railroad engineer J. J. Thomas explained. He is pictured here with his wife, Rose, on their wedding day. *J. J. Thomas,* Fifty Years on the Rail *(1912), Hagley Museum and Library.*

J. J. Thomas, the Alabama trainman who so distrusted air brakes, exemplified the kind of railroad man who prospered under this system. In the conclusion of his memoir, he pronounced his career a great success. "Duty has been my watchword," he explained. *Duty* is a capacious word, but clearly one of its meanings for Thomas was careful attention to rules and responsibilities and hence to safety. In the same paragraph Thomas defended his safety record by writing, "I was called into the Superintendent's office once for running a railroad crossing, but explained it satisfactorily," implying that this had been the extent of rule breaking and risk taking in his career.[58]

The litany of accidents in Thomas's memoir suggests a more complicated and ambiguous story. He experienced his first serious train wreck about a year after he began working as an engineer, around 1860. In his memoir, he uses the passive voice to explain what had happened: "My engine was run off the track, and I was hurt so badly as to disable me for work for some time."[59] His next job lasted until a bridge collapsed under his locomotive. He walked away from the accident but heard through the employee grapevine that a railroad superintendent intended to have him conscripted into the Confederate army for "putting that engine in the creek." Soon thereafter, he took preemptive action, abandoning a locomotive on a siding and boarding a passenger train for Meridian, Mississippi, where he obtained an engine driving job with another railroad company.[60] After the war, he survived additional accidents that left him with a mashed hand and severe burns from the steam escaping a wrecked locomotive.[61] The accident he regretted most was running over his wife's friend as she crossed the tracks after a visit to the Thomas's house, lethally wounding her.[62]

With the exception of the bridge collapse, which he blamed on his supervisor, Thomas's account is vague about the causes of his accidents. Nowhere does he directly admit responsibility. Nor does he blame his employers or other workers. He gives the impression that he saw accidents as an inevitable part of the job. He drew no broader set of conclusions or safety prescriptions from them, except perhaps that inhaling steam is extremely painful and to be avoided even if it meant jumping from a moving train.

Thomas also took pride in certain displays of masculine bravado widespread in the work culture of engineers. Indisputably, it took a great deal of courage and skill to send a steam locomotive hurtling through the dark at sixty or more miles an hour, as passenger train drivers routinely did. Men who were paralyzed with fear posed a threat to themselves and others because they could not be relied on to make split-second decisions. But engineers also took it upon themselves to push beyond the risk taking required by their employers to demonstrate their abilities, to make up lost time, and for a bit of excitement to punctuate an often-tedious job.

Risk taking as a demonstration of skill and bravado and a way of having fun was epitomized in the practice of engineers racing their trains side by side on stretches of doubled track. In his memoir, Thomas expressed disappointment that the sale of the railroad he worked for put an end to informal train races, "which at times," he said, "were quite exciting."[63] Displays of masculine bravado also constituted a part of engineers' leisure culture. Many of these men drank heavily and spent the hours between runs buying each other beers in saloons. Since engineers were often on call to work on short notice, this practice sometimes resulted in episodes of drunk driving.[64] Thomas did not drink. Instead, he regularly got into fistfights—sometimes to defend his honor and sometimes to defend his reputation as a skillful fighter. From the perspective of our own times, it is hard to imagine why Thomas would brag about taking chances or behaving badly, seemingly contradicting his claims to attention to duty. But these themes are ubiquitous in accounts by and about nineteenth- and early twentieth-century railroad engineers, suggesting that risk taking was an accepted part of railroad culture.

Thomas's pride in his own bravado also illustrates one of the central contradictions of nineteenth-century American vernacular risk culture. White, native-born men were widely seen as the only members of society who could be entrusted with the safety of the public and with the control over powerful, potentially destructive machinery. But they were also allowed and encouraged to engage in risk-taking behavior that would have been considered evidence of irresponsibility and moral weakness if performed by a woman, a recent immigrant, or an African American. This risk taking was considered indicative of courage and independence—two personal qualities that, along with intelligence, qualified a man to be responsible for the safety of others.

These men defended their monopoly on driving trains by claiming that women and minorities lacked the innate qualities necessary to do the job. Well beyond the nineteenth century, male railroad workers did not have to worry about women challenging their competence and competing for their jobs. Labor shortages and railroads' constant efforts to economize on labor costs did mean they had to be concerned about African Americans. In the antebellum South, engineers like Thomas had worked side by side with black firemen who performed the hard work of heaving coal or wood into the engines, picking up the knowledge needed to drive in the process. In some Southern states, legislators worried enough about these workers to pass laws preventing them from working as engine drivers. In the aftermath of the war, blacks sometimes competed directly with white workers for jobs. Whites responded by arguing vociferously that whiteness was a marker for competence and safety, sometimes in the face of obvious evidence to the contrary. For instance, white workers on the Houston and Texas petitioned the railroad's president against the employment of blacks in one of the railroad's switching yards. They suggested that switching work "should not be placed in the hands of ignorant Negroes when sober, industrious, reliable and experienced white men may be employed." Mr. Lovett, the president, replied that the statistics showed that "the cost of repairing cars in the Houston yards was only one fifth as much per car as it was in the other yards of the company," presumably because black workers were much more careful with the company's property. There were also fewer employee injuries in the Houston yard.[65]

This kind of direct challenge to white male workers' claims was very rare. In general, the culture of railroad men and railroad management worked against close scrutiny. Codes of masculine behavior proscribed self-recrimination and compelled men to cover for each other or to at least keep silent about misdeeds. Railroad rule books required engine drivers and conductors to check each other's work, paying particular attention to rules and protocols. They were supposed to admonish each other for failing to carry out work properly and, as a last resort, report infractions to a supervisor. In practice, however, they were often unwilling to risk making an enemy of a coworker or getting a reputation as disloyal to a "brother" railroad man.[66] R. L. Caincross, a conductor on the Gulf, Colorado, and Santa Fe, claimed that when workers talked about accidents in lodge halls and brotherhood magazines, it was almost always in the vein of extending "sympathy to those at fault." Few thought it appropriate to publicly admit any responsibility for accidents or to discuss what could be learned from mistakes.[67]

Ultimately, responsibility for enforcement fell to local supervisors. Some commentators in the railroad press felt that management practices and managers' attitudes contributed significantly to failures of discipline. The problem began with hiring. Railroad managers generally preferred internally promoting men who were a known quantity. But this practice created its own opportunities for favoritism and nepotism.[68] Moreover, in the last decades of the century, railroad workers became increasingly mobile, pushed from job to job by an unstable economy and pulled by potential economic opportunities.[69] Faced with hiring men about whom they knew little or nothing, managers often fell back on their own ability to make subjective judgments about native intelligence and whether a potential employee was telling the truth about his previous work history. In this "generation of boomers," workers could easily move, erasing unfortunate work histories in the process. Until the 1890s, most railroads did not keep detailed records of individual employment histories. Supervisors needed to remember whether an employee had done the same thing before and to make judgments about whether he was a "good employee" who had just made a mistake or a "bad" employee who was insubordinate, careless, or worse. In the highly mobile world of railroad work, memories were often short and reputations made on good first impressions and the ability to cover one's tracks.[70]

The long-standing practices that had worked well when railroads were small became increasingly problematic as rail networks and managerial bureaucracies grew. Some observers believed that smaller railroads exercised the most effective discipline because managers spent more time out on the road and less time doing paperwork.[71] On big bureaucratized roads like the Pennsylvania, competent local supervisors often did not stay long enough in one job to get to know their employees or to gain a good understanding of particular local problems.[72] These organizations also placed heavy emphasis on budgets. Faced with tight payroll allocations, they hired younger, inexperienced workers who would labor for less than seasoned men.[73]

Supervisors were also not immune to the hidden codes of masculinity that shaped worker behavior. It is somewhat surprising to discover that, given how much power employers had in this period, they still went out of their way not to offend trainmen's pride. In doling out punishments and corrections, many supervisors dreaded what the *Railway Gazette* described as "narrowness and touchiness which makes men restive under instruction." "Giving elementary instruction to a man 30 or 40 years old is a delicate business," the author told his readers, probably unnecessarily. He thought that engineers were most likely to overreact because, as he put it, "they are the men in whom the superiority of their natural over their acquired qualification is most marked (in other words, they thought they had a God-given talent for driving trains)."[74] In an era of intense labor conflict, managers also knew that ignoring worker expectations about how discipline would be carried out might set off larger battles. As the railroad brotherhoods grew in influence, the accusation also circulated that some managers avoided disciplining or terminating dangerous or inefficient men because they feared antagonizing these organizations.[75]

The bargain establishing railroad workers' independence had always been open to gradual renegotiation, mostly unilaterally from the corporations' side. In the decades after the Civil War, many railroad companies made piecemeal efforts to improve the existing system of rules and punishments. Accidents prompted superintendents to rewrite existing rules and add new ones. Railroad disasters stimulated a broader conversation about standardizing rules, eventually resulting in the 1887 publication of the first set of "universal" rules by the American Railway Association. But better rule books did not necessarily make for safer railroads.

"It is one thing to give an order and another thing to see that it is enforced," H. S. Haines rightly pointed out in 1893.[76] Railroad managers found it notoriously difficult to know whether trainmen were obeying rules until an accident revealed the breach in discipline. Techniques of supervision and surveillance used in other industrial workplaces did not transfer to the mobile workplace of a train or the narrow confines of a locomotive cab. In the late 1880s, as railroads began to focus on discipline, some hired private spies or "spotters" who traveled incognito on passenger trains to catch conductors pocketing fares. They also hung around saloons, hoping to surreptitiously photograph on-call engineers with their hands wrapped around a shot and a beer.[77] Other companies experimented with employing inspectors who periodically rode along in the cab checking to see that speed limits and signals were duly observed. So-called decoy signs were also planted along the right of way so that managers could test whether engineers were paying attention. Such measures undoubtedly revealed carelessness. Workers also bitterly resented them as a violation of trust implicit in the unspoken bargain that railroads would not enforce rules to the letter as long as trains ran on time and without major accidents. One Pennsylvania Railroad manager told his superiors that he avoided such techniques because they had been "unfavorably commented upon by the employees," who thought spying "descended to meanness in the way of Kodak pictures entering and leaving saloons, &c."[78]

Railroads also began trying to impose some oversight on the process of hiring, retention, and promotion by testing employees for mental and physical competence. In the late 1870s, the Illinois Central pioneered compulsory eye exams for trainmen, including tests for color blindness. In the next decade, many other lines followed suit.[79] Increasing numbers of railroads also instituted written tests, particularly for employees seeking promotion. In the 1890s, the Pennsylvania required all firemen to take the promotion exam even if they had no intention of moving up as a means of forcing out the incompetent and unmotivated.

None of these measures addressed criticisms that the railroads had no formal mechanism for inculcating good judgment. The existing system largely relied on the employee to make the connection between the act and the punishment and to change their behavior accordingly. Wise managers tried to correct misconceptions about risk as part of the disciplining process, but managerial inconsistency and trainmen's belief that they knew better constituted formidable barriers to learning. Many others believed that punishment should be an end in itself—that erring employees should, in one way or another, pay the railroad back for the cost of their mistakes. The point was controversial enough that the Pennsylvania Railroad's Association of Transportation Officers began their discussion of the discipline problem by assenting to the General Supervisors' proposition that "the object of discipline is to prevent a recurrence of the thing for which discipline is applied," that it is for "corrective and not punitive purposes."[80] This was a new idea.

Inculcating Good Judgment

Reformers placed their greatest hopes in the adoption of what was variously called the "Brown System," "discipline by record," or "discipline without suspension." As the names suggest, the essence of the system was the accumulation of demerit points or "brownies" (probably the origin of the colloquial expression "brownie points"). Rather than suspend an employee for each infraction and allow him to return to work with a clean slate, points built up as part of a written record. After a given number of points had been accumulated, the employee was called in for a conference and either dismissed or otherwise punished. Some railroads also gave credits or even cash bonuses for meritorious behavior. Managers reserved the right to summarily terminate employees for drunkenness, insubordination, and other gross misconduct.[81]

The Brown system was first introduced by George R. Brown on the Fall Brook Railway in 1883. For nearly a decade, it was viewed within the industry as the idiosyncratic innovation of a small New England railroad. In 1894, it was put into effect on the Indianapolis Division of the Pittsburgh, Cincinnati, Chicago & Saint Louis Railway, setting off a chain of adoptions. Three years later, thirty-two railroads were operating under some form of the system; within a decade, the majority of major roads had some version of it.

Proponents imagined that the system would be useful in every railroad department, but railroads typically applied it first in train operations because they were seen as the nexus of the discipline problem. Managers targeted conductors and particularly engine drivers because these individuals most often made decisions that led directly to accidents (particularly collisions) and because they were difficult to supervise in more traditional ways. Managers were attracted to the system because it seemed to solve many problems associated with traditional discipline. Aware of demerits building up in their files, employees would make a greater effort to avoid repeating the same behaviors. Files were supposedly open at all times for employee inspection to avoid the impression that supervisors dished out demerits on the basis of personal enmities. Some railroads also adopted the practice of "bulletining" examples of behaviors that had led to accidents. Anticipating a technique favored by safety-first advocates, they wrote up didactic accounts, carefully removing any identifying features that would embarrass individual employees. These were circulated in the form of newsletters or posted in station houses. All these efforts were directed toward cajoling employees into internalizing discipline and adopting managers' ways of thinking about risk.[82]

Railroads also adopted the Brown system for reasons that had nothing to do with preventing accidents. The practice of suspending employees forced managers to find replacements. When labor markets were tight, supervisors were tempted to ignore breeches of discipline rather than lose an essential employee for a week or a month.[83] Advocates also responded to contemporary discussions about the causes of poverty by arguing that this system was more humane than fining or firing workers because it did not impoverish innocent women and children who were dependent on a male breadwinner.

Written records also functioned as a way of disciplining local managers. The cumulative nature of the Brown system forced them to create employment records for workers and to justify punishments in writing. Before a particular railroad introduced the system, a circular was sent out to managers and employees explaining it. Some roads also issued very elaborate notices to operating officers "outlining what was required of them, and describing the system . . . specifying in great detail what system of debits and credits would be used and the number of credit marks to be allowed for perfect service for given periods of time."[84] Directives and accountability in writing also made punishments more consistent from district to district and employee to employee. In the 1890s, many railroads were very concerned about employee perceptions about fairness because seemingly arbitrary punishments were a major source of small-scale labor conflict. Independent of the Brown system, some roads instituted a process of appeal that gave higher-level managers more control over their local subordinates.

Some contemporaries remained unconvinced that the Brown system improved discipline and prevented accidents. One of the most scathing critiques came from a former railroad signalman, F. O. Fagan. In 1908, he published an inflammatory little book, *Confessions of a Railroad Signalman*. The author claimed to be a longtime railroad employee who felt compelled to tell the public why so many railroad accidents occurred and what could be done about it. Like many others, Fagan argued that the American people could not afford to wait for technological fixes. They needed to have better discipline on railroads. Fagan thought that the Brown system was not the answer. He argued that its main purpose was to "promote good feeling between the men and the management."[85] The "Brown system has abolished publicity [of wrongdoing] and done away with pecuniary loss," he wrote. "The employee is now aware that no one can touch his pocketbook, no one can wound his pride, or hold him up as an example to his fellows."[86] "Harmony [between management and employees] is the altar upon which the interests of the traveling public are continually being sacrificed."[87]

Fagan advocated the use of safety committees through which men in the operating departments could dedicate themselves to the problem of safety and point out safety issues to management. He had apparently suggested the establishment of a "safety league" but had been told that upper-level management would veto the idea because "they have always frowned upon any such democratic relationship between men and management, such as a Safety League would initiate."[88] The idea did not exactly originate with Fagan. By 1908, the industrial safety movement had already accepted this practice. Some of the largest steel and chemical manufacturers had begun encouraging their employees to meet together to talk about safety and to point out problems in factories to their employers. Like the Brown system, these meetings helped workers internalize their bosses' ways of thinking about risk, or at least that was the theory.

AS THE RAILROAD INDUSTRY MATURED, the fundamental nature of railroad risk remained the same: human error and technological failure undermined predictability and led to accidents. In the 1910s, increasing numbers of railroads began to implement more broad-based efforts to improve railroad safety involving internal safety committees and even safety departments intended to both identify problems through the eyes of those who actually operated trains and change the culture of railroading to bring it more in line with the methods and expectations of an emerging safety movement in industry. Both railroad managers and regulators also continued to focus on improving the reliability of technology and finding both supplements and substitutes for human judgment.[89]

In some ways, efforts to manage railroad risk helped pave the way for the early twentieth-century safety movement. Railroads also provided a template, not always consciously acknowledged or applied, for a host of other transportation systems, including highways and air traffic. On the other hand, the peculiar character of railroads as systems initially kept railway safety experts from seeing how their practices might translate into other contexts. Those insights came instead from the less systemic environment of factories.

6

The Professionalization of Safety

In the last decades of the nineteenth century, a small group of men and women created a new profession: the safety expert. This was the era of professionalization, a time when practitioners of older occupations, such as medicine and law, first created professional organizations, and new occupations, such as social work, claimed that their specialized methods offered better solutions than common sense.[1] The professionalization of safety encompassed both these patterns. This new group of professionals defined safety (or, more precisely, the management of risk) as a distinct body of knowledge, separate from the actual ability to drive a train or ride a horse. By doing so, they ruptured the vernacular culture's linkage of safety and skill. To equip themselves, safety experts gathered together the loose conglomeration of risk-managing techniques developed since the eighteenth century. Their tool kit included quantification, safety devices, regulation, and training. Professional organizations became an important means through which they shared information with each other and defined professional norms.

By the 1910s, bigger ambitions colored safety experts' profession building. They hoped to create a social movement that would convince the public to put "safety first." As part of this effort, these men and women claimed ultimate authority in defining unacceptable risk and deciding what should be done about it. That claim proved easier to sell to legislators and managers than people who actually worked in factories or bought the first automobiles.

A growing sense of crisis around the problem of industrial labor provided the context for the professionalization of safety. The first safety professionals were government employees who inspected industrial workplaces, including mines, railroads, and factories; collected accident statistics; and recommended corrective measures to employers and legislators.[2] Of these three kinds of specialists, factory inspectors and their privately employed counterparts played the most important role. They carried the message of safety first out of industrial workplaces into streets, schools, playgrounds, and that stronghold of the vernacular—the private home.

Recognizing and Quantifying the Problem

The push for factory safety originated with working people's organizations in the years after the Civil War. By the 1880s, it was dominated by men and women connected in various ways to the Knights of Labor, the most powerful labor organization of the era. Members of the labor movement and their allies successfully fought for state intervention in industrial workplaces. Legislatures created bureaus of labor statistics and factory inspection to collect data, enforce regulations, and recommend changes to machinery, buildings, and work practices.[3] They drafted new laws requiring employers to supply information to the state about workplace wages, hours, and conditions (including accidents). Other statutes placed restrictions on the use of child labor and set requirements for machine guarding.

Rapid, often-chaotic industrialization, resulting in enormous collateral damage—tens of thousands of broken bodies and families impoverished by the loss of breadwinners to accidents—provided the motivation for reform. As the cataclysm of the Civil War drew to a close, thousands of soldiers limped home to the factories, workshops, mines, and other industrial places of employment. In the next half century, they were joined by waves of new recruits: men, women, and children, who left farms, burgs, shtetls, and villages for life of wage labor in New York, Cleveland, Buffalo, Chicago, and hundreds of other cities and towns across the nation. The term "factory" hardly begins to describe the diversity of places where they found work. In tiny tenement workshops, women and children not only made clothing but also picked nuts and assembled artificial flowers for ladies' hats. Small factories, many with only a handful of employees, turned out coffins and horse brushes, laundered clothing, and cut metal. On the other end of the scale, the first large corporations employed thousands of workers to feed the nation's growing demand for iron and steel, petroleum, and meat.[4]

The risks of industrial labor were not unknown in 1865. Railroads, of course, were already notorious, and textile mill workers, canal diggers, and journeymen in antebellum machine shops had scores of stories to explain their missing body parts. The post–Civil War change was in scale and degree. More and more workplaces replaced tools with increasingly powerful machinery. New chemicals—a hallmark of this industrial revolution—ate away at workers' bodies and sapped their vitality. The paternalism and sharing of risk that characterized antebellum iron plantations and gunpowder factories disappeared as proprietorship was replaced by incorporation. Constant novelty and an influx of workers who had not grown up around machinery and who were not fluent in English rendered the older ways of learning about the risks of particular trades increasingly useless. Greed, ignorance, and desperation fueled this already volatile mixture.[5]

Long after the dangers of railroads began to attract public attention and reform efforts, most Americans remained unconcerned about industrial risk. Farmers, legislators, and members of the growing urban middle class did not share the dangers of molten steel or unguarded machinery in the same way that they shared the risks of train wrecks. Nor did they have any opportunity to witness directly the events that crippled and killed industrial workers.[6] Especially in the aftermath of a war, it was easy to impute a different cause to empty sleeves and wooden legs. Even the well-intentioned found it difficult to disaggregate the social costs of industrial accidents, particularly cascading effects on children, spouses, and communities, from other causes of poverty.

Inspired by the British Parliament's passage of a series of Factory and Workshop Acts, a few antebellum state legislatures had conducted half-hearted investigations into the problem of child labor. Little resulted from their labors in terms of either information or legislation.[7] A century and a half after Cotton Mather and others had carefully counted Boston's smallpox victims and quantified the risks of inoculation, no one knew how many victims industrial employment had claimed. Nor were many objective facts available to evaluate an increasingly vocal labor movement's claims about exploitative wages, hours, and conditions. After the war, legislators became much less reticent.

In response to a growing clamor from labor leaders and reformers that government officials do something about the "labor problem" and other ancillary social effects of rapid industrialization, an increasing number of states formed a distinctively American entity, bureaus of labor statistics, to gather information. Massachusetts was first, establishing a bureau in 1869. Over the next quarter century, thirty-two states created their own bureaus of labor statistics or labor bureaus (incorporating both statistical collection and enforcement). State-sponsored bureaus of labor statistics provided the most important means for generating what contemporaries simply called "facts." The term *statistics* in the name of these agencies says much. In an era increasingly infatuated with "social physics," cold, hard numbers provided the seemingly objective proof legislators needed to agree that America had a problem. Linked to carefully chosen case studies that gave those numbers a human face, sociological data helped propel the passage of more and more factory legislation.[8]

Bureaus of labor statistics belonged to the first wave of modern regulatory activity that also included public health departments and railroad commissions. For a brief period in the early 1870s, Charles Francis Adams, then chair of the Massachusetts Railroad Commission, and Carroll Wright, the first Massachusetts labor statistics chief (and later legendary head of the United States Bureau of Labor Statistics), might have crossed paths in the State House on their way to presenting compilations of data to lawmakers.[9] But Wright's mission was somewhat different from that of Adams. Risk was an important element in the investigations of both organizations, but because Wright's subject was a social group rather than a specific technology, possible lines of investigation were much broader and more amorphous. Especially in the earliest years, labor statisticians did studies of everything from accidents to wages in an effort to get an analytical understanding of why working-class people suffered from so much misery and frustration.[10]

Railroad commissions and labor statistics bureaus also targeted somewhat different audiences. Adams explicitly endeavored to inflame public opinion so that the public would pressure railroads to change their operating procedures. In the 1860s and 1870s, labor leaders held similar hopes for labor statistics. However, by the 1880s, most state commissioners tried to avoid the appearance of collecting data just to prove the labor movement's claims about low wages and terrible conditions. Public officials feared that more strikes and riots would result if they seemed to intentionally provide quantitative provocation for working-class revolt. Wright took on the public role as spokesperson for statisticians as objective experts whose "scientific" findings might do better than common knowledge or partisan opinion.[11] In his estimation, the facts he and his fellow commissioners collected were first and foremost for the use of legislators and other government officials; if they also proved useful to labor leaders, so be it.

Representative Martin Foran, a prominent Knight and former president of the Cooper's International Union, echoed Wright when he stood before his congressional colleagues in 1884 to argue for a Federal Bureau of Labor Statistics. Foran explained that "a very considerable portion of the imperfect and vicious legislation of which the people justly complain may be traced to a lack of reliable data and accurate knowledge upon the part of the average lawmaker." Statesmen required social science "based on statistics which collects, collates, arranges and compares facts," without which they "cannot ascertain those great principles in accordance with which the state must act if it would promote and foster the well-being and happiness of its citizens."[12]

The work of labor statisticians had the desired effect, supporting the passage of laws regulating the conditions of labor. Legislators initially focused on women and children because courts considered them to be "dependent classes," unable to freely contract the terms of their labor in the same way as grown men.[13] These efforts were also motivated by a growing spirit of humanitarianism and the labor movement's fear that children's low wages would undermine the labor market.[14] Beginning with the most industrialized states—Massachusetts, New York, New Jersey—lawmakers laid down a patchwork of statutes requiring employers to provide sanitary facilities, guard machinery, and forego employing very young children.

It rapidly became clear to reformers that many employers had no intention of voluntarily complying at the risk of losing competitive advantage (if they knew anything about the laws at all, which many apparently did not). Moreover, laws did not enforce themselves. European examples suggested that compliance rested on factory inspectors armed with expertise and police power. A rough template emerged. Labor statisticians identified and documented problems. Legislatures responded by passing laws to remedy those ills (within the constraints of constitutionality and convention). It then fell to factory inspectors to enforce statutes.[15]

Actual practice proved messier and full of greater variation. Not every state had both statisticians and inspectors. Most depended on inspectors to gather information as well as to enforce laws—placing them in an inherently conflicted position and stretching their base of expertise. Although European practices offered possible models, most of the men and women who worked for the new bureaus in their earliest years had little detailed knowledge of those practices. The profession of American factory inspector and the new kind of expertise it entailed emerged out of a mix of borrowing, inventing, and tailoring to the realities of local politics. Henry Dorn, a German immigrant machinist who served as Ohio's first chief factory inspector, both exemplified and played a central role in that process.

Inventing the American Factory Inspector: Henry Dorn

Ohio, promised land of antebellum western migration, became a different kind of frontier after the war. Among the first of the Midwestern states to industrialize, by 1880 it ranked fifth after New York, Massachusetts, Pennsylvania, and Illinois in the value of goods being produced.[16] Like many other immigrants, Henry Dorn came to Cleveland in the 1870s seeking new opportunities. This raw, young city was the last stop on a long odyssey that carried the native of Frankfurt am Main through the emerging industrial landscape of Europe and America. Born into the working class, Dorn's career began as a fourteen-year-old with an apprenticeship in a machine shop. Determined to be a designer as well as a maker of mechanical devices, he took night school classes in drafting and other "scientific" subjects. "In this manner," one of his colleagues later explained, "he became a master of mechanical engineering." Four years in Paris working for the Northern Railroad Company apparently did not satisfy his ambition or his wanderlust. He crossed the Atlantic in 1869, gaining employment in one of the meccas of America's emerging engineering culture, William Seller's Philadelphia machine shops. At some point afterward, Dorn tried without success to make the leap from employee to capitalist by establishing a cigar factory, before settling back into the working class and the machinist's trade. In 1881, his days cutting metal ended when he was partially (though not permanently) paralyzed by a shop floor accident at the H. P. Wire Nail Co. in Cleveland.[17]

Like many other highly skilled tradesmen of his era, Dorn actively participated in the labor movement, serving as president of the Cleveland Machinists' and Blacksmiths' Union. He joined the Knights of Labor in 1880 and was soon elected as an officer in various Trades and Labor assemblies. An unsuccessful run for the office of state senator and membership in no less than fifty-five organizations, including the Odd Fellows and the Masons, rounded out his resume.[18]

Dorn's prominence in these organizations probably figured in the governor's decision to appoint him as Ohio's first inspector of factories in 1884. But from the standpoint of experience, Dorn was an excellent choice.[19] He personified the kind of "practical mechanic" many legislatures thought appropriate for the job. He also proved himself an effective, if sometimes impatient, visionary and organizer. He played an important role in inventing an American version of the European factory inspector and propagated that vision by organizing the National (later International) Association of Factory Inspectors. Finally, he was the prototype of what would later be called a "safety engineer"—someone who specialized in redesigning existing technology to make it safer and creating specialized devices (in Dorn's case, fire escapes) with the sole function of controlling risk.

Dorn took office at a moment when the Knights of Labor stood at the peak of their influence, with an estimated seven hundred thousand members nationwide.[20] Moreover, in Ohio, the Democratic Party was eagerly cultivating the "workingman's vote" that had brought it back to power after a thirty-year hiatus. Flush with good intentions, newly elected legislators not only voted to establish a state office of factory inspection but also endowed it with powers far exceeding those in most other states (where inspectors were limited to regulating the employment of women and children).[21] The inspector had the legal right to "visit all factories and shops where ten or more persons are employed." If he (and later, she) did not approve of conditions, they could require employers to make specific changes in sanitary facilities, including toilets and urinals, lighting and ventilation, means of exit "in case of fire or other disaster," and "all belting, shafting gearing elevators, drums and machinery of every kind and description."[22]

On paper, authorizing factory inspectors to enter private property and to force employers to invest money in physical changes to the plant and productive machinery represented a significant expansion of the power of the state. Along with laws mandating air brakes and automatic couplers on trains, it marked the beginning of a long process in which government agencies would become increasingly empowered to require safety devices on various kinds of everyday technology. However, typical of state-sponsored reform in this era, the mandate was poorly funded and rested on a shaky political foundation. Like other pioneering inspectors, Dorn struggled against an overwhelming mountain of impediments, mostly ensuing from the contradictions and hypocrisies of gilded age politics. He held a political appointment, dependent on the good will of the governor and shifting tides of partisan politics for his job. As he gradually discovered, politicians were also more interested in the grand gesture of passing laws than the harder problem of finding money to fund their enforcement.

Zealous and initially optimistic, Dorn did everything he could to overcome these obstacles, beginning with trying to establish the legal extent of his powers. Although he was an engineer, he understood that his ultimate authority came from the law. He also recognized that words in the statute books could be interpreted in a variety of ways that might help or hinder his mission. In their daily rounds, inspectors trod on disputed ground. Dorn had perhaps also gotten wind of a troubling weakness in the system—prosecutors who simply refused to bring violators to court. Dorn tried, ultimately unsuccessfully, to get Ohio attorney general Lawrence James on record by asking him for a definitive statement about "what powers I can exercise under the law creating the office." He also queried James about his legal options in specific scenarios, for instance, what he could do if a proprietor refused entrance to a factory and what was the exact process through which violations of the law should be prosecuted.[23]

Initially, the legislature provided Dorn with little more than a title and a small budget for expenses. He made up for a lack of resources with extraordinarily hard work. During his first year in office, he ran the inspectorate out of a room in his house, keeping his own records and correspondence while visiting nearly 487 establishments.[24] By 1888, the legislature had found the money to give Dorn an office in the statehouse and three assistants. He stocked the office with a reference library and opened the door to the "working class of people" who, he said, visited frequently.[25]

Still, the pace of industrialization far outran the growth of the inspectorate. Day by day, Dorn and his tiny band of assistants trudged through the streets of Cleveland, Cincinnati, and a host of smaller towns giving what attention they could. In 1885, Dorn spent a few days visiting businesses along the C&P railroad in Cleveland. He found the Forest City Oil Works and Woodland Oil Company both in "good condition." In contrast, the Edson Type Case Company was ordered to install a fire escape, and the M. B. Sanders Company was directed to tear down a shed used as a water closet by its ten employees and to replace it with a "vault" that could be cleaned. He discovered that other buildings along the route needed banisters on stairs, guards around belting, and better ventilation.[26]

Dorn's mandate was so broad as to be potentially overwhelming. He seems to have concentrated on changes that would improve the safety and comfort of workers without a great deal of expense or inconvenience to employers. Later generations of safety experts would depend more and more on statistics to guide their judgments about what constituted an unacceptable risk. Dorn largely relied on his own experiences, observations, and a list of pet fixes. He understood that employers might be resistant to the costs of guarding machines and so repeatedly emphasized that ultimately attention to safety would save them money by preventing work stoppages, labor conflicts, and lawsuits.

Like other early inspectors, he was most effective in bringing an end to the very worst practices rather than raising the general standard.[27] He made a point of targeting factories governed not just by common, if dangerous, practices, but by the improvisation and ignorance of proprietors with no prior experience of industrial production. In the headlong rush to go into business, would-be entrepreneurs grabbed any space available and then added on to it piecemeal. For his fellow inspectors, Dorn laid out a few scenarios: men who "'grew up in the country' seizing the opportunity to start in business for themselves" set up "a little business in some little office or the corner of a loft." He fails to notice the absence of banisters on stairs and guardrails on balconies and walkways until someone falls to their death. Another hypothetical employer is "wrapped up in the selling end of the business," leaving management to his foreman, "who is of a careless disposition" and neglects to guard the freight elevator. "One day there is an accident. . . . If there is a suit for damages he fights it, and labor, in a way, has made an enemy."[28]

Engineering, in the form of analysis, redesign of existing technology, and invention of new safety devices played a central role in Dorn's approach to factory safety. Few workshops that he visited actually had extensive productive machinery. As a consequence, Dorn spent a great deal of effort thinking about how to make the buildings themselves safer, primarily by educating employers about simple devices that ensuing generations would take for granted as minimal even in the most dismal workplaces. Elevators proved a magnet for Dorn's attention. Builders installed them and employers used them with startling casualness. Many were little more than boxes suspended by a rope or chain from an engine (or hauled up by hand with a pulley system). Most lacked a mechanism to stop the free fall of the compartment if the rope broke. Doors or fences to prevent the unsuspecting from stepping into an open shaft were an afterthought. Dorn had to plead with the legislature to pass laws requiring features such as safety gates.[29]

The possibility of being trapped in a fire was perhaps the riskiest part of working in many of the fourth- and fifth-floor workshops Dorn inspected. During his first year in office, sixteen people died in a dramatic fire in W. Dreman & Co.'s Cincinnati rag warehouse and rope and twine factory. Only the day before, the proprietor used a legal loophole successfully to fend off a lawsuit involving victims of a similar fire two years earlier. As in the more famous Triangle Shirtwaist Fire twenty years later, horrified spectators watched helplessly as a number of young women jumped to their death from the top floor. Dorn used this example to argue for a law requiring fire escapes. This device constituted one of his favorite technological remedies. During his time in office, he used his engineering knowledge to design a series of elaborate fire escape systems.[30]

Technological fixes suited Dorn's talents and interests, but they also made sense given the constraints under which he and his fellow inspectors operated. Later safety experts spent extensive amounts of time on the shop floor trying to alter workers' risky behavior, but Dorn and his lieutenants had only a brief opportunity to change the conditions in the factories they visited. Guards on and around machines, on stairwells, and on elevators could function as their mechanical surrogates—a constant, silent reminder about the dangers of ironing machines and saw blades.

In his earliest years in office, Dorn created his profession mostly through improvisation and his own memories and reading about European factory inspection systems. But as the idea of creating bureaus of labor statistics and factory inspection caught on in a growing number of states, he realized that much could be gained in both prestige and knowledge by joining together with others in a professional organization.

Establishing Professional Networks

In the fall of 1885, Dorn somehow found time to begin a correspondence with the four other chief inspectors in the United States: Rufus Wade from Massachusetts, C. T. Fell from New Jersey, James Connolly from New York, and Henry Siebers from Wisconsin. He suggested that they and their staffs should meet to share ideas and discuss their experiences, with an eye to achieving greater uniformity in both labor laws and inspection practices. According to Wade, convincing these pioneer factory inspectors to talk to each other took "considerable time." Because the practice of American factory inspection had grown in a fragmentary way, "each inspector entertained views peculiar to himself on the subject, and these conflicting ideas had to be harmonized." Dorn also picked the site, the "city of brotherly love," where he had first landed in America. A dint of further letter writing convinced Philadelphia's mayor, Edwin Fitler, to let the sixteen participants (chief inspectors and their lieutenants) use the Common Council Chamber in City Hall the following June for a meeting of the soon to be grandly titled International Association of Factory Inspectors.[31]

In late nineteenth-century America, conventions and the written proceedings that resulted from them played an extraordinarily important role not only in forging collective professional identities but also in creating pathways for the exchange of information. Newly minted factory inspectors were more likely to accept information propagated by professional organizations as authoritative in part because it came from colleagues who were also professional friends. The men who attended the association's meetings told each other what it meant to be a factory inspector and shared ideas about how to do the job. As new members joined, they watched and listened and were, in turn, acculturated. These meetings and the resulting publications helped create something new—a national and international community of people who thought of themselves as professional safety experts.

New York's James Connolly must have been impressed by Dorn's presentations. After the Philadelphia meeting, he set off on a campaign to force New York employers to install fire escapes of Dorn's design. Connolly's report to the legislature featured long passages from Dorn's report at the annual meeting, as well as his carefully rendered drawings.[32] Connolly and Dorn also cooperated to investigate a New York orphanage that was shipping its charges to Ohio to work as underage minions in the glass factories.[33]

A year later, Ohio had a new Chief Factory Inspector. No evidence survives to tell us whether Dorn quit in disgust or was ousted, but the common pattern in this period, decried by commentators, was for governors to replace inspectors when they needed to reward a political crony with a government job. If individual factory inspectors did not often survive long in their positions, the existence of laws, agencies, and professional organizations could not be undone so easily. The profession continued to evolve and expand independently of people like Henry Dorn.

Settlement House Sociology and Industrial Risk: Florence Kelley

In the last decade of the nineteenth century, a growing number of social investigators, muckraking journalists, and reform-minded politicians, known collectively as progressives, discovered the problem of workplace risk. With their entrance, the influence of working-class people on the shape of the safety movement began to wane. Named as Illinois's first chief factory inspector in 1893, Florence Kelley represents a transitional figure bridging earlier working-class-centered activism about industrial risk and the entry of those who worked to redefine the methods and objectives of the movement after 1900. Her appointment to the position by progressive governor John Altgeld marked a departure from the pattern in other states of either giving the job to members of the labor movement or making patronage appointments.

Unlike Henry Dorn, Florence Kelley could make no claim to being a "practical mechanic." Instead, she exemplified the social science movement's ambitions to combine social research (sociology) with hands-on activism to create a "science of reform" implemented by professional social workers.[34] Armed with a degree from Cornell, where she had written a thesis on child labor, she could claim some mastery in the emerging discipline of sociology, including the ability to collect and analyze data. Recognizing the growing importance of legal expertise and credentialing in both creating legislation and litigating violations, she studied for a law degree while working as chief inspector—becoming one of the increasing number of lawyers in our story as it enters the twentieth century.[35]

Kelley's academic expertise on child labor, reinforced by her gender, played an important role in qualifying her for the job of Illinois Chief Inspector. For several decades, restricting the employment of children had been an entering wedge through which state legislatures began to regulate the terms and conditions of employment. Progressive reformers, especially social workers, shared these concerns but also had a grander agenda. Shaping the lives of children offered an important way to improve society as a whole. Kelley believed that widescale industrial employment of children threatened their health, education, and morals. Controlling these risks was a way of investing in the future. In the next few decades, safety professionals increasingly challenged the assumption, built into law and custom, that parents and others acting *in loco parentis* should have the right to expose children to whatever risks they deemed acceptable. They also promoted the more radical corollary that experts, rather than parents, should decide what children should learn about how to manage everyday risk. They helped to construct the notion of a sheltered childhood in which common risks such as accidents and disease were viewed as preventable aberrations.[36]

In the 1910s, Chicago gradually became the epicenter of the safety movement, labor activism, investigation, and reform. But the city Kelley came to know was a volatile mess of unresolved tensions around industrialization. Chicago was home to not only a number of the largest and most notorious industrial employers in the country—meat packers and steel producers—but also hundreds of tiny sweatshops and middling factories; an active, if internally divided, labor movement; and some of the most radical and influential social reformers in America, including Jane Addams and Henry Demarest Lloyd. The Haymarket riots and their shameful aftermath were still a fresh memory among city residents. The newly erected lathe and plaster edifices of the World's Columbian Exposition stood in stark contrast to nearby tenements and stockyards. The city and surrounding Cook County sat like a smoky island on the northern border of a large and overwhelmingly rural state. Many rural voters and their representatives felt little obligation to involve themselves in the metropolis's problems, let alone underwrite the costs of ameliorating them.[37]

Kelley's entry into the world of factory inspection actually began in her home state of Pennsylvania. In the company of Knights of Labor member Leonora Barry, she worked the hallways of the state capitol lobbying for the passage of Pennsylvania's first factory inspection law. The bill passed into law in 1889, and Barry was appointed as one of two female inspectors.[38] Kelley moved on to Illinois, taking up residence at the Hull House settlement, which had been founded by Jane Addams and Ellen Gates Starr two years earlier.[39]

Once ensconced in Addams's inner circle, Kelley benefited from Hull House's growing importance both as a gathering place for experts on working-class problems and, along with the neighborhood surrounding it, as a kind of laboratory for gathering data and testing reform strategies.[40] U.S. Commissioner of Labor

Statistics Carroll Wright hired her to direct a survey of sweatshops in the neighborhood around the settlement house. He had met Kelley at a conference where she gave a paper based on her college thesis, "The Law and the Child." As a result, she rapidly became the person politicians and journalists sought out when they wanted information or a guide into the warren of tenement rooms and tiny workshops. In these surroundings they discovered women and children (or even whole families) laboring often in the same space in which they ate and slept.[41]

As a consequence of her experience, Kelley became a passionate advocate of regulating the employment of women and children, particularly in what were called the "sweated industries"—small workshops located in tenements and lofts where clothing, artificial flowers, and other goods were made, often in grim surroundings. She traveled down to Springfield, the state capital, in early 1893 to testify about the need for "stringent legislation," particularly of sweated labor. Soon thereafter, the legislature voted into law Illinois's first Factory and Workshop Act, which Kelley probably drafted. Initially, Governor Altgeld offered the job of Chief Factory Inspector to Kelley's friend and mentor, Henry Demarest Lloyd, elder statesman of Chicago's socialist reform community. When Lloyd declined, Altgeld presented the job to Kelley. She accepted, quickly assembling a staff composed of Hull House friends, including Alzina Stephens—a former Knight, whom she had known from her Pennsylvania sojourn.

The scope of the Illinois law differed greatly from the one within which Dorn operated. The legislation reflected the particular concerns of Kelley and her fellow Hull House reformers, as well as what the lawmakers would approve. The statute focused on female and child labor and on the sweating system's threat to the health of both workers and consumers through the transmission of infectious disease. Kelley was most proud of the clause limiting the number of hours women could work each day.[42] The law did not include any specific requirements for sanitation and the guarding of machines—omissions that Kelley came to regret.[43] The inspector's office had almost no power at all over the conditions under which adult, male laborers worked in steel mills, stockyards, and on Illinois's dense railroad system.

Looking back from our own time requires a leap of imagination to understand the opposition against Kelley's efforts to curtail child labor. Employers, parents, and children cooperated in resisting regulation. Kelley found that sometimes even school officials became complicit, ejecting "incorrigible" pupils on the premise that they would be better off working.[44] If pressed, defenders of child labor presented a number of arguments with which many nineteenth-century Americans would have agreed. Parents were entitled to their children's wages. A boy or girl's earning power could keep families out of abject poverty, particularly when one parent was absent, injured, or unemployed. Employment prepared working-class children for adulthood. Physical labor was not necessarily harmful to children; it provided a socially productive form of discipline for children who might otherwise, by dint of parentage and nature, be a burden to society. Less vehemently, employers also implied that industry should not be held responsible if child workers ended up crippled in adulthood; in a free labor system, the ability to work was a commodity to be bought, sold, and sometimes used up sooner rather than later. Sounding a familiar theme, one employer excused the lack of machine guards by telling a member of Kelley's staff that "children never get hurt till they get careless."[45]

Kelley argued back. First, she declared that children could not be expected to have the same understanding of risk as adults. After a teenage boy was killed in a can-making factory, the employer defended his safety practices and unwillingness to buy safer machinery by pointing out that the company required employees to carry a card with instructions on it while working. Kelley opined, "For middle-aged men, self-possessed and cautious, able to read these rules and ponder them, it would be a gruesome thought that the penalty of violation may be instant death." Growing boys might take a different message: "Unable to read, and all at the age when risk is enticing and the most urgent warning is often a stimulus to wayward acts, what excuse can be offered for using machinery lacking any essential of safeguard?" She thought that the company had created the cards primarily as a means for limiting its own liability if a worker was injured.[46] According to Kelley, the fault did not lie entirely with employers. Parents were also sometimes complicit in endangering their own children. Her staff reported that in one of the meat packing plants, "a boy has been found at work at a dangerous machine, *because his father had been disabled by it.*" "Keeping the place pending recovery depended upon the boy's doing the work during the father's absence," she added, condemning a practice that would have gone unquestioned if the injury had taken place on a farm and the father had been injured by an animal or piece of equipment.[47]

Kelley singled out parents whose children worked in the glass industry for scrutiny and condemnation because, for progressive reformers, the industry epitomized its most abhorrent scenario: parents who intentionally lived off their children's labor, in the process depriving them of both a childhood and a future. In Illinois, glass factories could mostly be found in Alton—a Mississippi River town, far south of Chicago. There, the Alton Glass Works employed over six hundred children, making it the single largest child employer in the state.[48] In the fall, families looking for employment floated down the Mississippi on rafts or in makeshift boats. Upon reaching Alton, they erected tents on the riverbank or railroad sidings or on the rafts themselves, which gradually froze in place. The result was a filthy shanty town, filled with the smoke of wood fires, without running water or sanitation, from which children as young as seven trudged off to work six days a week.[49] Kelley made a point of investigating and challenging glassmakers' claims that their glass boys worked primarily to support widowed mothers. She claimed that she could find only three "widows" (a word she consistently put in quotation marks) among the parents of the six hundred glass boys. Instead of keeping unfortunate women from penury, the glassmakers' willingness to violate child labor laws encouraged "undesirable people" to put their children to work, thus degrading the whole family.[50]

Inside the factories, one small fellow slowed down long enough to tell Kelley, "I am going to be 8 next summer." He and the other boys feared the anger of adult male workers, for whom they fetched and carried, too much to stop and talk. Kelley described a cluttered workspace with shards of broken glass on the floor. "Young children, with heads and hands bandaged, where they have received burns from melting glass or red-hot swinging rods," dodged in all directions to escape running into each other or the glassblowers' long pipes. Debilitating injuries and deaths inevitably resulted.[51]

In September 1893, International Factory Inspectors gathered in Chicago for their annual meeting. Kelley gave her paper on child labor but was a bystander to conversations about the more technical side of factory inspection. Dorn's intention that the conference would both share information and establish professional norms apparently swayed Kelley, who soon became an ardent advocate for mechanical safeguarding. Later that year she rebuked the legislature, "In the matter of legislative provisions for the safety of life and limb of employees, Illinois is one of the least progressive of the manufacturing states."[52] She could not even expand her mandate regarding child workers to include safety because the law "affords them no safety guards against falling down elevator shafts, burning up for want of fire escapes, being mangled in unguarded belting and shafting, or mutilated by uncovered saws and unprotected stamps."[53]

By 1895, she had become even more convinced that the statute needed to be reconceived. In her report for that year, she complained that the legislation revealed "a fundamental misconception of the scope and value of factory inspectors." Her deputies spent too much time "keeping children under fourteen years of age out of factories"—a task she thought would be better accomplished

Dangerous and unhealthy child labor in glass factories continued to preoccupy reformers after the turn of the century. Lewis Hines took this photograph in 1909. *Library of Congress.*

through the passage of a compulsory schooling law and the use of truant officers to enforce it. "More enlightened industrial communities," she argued, use factory inspection for not only the enforcement of child labor laws "but also the supervision of sanitary arrangements and safeguards of life and limb."[54]

Like Dorn, Kelley's tenure as a factory inspector was short. When Altgeld was voted out of office, his successor quickly replaced her with a new chief inspector with close ties to the glass manufacturers. Both Dorn and Kelley were extraordinarily dedicated and capable people, but they were fighting an uphill battle against both the politics of their age and the limitations of state-sponsored regulation. Optimism about how much factory inspectors could change the conditions of industrial labor waned as the corrupting influence of partisan politics and the spoils system became clear. On the other hand, the first generations of labor statisticians and factory inspectors firmly established a combination of social investigation, quantification, technical and medical expertise, and intervention as paradigmatic. They also laid the groundwork for the professionalization of safety work.

Advocates for the working class were not the sole progressives thinking about remaking industrial capitalism. As the twentieth century began, "efficiency" emerged as an important watchword for the men (and more rarely women) who created a distinctly American solution to the problem of workplace safety: private, in-house safety programs run by lawyers and engineers.

Privately Sponsored Safety Programs

The Illinois Steel Company's South Chicago works stretched like a vast fortress along two miles of Lake Michigan shoreline. Behind its walls and fences, more than six thousand men bent their backs to the business of making rails for American railroads. The Illinois mill, like other similar plants, could be read as a virtual encyclopedia of late nineteenth-century workplace risks. The muckraking journalist William Hard reported that, in 1906, forty-six men were killed while working in the plant. A dozen perished in dramatic blast furnace accidents—cooked by a torrent of hot metal. Others died more prosaically—electrocuted, suffocated while cleaning boilers, buried alive under mountains of ore, or run over by the railroad cars while traversing the yards. Although no one kept track of nonfatal casualties, Hard calculated that another 552 workers were temporarily or permanently disabled each year.[55] Steel mills provided seemingly endless ways to get hurt. Men broke arms and legs falling from catwalks and cranes. They smashed or sliced off parts of themselves in machine shops. Heat prostration and respiratory disease also took their toll.[56]

According to Hard, by the standards of the industry, the Illinois facility was not a particularly dangerous steel mill. "If it were, there would be no use in writing about it," he told his readers. "The exception proves nothing."[57] Moreover, he had probably obtained access to the plant through the aegis of a safety manager (after signing a waiver that he would not sue the company if injured while on the property).[58] The plant also received regular visits from representatives of its insurance company looking for hazards that might result in expensive claims for damaged equipment or employee lawsuits.[59] The problem, according to Hard, was that the company's idea of an acceptable level of risk included a substantial number of fatalities and permanently disabling injuries.

Making metal had always been dangerous, but certain characteristics of late nineteenth-century industry exacerbated its risks.[60] Like railroads, the steel business was characterized by huge machines, enormous workforces, and layers of management. At the very top, managers struggled with cutthroat competition and a volatile market. They worked to increase profit margins by continually investing in new technology while trying to shave labor costs. As a consequence, a largely immigrant labor force alternated between working twelve-hour days in physically exhausting conditions and months of forced rest when orders stopped coming, prompting layoffs. The mill was still the "foreman's empire," where groups of men worked under the supervision of someone who had risen from their ranks.[61] Shop floor workers and their foremen mostly made their own decisions about risk, based on common practice and individual experience, tinctured by the kind of masculine risk culture that accepted and sometimes even welcomed danger.[62]

Florence Kelley and the inspectors who followed her into the Illinois state office probably never ventured very far inside the complex on Lake Michigan. The steel industry was notorious for its various abuses of labor, but the employment of children was not one of its major sins. Consequently, given the limited scope of the Illinois factory laws, state inspectors could do little about what went on there. The story was not much different in states with stronger labor laws. In 1906, Kelley looked into the Pennsylvania factory inspector's relationship with the Pittsburgh steel mills as part of a monumental study of working-class life, the Pittsburgh Survey. Kelley pronounced the situation "disgraceful." Five inspectors covered the whole of the Pittsburgh district, which comprised 250,000 wage earners, including "seventy thousand in the steel mills, 20,000 in the mines,

50,000 on the railroads."[63] The inspector she followed through a factory paid no attention at all to dangerous machinery but "confined himself to children and their certificates." The chief inspector's annual report gave no evidence of any prosecutions.[64]

Following Kelley's analysis, fellow investigator Crystal Eastman pointed out that even if the law was consistently enforced, it was of little effectiveness in mandating specific safety devices and modifications of machinery. Each industry had its own "special problems" requiring specific technical knowledge and regular attention.[65] Any real change would have to come from within the companies themselves.

That change was already in the making. In 1901, Illinois Steel and the Carnegie-owned Pittsburgh mills became a subsidiary of U.S. Steel as part of a series of mergers, creating the largest corporation in America. This bold move on the part of J. P. Morgan and a group of investors helped change how top managers assessed the cost of workplace accidents. In the old order, accident prevention was viewed as a paternalistic measure that added to labor costs. Because U.S. Steel could now set prices within the industry, shaving pennies was much less important. Monopoly virtually ensured a profit. On the other hand, the new corporation found itself under constant investigation for practices that might violate the Sherman Anti-Trust Act. Public opinion and, more importantly, the opinion of members of Congress became a great deal more significant. U.S. Steel offered muckraking journalists and crusading reformers of various stripes a huge target.[66]

Legal changes concerning corporate liability also figured in the company's calculations. By the end of the century, juries increasingly ignored legal rules that had long protected employers from being sued by injured employees and their families. The Fellow Servant Rule and the Doctrine of Assumed Risk gradually crumbled with each additional award. Reformers also strongly advocated for state-mandated accident insurance (what would come to be called workmen's compensation insurance). They hoped it would function not only as a social safety net but also as an incentive for employers to lower their premiums by engaging in accident prevention.[67]

Not inconsequentially, the ultimate power to make decisions at U.S. Steel now rested with a corporate lawyer rather than with someone from a manufacturing background. Eldred "Judge" Gary had risen to his lofty position because of his ability to cast a dispassionate eye on all the factors in a business situation without getting bogged down in custom and precedent. Because Gary had not come into the business from the shop floor, he had no personal stake in holding on to the existing, shop-floor-based risk culture of steel mills. He also understood the liability issues around accidents. As a practicing lawyer in the 1880s, he had taken on personal injury cases involving accidents in factories and railroads. He told his biographer, Ida Tarbell, that he had represented both corporations and injured individuals in those lawsuits. Gary had also acted as general counsel to Illinois Steel in the 1890s and was therefore in a position to know about the company's early innovations in workplace safety.[68]

In 1906, Gary began actively working to standardize labor practices. To address the problem of accidents, he called a meeting of "casualty managers" from various subsidiaries. He asked them to create a central committee for safety.[69] Out of these emerged the idea of creating internal safety campaigns orchestrated by safety professionals. Gary probably adopted the idea from a program already in place at the South Chicago works of Illinois Steel—the plant that Hard had written about in such scathing terms.[70]

The casualty managers Gary brought together were mostly more accustomed to managing risk through the legal system than through safety engineering or education. As in railroads, casualty (or "General Claims," as it was called at Illinois Steel) departments existed primarily to protect steel companies from liability. At Illinois Steel, a "safety investigator" would routinely bring a photographer with him to the scene of an accident. A picture was taken of the victim to prevent later fraudulent claims.[71] As with railroads, claims agents typically offered a cash payment to the worker or his family in exchange for signing a promise not to sue. If an injured worker did take the company to court, a lawyer from the casualty department would represent the company.

On at least one occasion, the practice of documenting accidents paid off when a man calling himself "Nic Halic" sued the company for twenty-five thousand dollars for an 1892 accident in which a wheelbarrow fell off a runway, fracturing his skull. The company's lawyers provided a photograph to prove that the seemingly brain-damaged plaintiff who showed up in court was not Nic Halic. In fact, their investigators discovered he was Halic's "half idiot" cousin, Dave Haralovic, who had been recruited by the family to perpetuate a fraud because Halic was still angry about the measly two hundred dollars he had received at the time of the accident. According to witnesses, his cousin was not debilitated by industry but had been born with his disability.[72]

By 1906, some of these casualty managers spent their time not only fending off lawsuits but also, as a U.S. Steel executive later said, giving "some study to the subject of accident prevention." Gary's mandate to create a central safety committee as well as plant-based committees transformed the nature of casualty departments.[73] The U.S. Steel program included many features that would have been familiar to Dorn, Kelley, and their contemporaries—most notably technological fixes such as guards on machines, governors and shutoff switches on engines, regular inspections of boilers, and warning signs and guards at railroad crossings.[74] U.S. Steel's new safety campaign had several effects. Like the Factory Inspector's Association, it provided a means of exchanging information and professionalizing individual safety managers. For the first time, the company systematically collected and analyzed data about the causes of accidents. The corporation also pumped large sums of money into making plants safer. For instance, by replacing old furnaces with thicker-walled models, they virtually eliminated the problem of molten metal breaking out and burning workers to death. The corporation also employed safety inspectors to visit various plants. Like state-employed factory inspectors, they prepared reports with "suggestions for improvement." Their reports went, however, to the central committee rather than a legislature or a judge.[75]

The central committee also added a notable feature that, in various guises, would become characteristic of the safety movement. Plants created foremen's and workers' committees. Each met weekly and submitted reports and suggestions for fixing hazards observed while on the job.[76] This technique had the dual function of harnessing craft knowledge and experience (without giving workers authority to change anything) while also teaching them to "think safety first"—to see their workplaces in terms of risk. The problem of the "human element" was also addressed through lectures by foremen and "sermonettes" printed on pay envelopes. The use of print propaganda anticipated the safety posters and other techniques that would characterize the safety-first movement, but the rhetoric echoed the ancient themes of carefulness and duty: "Always be careful and take no risks," and "The exercise of care to prevent accidents is a duty which you owe to yourself and your fellow employees."[77]

To many observers' surprise, U.S. Steel's new safety program actually paid off.[78] While Gary's public relations people were busy packaging it as a "humanitarian" effort, the accountants discovered that the estimated $750,000 the company spent yearly on safety had yielded an estimated $1.4 million in savings on injury payments and other costs.[79] The message was not lost on other large corporations. One by one they joined the safety campaign. Some, like the DuPont Corporation, had a long tradition of paternalism and of work practices aimed at reducing risk. Others were fresh converts.

U.S. Steel, with its huge cash reserves and public relations problems, remained something of a special case in a manufacturing economy still largely made up of smaller companies—the machine tool makers and glass manufacturers with whom Dorn and Kelley had negotiated. For small businesses, investment in internal safety programs often seemed like an unnecessary luxury. That calculation was changed by the rapid adoption of workmen's compensation insurance laws, the first in Washington in 1911, spreading to forty-two states by 1920. These laws established a system in which companies paid premiums to private insurance companies or into state insurance funds. If a worker was injured on the job, he or she could collect compensation regardless of who could be held responsible. In exchange, the individual forewent the possibility of suing for damages in civil court.[80] This "compensation bargain" largely bypassed the legal process of determining negligence. In theory, the resources companies had expended paying lawyers and claims agents to fend off lawsuits by blaming workers for causing their own injuries could be shifted into safety work and premiums. At least in some states, the premiums were structured to reward employers with good safety records.[81]

This rare convergence of interests also changed the nature of government-sponsored factory inspection. For the first time, large employers began making an active effort to cooperate. In 1912, the Wisconsin Industrial Commission, a state agency, hired Charles W. Price, a safety expert from International Harvester. His job was to "sell" safety to corporations and workers.[82] By the late 1910s, the line between corporate and state-employed safety experts increasingly blurred as professional identity became nearly as important as the reporting relationship. Many of the most important figures in the emerging safety movement moved between jobs in private industry, factory inspection, and insurance companies.[83]

Birth of the National Safety Council

Among the corporate converts to thinking safety first were a few energetic members of a somewhat arcane professional organization: the Association of Iron and Steel Electrical Engineers. These men first met together in 1907, the year after Gary's directive to institutionalize safety work in U.S. Steel. Although they presumably had much to discuss regarding the application of a relatively new power source to steelmaking, founding members later claimed that safety was a central concern from the beginning—so much so that by 1912 the safety-minded among them had decided to create a separate organization. Initially called the National Industrial Safety Council, it was soon renamed the National Safety Council (NSC) to reflect the broad ambitions of its founders.

One of the early members later remembered that important inspiration had come not just from Gary's directives as head of U.S. Steel or the casualty department lawyers, but also from a higher power. The conference began with an invocation from Reverend John McDonald, a one-armed former coal miner, who preached the "Gospel of Safety" to the conventioneers.[84] McDonald told them "to save life is the noblest of all purposes." Invoking one of the profession's favorite themes, he also noted that "it conserves the best assets of the nation."[85] Although neither the organization nor its leaders made a formal connection to the social gospel movement gaining national attention at the same time, they did conspicuously adopt the movement's language. Their work was described as a form of "applied Christianity."[86] They portrayed themselves as evangelists preaching the gospel of safety wherever someone could be made to listen. Safety campaigns became "crusades." The movement even adopted a quasi-Christian symbol, the green cross, as its emblem.[87]

Inspired by McDonald's talk and the rest of the proceedings, Lew Palmer, a Princeton-educated electrical engineer employed by Jones and Laughlin Steel Company, set out to make contact with the most important figures in industry, government, and insurance interested in the safety problem. The following September, the initial group met in Milwaukee, Wisconsin, for a "safety congress." There they created a committee charged with creating a permanent organization "devoted to the promotion of safety to human life in the industries of the United States." Palmer had done his homework and, with characteristic self-confidence, gone straight to the top. The committee brought together the U.S. Commissioner of Labor, the Director of the Bureau of Mines, representatives of the Interstate Commerce Commission, the Labor Department, Red Cross, National Association of Manufacturers, and Aetna Life Insurance Company. Not surprisingly, representatives of U.S. Steel in general, and Illinois Steel in particular, were also on the list and would soon play a central role in creating a corporate-sponsored safety movement.[88] Labor unions were noticeably absent.

As a result of the meeting, the organizers decided to create what they would later call a "clearinghouse" for safety information: the National Industrial Safety Council. This association was primarily a business organization—part of what Palmer called the "American Plan"—wherein workers and management pulled together toward a common goal without interference from labor leaders. At the council's annual meetings, subsections proliferated as increasing numbers of people identified their work as being about safety. At the same time, industry trade associations—Portland cement makers, laundry owners, textile manufacturers—created their own safety divisions.[89]

Attendees of the 1912 Milwaukee safety congress were still overwhelmingly government employees. Not surprisingly, they discussed regulation as fundamental to making workplaces safer. In succeeding years, this approach, along with the people who advocated it, was marginalized in safety council meetings. Instead, a new generation of privately employed safety professionals stressed the safety techniques pioneered at Illinois Steel. They emphasized education and coercion to change the behavior of those at risk and reserved the prerogative of ordering guards on machines and other material changes for themselves. They justified focusing on changing worker behavior with the often-repeated claims that 70 percent of all accidents were preventable, but that the technological limits of safeguarding had mostly been reached.

The "American Plan," as Lew Palmer called it, along with the movement's methods, symbols, and slogans, gradually became familiar to growing numbers of Americans. It was also America's first significant original contribution to the way the rest of the industrializing world dealt with risk. Over the next few decades, British workers would be admonished to "think safety first." The green cross and suitably translated slogans would appear in places as far away as machine shops in Brazil and railroads in Japan. The position of corporate safety expert (as opposed to that European invention, the factory inspector) was replicated in steel mills and textile factories.[90]

BY THE TIME THE NATIONAL INDUSTRIAL SAFETY COUNCIL was born, industrial safety had become a burgeoning profession, giving rise not only to safety departments in industry but also to a proliferation of academic programs, textbooks, professional organizations, and specialtics. The establishment of the American Society of Safety Engineers in 1911 signaled the solidification of this new professional identity, as did the establishment of a new trade journal, *Safety Engineering*. By the mid-1920s, a growing number of colleges and universities, including Penn State, Ohio State, and the University of Pittsburgh, had added safety to their industrial engineering curriculums.[91] In safety-minded corporations, experts like the DuPont Corporation's Louis DeBlois enjoyed increased influence and prestige.

For other reformers, including Lew Palmer and other early organizers of the first Industrial Safety Congress, the cause of industrial safety was already too small. In 1913, they decided to drop "Industrial" from the name of their two-year-old organization, rechristening it the National Safety Council. The likely catalyst was Illinois Steel Company's participation in the Public Safety Commission of Chicago (discussed in the next chapter), which resulted in an invitation to George Whittle, the commission's chairman, to give a paper at the Second

Safety Congress of the National Council for Industrial Safety. Whittle outlined the commission's plans for a safety campaign in Chicago and then concluded with an impassioned plea: "Let me urge upon you the importance of broadening your efforts; let this organization become instead of the National Council for Industrial Safety, the National Council for Public Safety."[92] The next summer, the NSC also embarked on a new mission for public safety by sending a list of suggestions for safer driving out to people on their mailing list, as well as a press release explaining that more people were "killed by reckless driving than on many battlefields."[93]

Whittle's use of a statistical comparison was telling. The growing body of quantitative evidence collected by government officials and private organizations revealed a counterintuitive truth: the risks of living in an industrial society were not restricted to or even primarily rooted in riding railroads or working in factories. In heeding Whittle's call to action, the founding members of the NSC joined a movement that was already taking shape under the sponsorship of a diverse group of actors.

7

The Safety-First Movement

"Will you please tell me when and how the 'safety first' movement originated?" a puzzled *New York Times* reader wrote to the newspaper's "Queries and Answers" column in 1915.[1] That year, the slogan suddenly seemed everywhere. "Think safety first," declared posters in factories, children's books, and public places. The movement's earnest admonitions quickly became a source of jokes and allusions—the sure sign of a popular fad. Princeton University junior F. Scott Fitzgerald contributed the lyrics to a satirical musical bearing that title. A few years later, Harold Lloyd's comic masterpiece, *Safety Last*, featured the indelible image of a hapless bumbler swinging from the hands of a giant clock in defiance of common sense and the spirit of the age.[2] Lloyd and Fitzgerald might have their fun, but spreading the message of safety first was serious business for a host of engineers, lawyers, doctors, statisticians, and safety professionals. Fired with the zeal of Progressive Era reform, they aimed to decrease substantially the overall number of accidents through education and engineering.

When and how did the safety movement originate? The historically myopic expert from the *New York Times* declared that it began with the founding of the cumbersomely titled Safety First Federation of America of New York a few months earlier.[3] A more accurate answer was, from lots of places, including railroads, streetcars, insurance companies, and the new challenges of living with automobiles.

The rise and fall of the public safety movement took place at a moment when the notion of reform was itself in flux. While the movement started among social reformers, its most avid proponents came from the emergent business class.

For middle-class reformers, thinking "safety first" promoted efficiency and conservation of resources. In a happy coincidence, safety promised to protect human lives while simultaneously saving money. This orientation toward the needs of corporations helped define the methods and goals of the public safety movement. Regarding government regulation of individual and corporate risk-making behavior with suspicion, public safety advocates preferred to believe that, properly persuaded, most Americans would voluntarily choose safety first.

As historian Barbara Welke astutely points out, safety professionals saw accidents as "the product of habits inconsistent with the modern world."[4] Consequently, changing those habits ranked high on their collective agenda. They acknowledged that technological fixes (or "safeguards" in the language of the day) had a place in guarding and disciplining the public. But as it turned out, systematic research into the use of passive protection for use outside industry lay mostly in the future. Instead, safety experts set about convincing men, women, and especially children to take responsibility for their own safety, teaching them the right way (meaning the safe way) to move through their daily lives. Many of their suggestions became not only part of school curricula and traffic laws but also part of our collective common sense, embodied in habits such as looking both ways before stepping into the street.

By the mid-1920s, millions of Americans had heard the message of safety first in one form or another. Yet, the promulgation of information did not lead to a steady decline in the number of accidents. Safety experts who had committed themselves to the cause absorbed a hard lesson: just because they literally thought about safety first, meaning before everything else, the public was not necessarily convinced to do likewise.

Pioneers of Public Safety

As spectacular as railroad crashes or steel mill explosions could be, they endangered limited numbers of people. The rapid spread of electric streetcars and interurban traction lines in the late nineteenth and early twentieth centuries, however, threatened anyone who stepped out onto city streets. Streetcar lines traversed dense urban environments characterized by a chaotic mixture of horses, pedestrians, and a growing number of automobiles. The typical streetcar accident involved passengers falling as they got on and off cars. Collisions with other vehicles and with pedestrians were less frequent, but still far more common than with trains traveling on dedicated right-of-ways.

Beginning around 1908, urban traction companies conducted the first large-scale, community-based safety campaigns.[5] The initiators of street railway public safety efforts belonged to a profession already familiar to us—claims agents who labored to protect their employers from the costs of accidents and more particularly from lawsuits. The first electric-powered street railways began to replace slower horse-drawn cars in the late 1880s at precisely the moment that transformations in liability law were changing how corporations thought about risk. The introduction of a contingency fee system allowed poor people to file lawsuits without worrying about paying lawyers' fees up front. At the same time, urban juries included more working-class men who were inclined to disregard legal precedent in favor of redistributive justice.[6] As a consequence, growing numbers of victims refused to be paid off on the spot, choosing instead to take their chances with a larger settlement in court.

Paying damages cost street railway companies a stunning amount of money, estimated by some agents to be as much as 5 percent of their employer's gross receipts.[7] This outlay also constituted one of the most unpredictable costs in running such a business. Moreover, given the legal climate, the industry worried that these costs would rise. Even though claims agents estimated that 80 percent of plaintiffs eventually lost, the streetcar companies still had to pay lawyers.[8] When claimants did win, the average payout, at least in New York, was $1,542—a lot of money in 1900.[9]

At the beginning of the twentieth century, being a successful claims agent meant fending off lawsuits and limiting payouts. Agents used legal tricks, intimidation, and a variety of other means to avoid paying damages. Like steel company claims agents, streetcar agents kept a particularly sharp eye out for what the industry called "fakirs"—people who pretended to be injured to dig into the traction companies' deep pockets. As E. F. Schneider, a recent convert to the safety-first movement, confessed, "I was brought up in the old school which followed the theory that almost every claimant should be considered a fraud." By 1910, however, Schneider had joined a growing number of agents convinced that preventing accidents offered a better means for lowering costs. In retrospect, he concluded, "If we had spent half of the money preventing accidents as we did fighting claims, our accident account would not be so great."[10] The following February, Schneider reported that his efforts to promote safety on behalf of a Cleveland streetcar company cut payments for damages by more than half.[11]

As in other parts of the safety movement, the collection of statistical data guided expert interventions. Schneider and his fellow claims agents already knew a surprising amount about what kinds of accidents led to claims—not only because of lawsuits but also because they had convinced motormen and conductors to fill out accident reports.[12] These data confirmed that the majority of injuries and deaths took place when passengers were boarding or leaving streetcars.[13]

But here safety-minded claims agents faced a challenge. As with early railroads, no one received formal instruction about how to ride a streetcar. Instead, this new technology had been folded into the practices of everyday life—parents instructed children, friends told friends, and most people learned by doing or by watching riders who appeared to be more experienced. Expedience and habit prevailed over caution. People hopped on and off cars without holding on to handles; if they slipped, they sometimes ended up under the wheels. Encumbered by long skirts, female passengers needed to exercise particular care. Exhaustion, distraction, and inebriation also led to accidents. At least one Connecticut company gave its drivers special instructions about how to eject drunks so that they would not be run over.[14]

The presence of streetcars and, increasingly, automobiles rendered being a pedestrian much more dangerous. Before the safety movement introduced the concept of jaywalking and painted crosswalks at street corners, early twentieth-century pedestrians plunged across city streets wherever they liked and whenever they saw a break in the traffic.[15] Automobile and wagon drivers, eager to be on their way, scooted around streetcars discharging passengers, sometimes crushing an unlucky pedestrian between two large vehicles. Children playing in the road presented a particularly heartbreaking potential for accidents. One 1920 Detroit study of 3,051 children found that 90 percent of them spent the daylight hours outside of school playing (or, for 7% of them, working) in the streets. Nine percent spent their leisure in private yards and vacant lots; a scant 1 percent made use of public playgrounds.[16] Inevitably, some children ended up under the wheels of the streetcars not just because they were not paying attention, but also because they took risks such as catching rides on the backs of cars. One author ascribed the latter behavior to "trolleyitis," a "highly contagious" and sometimes lethal disease "almost always confined to boys."[17] Although the law held that parents who let their children do so were legally negligent, claims agents dreaded visiting broken-hearted families and feared facing them in court.[18]

Traction companies used a wide variety of methods to educate the public about the dangers of trolleys. Sometimes their efforts were concentrated in "safety weeks," which provided a model for the safety campaigns later adopted by the NSC and affiliates. These campaigns called attention to the risks already so commonplace that their dangers had become invisible to the public. For safety experts, streetcars had the advantage of being rolling billboards. "Caution bulletins," posted in cars, instructed passengers on potentially dangerous practices.[19] Larger posters on the outside reached out to pedestrians and drivers. Some companies also took out newspaper ads, particularly during week-long safety drives. A Baltimore newspaper advertisement titled "Look Out for the Little Ones" reminded parents that, although accidents were inevitable and the price of modern transportation, they could protect their children by keeping them out of the streets.[20] In Chicago and Brooklyn, teenage girls received special instruction in "proper streetcar habits" in schools, sometimes using specially built platforms that allowed them to practice boarding and alighting.[21] The Cincinnati schools cooperated with the local traction company to publish a booklet containing a list of "don'ts" for accident prevention.[22]

People run over by streetcars (and automobiles) also interested Chicago coroner Peter Hoffman, but for different reasons—his office counted the bodies. Hoffman personified the kind of public official who played an early and important role in promoting the gospel of safety first to the general public. Traditionally, coroners investigated sudden deaths, primarily to determine whether criminal activity was involved. At the turn of the century, formal training in forensic pathology began to play an important role in professionalizing coroner's offices, making them part of the physician-directed public health movement.[23] Hoffman, in contrast, had none of the medical credentials that allowed other coroners to professionalize. He was the successful product of Chicago's highly developed patronage system, an elected official who bootstrapped himself into politics from a modest start as a grocer, followed by seventeen years in the offices of the Chicago and Northwestern Railroad.[24] His skills were social rather than scientific: a flair for publicity and a knack for cultivating a wide circle of influential friends and acquaintances.

Hoffman heard the gospel of safety first from a railroad man: Mr. R. C. Richards of the Chicago and Northwestern Railroad (Hoffman's former employer). Beginning around 1912, Richards made the rounds in Chicago, sermonizing to the Association of Commerce and other groups (at least one of Hoffman's friends was impressed enough to come away believing that Richards had personally coined the term "safety first").[25] A public safety campaign must have sounded to Hoffman like an ideal way to bring positive attention to himself and the coroner's office which might have the ancillary benefit of saving lives. Not only did a campaign have the potential to please the public, but it also had found favor with the large manufacturing and transportation firms Chicago politicians aimed to cultivate. Pursuing the noble goal of safety first also elevated the coroner's office to more than a repository for dead bodies. As Hoffman explained, "The Coro-

Brooklyn schoolgirls practiced climbing onto a streetcar as part of a public safety program. *Library of Congress.*

ner's greatest service to humanity will be in the classification and tabulation of all causes of death, in order that through this recorded experience of many catastrophes, we may learn to avoid them." The safety movement, he suggested, is based on "learning from experience"—collective, statistically described experience, rather than "the experiences and misfortunes connected with our own lives or those of our neighbors and friends."[26]

There were, however, structural limitations on Hoffman's ambitions. In reality, the coroner's office did not have any statutory power to create regulations or change public behavior. Public officials in Chicago and Cook County resisted sponsoring a public safety campaign because they viewed the costs involved as an unjustifiable use of public funds. Thus, in 1913 Hoffman created a privately

Chicago coroner Peter Hoffman demonstrated his skill in forensics for a newspaper reporter in 1913. *Courtesy, Chicago Daily News Negatives Collection, Chicago Historical Society.*

funded organization to implement his vision: the Public Safety Commission of Chicago and Cook County. He also set up a permanent headquarters staffed by a secretary, a statistician, and former newspaperman C. Koerner, who Hoffman deputized to take care of publicity.[27] As his first official act, Hoffman extended invitations to a wide variety of influential Chicagoans. He later bragged that the mayor, chief of the fire department, superintendent of police, and representatives of the judiciary, the churches, and the schools were among those who agreed to lend their names to the cause. So too were representatives of the "great railway express and teaming companies; the street railways and the automobile interests and club; the Illinois Steel Company and other great manufacturing concern," as well as "the Association of Commerce and the Federation of Labor."

The organization also received a large donation from Louis N. Hammerling, president of the American Association of Foreign Language Newspapers. Hammerling saw to it that select articles by Hoffman were translated and distributed to the 690 ethnic newspapers, where they could be read by new immigrants.[28]

Initially, the Chicago commission's efforts followed the lead of the streetcar companies, focusing on traffic safety. The ever-ambitious Hoffman, however, soon expanded his purview. After noting that hundreds of Chicagoans died each year of drowning, he asked the police department to increase their patrols of the beaches and to enforce fencing laws. An upsurge in the number of tetanus cases prompted him to write letters to physicians asking them to stock anti-tetanus serum. He asked the state legislature to require manufacturers of poisonous household substances to put them in distinctively shaped and textured "porcupine" bottles so that consumers could identify poison containers in the dark.[29] Like other big-city coroners, Hoffman also noted that the percentage of homicides and suicides involving firearms was rapidly increasing. In *What You Must Know for Safety*, a 1915 book describing and promoting the Chicago campaign, he pointed to the passage of a strict gun control law in New York as a potential model for Chicago.[30]

Hoffman was elected Cook County sheriff in 1922. The Commission for Public Safety had by then long since disappeared from view, as politicians and industrialists lent their names to other causes. Crime control became Hoffman's new enthusiasm—a not-so-surprising development given Chicagoans' increasing preoccupation with machine gun–toting gangsters. The streetcar companies, too, soon lost their role as the head of the safety movement, as their industry (and its dangers) was soon eclipsed by the spread of the automobile.[31] Nevertheless, the traffic safety campaigns organized by streetcar agents and professionals like Hoffman set the stage for a broader public safety movement in the 1910s.

The National Safety Council's Public Safety Campaigns

In October 1918, hundreds of NSC conventioneers checked into the brand-new Statler Hotel in St. Louis, Missouri. Although a war was on in Europe, you would not know it from the official conference proceedings. Instead, the big news was the NSC's first public safety campaign, held in Rochester, New York, over the previous summer.[32] Between May and September, the NSC's recently organized Public Safety Division, members of the Chamber of Commerce, and public officials cooperated to turn Rochester into a laboratory for applying the industrial safety movement's methods to an entire community. Mothers' clubs, Boy

Scouts, and even the chauffeur's union engaged in safety assemblies and formed safety councils modeled after factory practices. A complaint bureau, rather than a safety manager, received reports from zealous citizens. Safety posters hung in public buildings and spaces illustrated the consequences of carelessness, and slogans summarized the heuristics of correct behavior. The campaign's organizers collected and analyzed accident statistics in the hope of identifying hazards and measuring progress in mitigation. Even the clergy got involved. On "Safety Sundays," parishioners listened to the gospel of safety preached from a dozen pulpits in this former epicenter of evangelical Christianity's "burned-over district."[33]

City-wide safety campaigns were not new in 1919; they had originated with streetcar companies a decade earlier.[34] However, the Rochester campaign dwarfed most previous efforts in scope and duration. It also inspired the NSC and its affiliates to sponsor more than fifty city-wide events in the next half decade.[35] The idea behind the campaign was to bombard the public with ideas and information about safety to reach a collective threshold of safety awareness. Organizers theorized that this strategy would lead to self-sustaining local groups and changes in behavior that would last well after officially sponsored activities ended. It also represented the ultimate expression of two core ideas of the emergent twentieth-century approach to controlling risk: that it was the job of experts to identify unacceptable levels of risk, and that it was the public's role to internalize expert-generated ideas and to exercise self-discipline in implementing the safety movement's recommendations.

The Rochester campaign was the brainchild of Robert Campbell, an attorney for the Illinois Steel Company, chair of the company's Committee on Safety, and the first president of the NSC.[36] The plan for a large-scale urban safety campaign resided in the council's files for several years while Campbell raised five thousand dollars and found a community ready and able to cooperate. Rochester contained all the necessary elements. It was an industrial city dominated by the factories of the Eastman Kodak Company, an early and eager participant in the industrial safety movement. Boosters had already infiltrated the local public schools, organizing "junior safety patrols" and a safety club. The Automobile Club of Rochester busied itself posting warning signs on dangerous roads. Police Chief Quigley fashioned himself into a zealous and systematic collector of accident statistics needed to measure the campaign's effectiveness.[37] As in other cities that conducted similar campaigns, the local chamber of commerce became central to the NSC's efforts.[38] In Rochester, the chamber provided office space for Julian Harvey, the director of the campaign, and promised to continue various projects after the official campaign ended.[39] Harvey and chamber members organized committees. These were headed by the same kinds of people that had involved themselves in traffic safety over the previous decade: the chief of police, the coroner, a claims agent from the street railway company, the president of the auto club, and the superintendent of schools.[40]

Campbell's plan incorporated all the elements of the factory safety tool kit: quantification, safety devices, regulation, and training.[41] Developed in industrial workplaces such as Illinois Steel, these techniques had proved extremely effective in lowering the accident rate. However, translation into a public setting, even a community like Rochester dominated by corporate interests, turned out to be more complicated. In the first place, workplace-based methods of coercion and discipline were inappropriate for a public campaign. Engineering techniques known as "safeguarding" seemed to offer a more reliable approach. Unlike changes in personal behavior, safeguarding did not require active, day-to-day thought or cooperation from workers or managers. Instead, technological fixes automatically prevented individuals from endangering themselves. In Rochester, however, Campbell and his allies found very little physical safeguarding possible. Most changes to buildings and roads would require substantial capital investments and take time to execute. Nor was mandating safety devices on automobiles or implementing an inspection system a part of Campbell's plan, perhaps because auto club members, who had helped to sponsor the campaign, resisted being told what kinds of automobiles or aftermarket safety devices they should buy.

Faced with these limitations, safeguarding efforts in Rochester focused primarily on systematizing the layout and use of public streets to promote traffic safety. Campbell had been in touch with William P. Eno, the era's best-known expert on traffic, and had adopted some of his ideas.[42] Eno was a wealthy automobile enthusiast who, beginning around 1903, had begun promoting uniform rules of the road for automobile drivers, as well as a set of simple tools to promote the regular flow of traffic, such as painting lines down the center of streets and at crosswalks.[43] Business progressives (many of whom were early adopters of the automobile) found Eno's ideas attractive for a number of reasons. Safety was a by-product of Eno's larger goal—making traffic flow more efficiently. A uniform set of rules rendered the behavior of other drivers more predictable. In theory at least, predictability meant that drivers could go faster because they did not have to negotiate on the fly or leave a margin of error.

In industry, safety experts viewed "enforcement," by which they meant the ability to discipline workers through pay cuts or dismissal, as another one of their important tools. But these kinds of threats were not available to safety professionals working outside the wage nexus. Instead, legal sanctions provided the primary tool for disciplining drivers. To make enforcement of new traffic laws palatable, Campbell framed the process in terms of education. Chief Quigley helped set up a "speeders court" to which drivers were summoned and warned.[44] The presiding magistrate, Arthur G. Barry, explained the law to each driver and gave them pictures of automobiles wrecked by drivers behaving in a similar manner.[45] Campaign organizers did not go so far as to even suggest using regulation to make streetcars safer (although many other cities had already done so), probably because they feared the loss of financial support from streetcar companies.[46]

Wary of confronting capitalists directly and ideologically committed to volunteerism, public safety advocates placed heavy emphasis on public relations and safety education. Council leaders often justified this tactic by claiming that 70 percent of all accidents were caused by "carelessness" that could not be remedied through additional technological safeguards.[47] Like other progressive reformers, as well as the business community, they also strongly believed in the power of information, properly packaged and delivered, to shape behavior. If the techniques of mass marketing could be used to convince tens of thousands to buy Crisco or a particular brand of cigar, safety experts saw no reason why this approach could not be employed to curb reckless driving or convince pedestrians to look both ways before crossing the street. As a premium for joining the Safety Council, NSC leaders already mailed out pamphlets, posters, and other information and gave advice about creative ways to propagate the safety message. It required only a small adjustment in NSC tactics to hire publicists, commercial artists, and professional writers to target a wider audience. A blizzard of information ensued. Campbell and his team convinced the city's four dailies, as well as weekly German and Italian language newspapers, to run nearly five hundred stories over the course of the summer.[48] They drafted twelve thousand school children to go door-to-door distributing a window placard stating "We are Helping to Make Rochester Safe."[49] The Boy Scouts handed out stickers for automobile windshields.[50] The NSC also commissioned a series of short films. Crowds attracted by free showings of a Charlie Chaplin film first had to sit through "The Price of Thoughtlessness" and "Careless America." They even involved the public through a slogan contest resulting in the less-than-pithy "Make Rochester First in Safety First."[51]

Organizers also set up over 125 public meetings, estimating that they reached eighty-five thousand men, women, and children. They imagined that the numbers would have been even greater had the campaign taken place in the wintertime, when clubs and other organizations were more active.[52] Acculturated to turning out for war rallies, parades, and other public events, Americans in this period willingly gave their time to be part of a crowd. As long as the safety movement remained both a novelty and a popular cause, curiosity and community spirit also drew people in.

NSC leaders gave data gathering and analysis a central place in their campaign. During the summer, periodic press releases used statistics to tout successes. "Big Decrease in Accidents," proclaimed an August 6 article in the *Rochester Times-Union*. The overall number of accidents causing injurics had dropped 28 percent from the year before, according to a report submitted to Chief Inspector Quigley.[53] In the aftermath of the campaign's summer offensive, the committee tallied up the numbers and tried to assess their efforts. Accidents involving pedestrians were down, but collisions between vehicles increased by 2 percent. "Statistics are not necessarily conclusive," the director of the campaign stated, unself-consciously undermining one of the basic premises of the safety movement. "Personally, I would not have felt discouraged had we been unable to show a decrease in accidents in the first six month's work."[54] He thought that the real value of the Rochester experiment was in "the knowledge and experience obtained for use in other municipalities."[55]

The disappointing statistics brought home a bitter truth to safety advocates. Most adults had little reason or motivation to participate voluntarily in safety meetings or other organized activities week after week. They weighed the obligation of being careful against their own sense of personal freedom and convenience and chose the latter. Consequently, early on safety professionals targeted children as the recipients of their efforts. Not only were young people presumably more educable than adults, but schools, clubs, and other organizations like the Boy Scouts provided captive and potentially impressionable audiences. If the goal was to remake a culture, children were the obvious place to start.

Children First

Throughout time and across cultures, adults have taught children how to navigate through a potentially dangerous world without being killed or grievously injured. In the process, they have consciously or unconsciously inculcated their culture's ideas about risk. Both direct instruction and personal example teach children which risks are worth taking and which should be considered foolhardy or valueless. As we have seen in preceding chapters, these lessons were typically imparted by parents, masters, and other people in the community as an incidental part of learning other life skills. Hence, a mother might caution her daughter to pin up her skirts while tending the hearth; a father might hand the reins over to his son with a warning about runaways and a winking acknowledgment of the pleasures of speed. Adults expected children to be careful, but more importantly, they required them to be obedient. As the title of a mid-nineteenth-century children's book explained, "Duty is Safety."[56] Indoctrination, generation after generation, reproduced (with gradual variations) not only structures of authority but also rules, habits, and expectations.

By the time the NSC leadership set its sights on reaching children, a generation of social workers had already developed the habit of questioning whether working-class parents knew how to adequately protect their children. As we have already seen, child labor laws and other child-saving measures challenged not only the venality of employers but also the right of parents to expose children to risk. The NSC's efforts went farther, targeting not only working-class children and their parents but also middle-class families.

Safety professionals' ideas about how children should learn about risk and what they should do with that information undermined a central dynamic of the risk vernacular. They subverted parental and community authority by arguing that children should learn about risk from experts and, in keeping with educational theories of the time, from their own judgment and experience. Shaping the behavior of the next generation was only part of the safety-first agenda. They also viewed children as a means to reach inside the private domain of the home to transform adult behavior. Children would, in effect, discipline parents to think safety first through homework assignments asking them to document risks inside the home and through general self-righteous hectoring.[57] Boys were of particular interest for safety reformers. As we will see, the largely male leadership of the safety movement danced around the paradox of male risk taking; they acknowledged that boys had far more accidents than girls, but they also thought that boys should provide leadership in promoting public safety.[58]

Very early on, safety-first advocates realized that schools gave them a captive, impressionable audience for their ideas. The New York–based American Museum of Safety, for example, worked with streetcar companies to sponsor an enormous safety crusade in Brooklyn as early as 1910. By 1912, the museum had begun vacation schools, organized field trips to the museum, and training museum staffers to lecture in the public schools. By December of the following year, the museum claimed to have reached 239,000 schoolchildren with the message of safety, signaled by the red and green logo of the museum emblazoned on lapel pins that many students proudly sported.[59] In creating safety education for children, members of the safety movement had bigger ambitions than just keeping kids out from under the wheels of automobiles and streetcars. A school lecturer employed by a steel company explained to the *New York Times* that he and his employer hoped to instill "the necessary habits of caution not only that they may carry the message home to wage earners there, but in order that the next generation wage earners may have 'safety first' ingrained in their character."[60]

In reaching out to children, safety educators initially tried to duplicate the techniques that worked so well in industry. Most importantly, they organized "junior safety councils" where school children could meet to share information about hazards they observed at school, in the streets, and at home. With adult guidance, boys and girls were asked to devise procedures for mitigating those risks. In keeping with the emphasis on street safety, school "safety patrols" consisting of groups of children who acted as crossing guards were one of the most common activities of the safety councils. Instructors also gave out lists of dos and don'ts and invited children to put on safety plays or sing safety songs.

The NSC took on children as a constituency very early in its history mostly because of Lew Palmer, one of its founding members. Nothing in the surviving evidence directly explains why Palmer made it his special mission to bring the gospel of safety to America's children. Indeed, Palmer's official NSC biography, as well as his obituary, paints a picture of a very unlikely child educator. Both documents begin with two facts: in 1898, he earned a degree from Princeton in electrical engineering and, while there, became a star football player—one of two diminutive men known as the "pony" ends.[61] The game at which he excelled was characterized by massed players charging into each other in tight formations like the "flying wedge." These techniques regularly resulted in crippling injuries and enough deaths that in 1905 President Theodore Roosevelt brokered rule changes opening up the game to cut down on casualties.[62] This was a surprising start for such an ardent proselytizer for safety first, made even more surprising by its prominence in his later self-descriptions. Here was a man who ardently believed in safety first but was not, as the Victorians liked to say, a milksop.

Palmer was acutely aware that if children took the message of safety first literally—making safety a priority over everything else—society would eventually come to a grinding halt (beginning with the playing of football). He wanted children to view safety as an exciting challenge rather than a boring restriction on their ability to have fun and explore the world. Therefore, he and the educators and popularizers with whom he worked crafted a complex message about risk taking. As a 1920 article in the woman's magazine *Delineator* stated, children "should be taught the very wide difference between courage and carelessness; that it is a knightly act to rush before a car or a wagon to save another's life, but it is wasteful and cheap and stupid to rush before a car without looking or to save time."[63] Moreover, the pursuit of safety could actually provide a grand adventure. How were children to know the difference between courage and carelessness, between necessary and unnecessary risk taking? Apparently, through a combination of expert instruction and their own ability to reason.[64]

Palmer's first effort to reach children came in 1913, at the height of streetcar safety campaigns. Through Palmer, the NSC contracted with Roy Rutherford Bailey, a writer of children's stories (including the ongoing adventures of Captain Ticklemouse), to write a safety book for children. No record remains of what instructions Palmer gave Rutherford, but it is clear from the text itself that he must have provided the writer with a list of risks, as well as inappropriate and appropriate ways to respond to them. Rutherford also seems to have been briefed on Palmer's idea that safety should be portrayed as an exciting adventure. The result was titled *Sure Pop and the Safety Scouts*.

Sure Pop begins with two children discussing a letter from their Uncle Jack, who is engaged in the dangerous business of subduing the natives in an unidentified South American country. Bob longs for adventure, telling Betty, "It won't be many years now before I can be a scout and an explorer myself."[65] His desire is answered sooner rather than later by the appearance of a mysterious figure who pulls another friend, Jimmy West, out from under the wheels of a streetcar. Captain Sure Pop is "tiny, but straight as a ramrod in his natty khaki uniform." Confronting the milksop problem from the start, Sure Pop explains that now that there are no more Indians left to fight, the great adventure for twentieth-century boys and girls is safety.[66]

In addition to traffic safety, Sure Pop teaches Bob and Betty a series of other lessons. Some of these revolve around the misadventures of their friend Chance Carter, who is slow to learn what Sure Pop calls "the rules of the game" about the difference between carelessness and reasoned risk taking. Chance breaks his leg while exploring a burnt-out building, requiring Bob to come to his rescue.[67] Other chapters involve paying attention to specific everyday risks that preoccupied safety experts, including fire prevention, injuries from falls, and tetanus prevention. Sure Pop lectured children about not playing with matches, picking up banana peels so no one can slip on them, pounding in rusty nails to prevent puncture wounds, and not taking chances with electricity. Nor are the familiar risks of older industrial technologies neglected. Sure Pop takes the children for a ride in a locomotive. The engineer emotionally explains how it "wrecks an en-

Better be safe than sorry.

— SURE POP

Standing behind a podium decorated with the National Safety Council's emblem, the children's book character Sure Pop leads a meeting of junior safety scouts. *Roy Rutherford Bailey,* Sure Pop and the Safety Scouts *(1915), Author's Collection.*

gineer's nerves" to worry about boys riding their bicycles down the tracks, playing on the turntable, and throwing rocks at the windows of the passenger cars.[68] They also visit a lumber mill where the chief engineer has somehow failed to learn about guarding machinery and other safety measures, as well as a steel mill where safety first has been implemented to everyone's benefit.[69]

During the course of their adventures, Sure Pop makes frequent references to the practices of the Boy Scouts. For instance, Bob and Betty are instructed to turn their safety first buttons upside down until they have done their "One Day's Boost for Safety," in imitation of the Boy Scout practice of leaving their neckties outside their shirts until completing a good deed.[70] With a little prompting from Sure Pop, Bob comes up with the idea of organizing a troop of "safety scouts." "Why not begin by organizing in patrols and then in troops, just like the Boy Scouts?" he asks rhetorically.[71] These frequent references were not just an acknowledgement of the growing importance of scouting in American culture. They also reflected another one of Lew Palmer's strategies: a year after *Sure Pop*'s publication, Palmer convinced the head of the recently founded Boy Scouts of America that scouts should have the opportunity to earn a badge in safety. It became available a year later.[72] Palmer worked to have a number of prominent safety men on the organization's Executive Committee.[73] He also wrote a guide for Scouts aspiring to a merit badge in safety. In addition to showing a written mastery of "safety first principles" as applied to home, school, streets, and railroads, candidates for the badge were required to implement at least two safety principles in their own homes. Palmer's guide suggested that the zealous Scout might "construct neat and noticeable signs all of the same color, size, and lettering," bearing the words "take no chances." He could then affix the signs wherever an accident might take place: "on the door of the medicine closet, over the stove or the gas range; where the matches are kept; over the kerosene can, and so on throughout the house."[74]

Boy Scout troops also provided the NSC with a workforce for community-based safety campaigns, playing a conspicuous role by marching in parades and taking part in safety patrols. Empowered by their leaders and their uniforms to chastise careless adults, they handed out cards to jaywalking pedestrians in Connecticut and put forty thousand safety stickers on automobiles during Cleveland's 1919 Safety Week.[75] In Rochester, Scouts massed on busy downtown street corners during the holiday shopping season, linking arms to obstruct pedestrians trying to cross against the lights.[76] Palmer and other safety professionals involved with scouting had even imagined that the training Scouts received would allow them to take responsibility and act until firefighters, police, and medical personnel could arrive at the scene of an emergency.[77]

Palmer also used his influence with the Boy Scouts to promote what he termed "safety engineering" as a career. "The work of the safety engineer is to educate people on the dangers around them, in their homes, their occupations, in their pleasures, everywhere," he wrote in the primer for the merit badge. The safety engineer should also "provide or advocate ways for fighting these dangers," as well as strive to "invent methods for eliminating the danger altogether."[78]

The safety badge and related activities remained in the Boy Scout repertoire, but Lew Palmer moved on, looking for new challenges. After the war, he took a job with an insurance company and gradually stepped back from direct involvement with teaching children about safety. Other industrial safety professionals were also less likely to approach schools directly. They still thought that early education in safety was critical to their larger mission, but they had begun to hand the job over to professional educators.

In this increasingly specialized role, E. George Payne wielded particular influence. Payne was employed as the president of Harris Teacher's College in St. Louis, Missouri, when approached about implementing a safety program in the St. Louis city schools in preparation for a city-wide safety campaign.[79] Like many early safety advocates, Payne was a self-made man, fired by personal ambition and idealism. In promoting safety first, he saw an opportunity to serve both. Therefore, he tackled the assignment with gusto, announcing a goal of reaching each of the one hundred thousand children enrolled in St. Louis public schools.[80] Payne had climbed into his current job from a variety of positions (including teaching) in the Kentucky public and private schools. He understood that teachers and administrators experienced a great deal of pressure to cover a tightly packed curriculum. In response, he developed strategies for incorporating "safety lessons" into virtually every aspect of the curriculum from mathematics to drawing. This integrative approach became his trademark contribution to safety education.

Payne's 1919 *Education in Accident Prevention* summarized his ideas. This collection of essays, or "syllabus" as Payne preferred to call it, rapidly became the most important guidebook for would-be safety educators. Some of his concrete examples came from a safety campaign in St. Louis, others from Payne's fertile imagination. In the latter category, the book suggested attending a coroner's inquest as an appropriate activity for a middle-school English class. Afterward, students could write reports to be edited by their classmates and used as the basis for giving public speeches presented to their classmates. Payne provided an example written by a seventh-grade boy, "Jack Dierberger." In his essay, Dierberger recounted listening to the coroner detail the sad story of Cletus A. Duval, a thirteen-year-old boy dragged seven blocks to his death after hitching a ride on a streetcar by hanging on to the window bars. Cletus fractured his skull and had one arm and both legs cut off. Nevertheless, the injured boy managed to cling to life for two agony-filled hours after the accident.[81] Similar lessons in both skills and safety could be learned through class projects taught successively to each grade level. Payne suggested that third graders should collect pictures showing "the soldiers of carelessness as opposed to carefulness." Students could use these to create collages showing progress toward health and happiness.[82]

The NSC adopted Payne's safety plan and used it to launch a national program that was implemented by local safety councils and school boards. Payne's newfound prominence resulted in an invitation to join the faculty of education at New York University, where he began offering a course in safety instruction in 1922.[83] The council's commitment to safety education was further strengthened by the enthusiastic intervention of Albert W. Whitney, general manager of the National Bureau of Casualty and Surety Underwriters. In 1922, Whitney introduced a resolution to create a Safety Education Section of the NSC, supported by funds from the Underwriters. Whitney remained vice president of the section for sixteen years, helping to oversee the introduction of a new NSC magazine, *Safety Education*, in 1926.[84]

In the late 1920s, a special subcommittee of the White House Committee on Child Health and Protection, headed by Albert Whitney, sent out a questionnaire to 1,862 school districts surveying the status of safety programs in schools. The compiled responses revealed some form of safety education in 86 percent of the elementary schools and 56 percent of secondary schools that responded. In addition, the subcommittee estimated that more than a hundred thousand boys and girls participated in school safety patrols, while a smaller group joined junior safety councils. Teacher training programs, including those at Columbia and Rutgers, offered courses in safety pedagogy, as did some states (in the form of summer school courses), while in other cities, local automobile clubs or safety councils provided lesson plans to schools.[85]

The overall effectiveness of this effort remained a matter of debate. The subcommittee claimed that progress could be measured by the fact that the accident rate among children was increasing more slowly than among adults.[86] E. George Payne disagreed with Whitney's optimistic assessment. In 1937, he published a bitter diatribe entitled "Contemporary Accidents and Their Non Reduction," in which he asked why school safety programs had not achieved better results in reducing the overall accident rates. Payne singled out the NSC for failing to mount an "intelligent attack upon the problem of accidents and the methods of their prevention." According to Payne, the council presented the appearance of a "professional society or a public-service agency," but in reality it was "more nearly a trade association primarily interested in saving money for its members and incidentally in saving human lives."[87]

Payne's ire was prompted in part by the publication of *Man and the Motor Car*. Authored by Whitney, it represented the NSC safety division's effort at creating a textbook for use in a growing number of driver's education courses. The problem, Payne claimed, was that Whitney (and presumably the Underwriters) was "less familiar with educational theory and practice than the teachers themselves."[88] Payne offered particularly harsh criticism of a driver's education program for high schools devised and promoted by the Underwriters. He described it as a "publicity stunt." He thought that its implementation would "likely retard the development of a program that would get at the roots of the accident situation."[89]

IN THE FIRST THREE DECADES OF THE TWENTIETH CENTURY, safety professionals gradually expanded the scope of their efforts from factories to streetcars to schoolrooms. They allied themselves in these efforts with businessmen, bureaucrats, and other members of a burgeoning middle class. Although the National Safety Council did not create this movement, it eventually came to dominate it by gathering together experts, sponsoring safety campaigns, and publicizing the results. By the mid-1920s, safety experts and their way of seeing the world had become firmly entrenched. Americans might resist their efforts, but there was no going back to a world without warning signs, accident statistics, and safety devices. However, the dream of a society volunteering to think safety first through every moment of their waking lives was just that, a dream.

A consensus had emerged about at least one issue: the automobile was rapidly becoming the most significant cause of accidental death, eclipsing factories, railroads, and streetcars. Numbers told the story. In 1936, Payne told his readers, the automobile accounted for "approximately 38,500 deaths and more than a million injuries . . . an increase from four deaths in 1907." These statistics, Payne claimed, spoke for themselves, making "our so-called civilization a mockery." More temperate observers might have put it differently. Clearly, millions of Americans had heard the gospel of safety. But the message they took away had not produced the intended effect. Put in the language of evangelical Christianity favored by safety men, the public had been educated on the nature of sin—carelessness, heedlessness, selfish disregard for the safety of others—but that did not necessarily stop the sinning. The public might be convinced about the desirability of thinking "safety first" in theory, but in practice, it often seemed an inconvenient and dull way to live one's life. In day-to-day practice, safety first was a syncretic faith, open to interpretation, manipulation, and outright heresy.

PART III

RISK IN A CONSUMER SOCIETY

8

Negotiating Automobile Risk

The automobile hurtled into American life nearly simultaneously with the gospel of safety first. Introduced as a toy for the wealthy in the last hours of the gilded age and as an alternative to the fast horses preoccupying the time and wallets of the wealthy, its transformation into transportation for the masses began with Henry Ford's 1908 introduction of the Model T. By 1919, the year the NSC staged its first urban safety campaign, an estimated 7.5 million motorized vehicles sped along the country's roads. In the next half decade, the number of cars, trucks, and buses reached almost 20 million. Not surprisingly, accidents followed. In 1920, 12,155 men, women, and children died on the roads (already ten times the fatalities of 1910). A decade later, the death toll reached 31,000, as motor vehicle accidents became an increasingly familiar part of everyday life, as well as the single most statistically significant cause of accidental death.[1]

Numbers made an impression, not only on safety professionals, but also on politicians, educators, and community leaders.[2] Automobile safety advocates pressed for sweeping changes in how Americans used the roads. They applied the safety movement's accumulated tool kit to transform unorganized public spaces into well-ordered, predictable systems. Their tools included creating and enforcing uniform traffic laws; educating, testing, and licensing drivers; and reengineering roads to make them more predictable. Experts and policy makers gradually recognized that significant numbers of accidents were, and would probably remain, unavoidable. A socio-technological system in which consumers had significant agency could never be controlled to the same extent as a rail-

This woodcut from a driver's education manual portrays the events following an automobile accident. *Albert Whitney,* Man and the Motorcar *(1936), Hagley Museum and Library.*

road or a steel mill. Acceptance of inevitability led to the safety movement's last great conceptual innovation: reengineering not to prevent accidents, but rather to lessen the severity of injuries. The enormous scale of the automobile accident problem also redefined and invigorated insurance and tort law as means for redistributing costs, binding them into a symbiotic relationship. Simultaneously the epicenter of an emergent consumer culture and a revolution in transportation, traffic safety rapidly became the twentieth century's most important locus for public deliberations and negotiations about how to protect the common good from the misguided or self-serving actions of individuals and corporations. Educators and automobile clubs, Herbert Hoover and Henry Ford all contributed to the discussion.[3]

At the same time, a robust vernacular culture took shape around the public's perception and use of motor vehicles. Because cars and trucks arrived into a society already in the midst of adopting modern, expert-generated ways of managing risk, the automobile vernacular developed as part of a dialogue. Drivers, passengers, and pedestrians selectively accepted, reinterpreted, rejected, or simply ignored expert-prescribed rules and technologies meant to keep them safe. Traffic signs and symbols, once novelties needing explanation, became naturalized. Habits and heuristics, which had first been taught in schools by safety professionals—look left, look right before stepping out into the street—came to be inculcated as a normal part of child raising. In other circumstances, a familiar pattern of trade-offs including cost, convenience, and the preservation of individual autonomy led to the rejection of safety measures. The persistent automobile safety problem also revealed the gap in risk perception between experts and the public, reinforced by the automobile's potent image as a consumer object in control of the user. Despite extensive efforts at education, many people continued to believe that skill and luck would protect them from becoming a statistic.[4] And finally, on the roads and in the offices of insurance agents and the chambers of traffic judges, clear distinctions between the two cultures seemed to disappear. Somewhere in the middle a kind of hybrid took shape.

Accidents Happen

According to police detectives, this was the sequence of events: late on Easter Sunday, twenty-one-year-old clerk Harry Oescheger and his wife drove back into Philadelphia after spending the day in Trenton, New Jersey. At the intersection of Third Street and Oregon Avenue, Oescheger's car hit a pedestrian. Oescheger then swerved, hitting a second bystander before fleeing the scene. Investigators speculated that the car had been traveling at "terrific speed" because both victims were lifted out of their "bedroom slippers" and deposited nearly twenty feet away. The first man, an African American whose name was never discovered, died on the way to the hospital. The second, a thirty-five-year-old pipe fitter named Jacob Herther, failed to regain consciousness and succumbed a week later. Before that spring, Oescheger might havc escaped with little more than a guilty conscience and a broken headlight. But at the beginning of 1926, Philadelphia's director of public safety, George W. Elliott, declared "war" on hit-and-run drivers. Two detectives tracked down Oescheger's car and matched broken glass found at the scene to its mangled headlight. They also discovered that the brakes on Oescheger's vehicle were not working. He had driven back from Trenton using only the emergency brake.[5]

What was Oescheger thinking? Obviously not about safety first. And why would he? After all, he had purchased his car as a form of speedy and convenient transportation, not as a murder weapon. If at all typical, very little in his interactions with automobiles, automobile culture, or indeed his life in general would have planted such a scenario in his imagination. What automobile salesperson in his right mind would make a point about the lethal potential of automobiles as a part of a sales pitch? Advertisements certainly did not. They focused on speed, power, styling, price, and perhaps reliability.[6] If perchance Oescheger had encountered the gospel of safety first, there was no guarantee (as the NSC had learned) that he would see the message as relevant to his own life. After all, since Oescheger was neither a factory worker nor a railroad engineer, there were probably no other circumstances in his life in which his actions would so directly endanger others.

Where else might Oescheger have learned about the risks of driving? He likely received no formal driving lessons beyond a few basic instructions from a friend or car salesman focused on operating the clutch and the throttle. Although Pennsylvania was one of eighteen states that required drivers to be licensed, the legislature gave the commissioner of highways broad discretion in deciding how to ascertain driver competence. Oescheger may or may not have ever taken a written test on the rules of the road or had his eyes examined.[7] If he hewed to the gender conventions of his era, he likely believed (despite all the evidence to the contrary) that white men such as he were by definition the most competent and safe of drivers.[8]

Undoubtedly, Oescheger thought about his failed brakes. But he felt enough confidence in his own skills to drive thirty miles of road using the emergency brake. He may also have weighed the risks against the consequences and inconvenience of not being able to drive home that night. Drivers probably made such decisions with some frequency because early automobiles (like most of the machines of the age) were mechanically unreliable. Investigators who conducted roadside tests in Michigan a decade later found that 40 percent of all cars they examined had defective brakes.[9] To twenty-first-century sensibilities, these shortcomings seem unacceptably dangerous. But in the 1920s and 1930s, the press and manufacturers repeatedly asserted that automobiles had improved enormously in function and reliability (which was true). Moreover, the general consensus held that good drivers could compensate for the shortcomings of automotive engineering. Drivers learned to work around frequent mechanical failures or they did not drive at all. Being able to nurse an ailing vehicle home was not only a necessary skill but also a source of pride, especially for young, cash-strapped male motorists.[10]

The speed at which Oescheger allegedly barreled down Philadelphia's narrow streets might have provoked more questions. General consensus held that excessive speed caused most road accidents. But what constituted a reasonable rate of speed remained the source of much conflict and personal interpretation. In the first decades of automobility, many municipalities adopted or retained speed limits of ten to fifteen miles per hour oriented toward protecting horse-drawn and pedestrian traffic. By the mid-1920s, some experts recommended raising the limit to a blistering thirty-five miles per hour on the open road and twenty to twenty-five miles per hour in "business districts."[11] Speed governors—devices that restricted how fast a car could go—offered a technically viable solution to limiting velocity while retaining enough power to climb hills, but consumers flatly rejected them (except for some employers who used them as a means of controlling risk in commercial vehicles). Car manufacturers continued to design their products to go much faster than the law allowed. Therefore, the temptation to speed was perpetual. Some experts counseled common sense, but, as most people would have guessed, the vernacular measure of a reasonable rate of speed proved highly variable.[12]

Another set of troubling questions remained: How much responsibility did the two victims bear for their own fate? What inspired them to stand in the street, after dark, on a Sunday night? Why were they wearing bedroom slippers? Might they have been drinking? Ironically, Oescheger might have avoided prosecution if he had not fled the scene. Elliott and the city's attorney general resorted to indicting most hit-and-run drivers under a 1917 law requiring them to give aid in the event of a crash.[13] In general, the courts resisted criminalizing the lethal consequences of automobile accidents. Although prosecutors sometimes filed "aggravated assault and battery" or even manslaughter charges, juries often proved unwilling to convict because they had trouble thinking of car accidents as intentional criminal acts. Moreover, both the law and common sense prompted the question of contributory negligence on the part of victims.[14]

In the 1920s, public and professional perceptions about automobile risk diverged as the latter group came to rely on accident statistics as an indicator of potential dangers. Most automobile users had difficulty imagining that these numbers could predict whether or not their risky choices would lead to accidents. Moreover, unlike the great train wrecks of the preceding decades, the cost of automobile accidents was metered out in a cascade of small events. No single moment of crisis, no disastrous event prompted a public outcry for structural change. Instead, the Philadelphia police added Oescheger's two victims to a growing pile of data that meant little to ordinary drivers, but a great deal to the kind of policy makers who increasingly paid attention to the question of how the risks of everyday life should be managed. By the mid-1920s, those data had become part of an increasingly well organized effort to use the police power of the state to prevent accidents by licensing and disciplining drivers like Harry Oescheger.

Expanding State Power to Discipline Drivers

Herbert Hoover learned to drive in the Australian outback, grinding across the empty, red desert between mining camps in the first years of the century.[15] Perhaps those huge distances bred a certain impatience and bravado in the young mining engineer. His close friend, Edgar Rickard, later warned a mutual acquaintance to "hold tight" when Hoover took the wheel of his Cadillac roadster "for he is a reckless and speedy driver."[16] By the 1920s, Hoover had become a wealthy and influential man who drove primarily for recreation, otherwise relying on a chauffeur. After Warren Harding named him secretary of commerce in 1921, the former engineer helped put traffic safety on the national agenda. Typical of the many reformers who tackled the automobile safety problem, Hoover never publically related his experiences as a driver to his reform efforts. His work promoting highway safety signified not the penances of a reformed sinner, but rather the actions of a public servant motivated by statistics. "There are 15,000,000 drivers of all shades of responsibility and irresponsibility," he told a reporter in 1924. "It does not take a statistician to calculate that there will be 20,000,000 cars in use by 1935 and, if present conditions are allowed to continue, this means 50,000 fatalities." As commerce secretary, he believed that the social and economic costs of so many accidents were unacceptable.[17]

Hoover is remembered by historians for many things, particularly presiding over the onset of the worst depression in American history. Historians of automobile safety recognize him for organizing a series of conferences on street and highway safety that brought together many of the major experts and institutional stakeholders of the emerging world of motor vehicle safety.[18] These conferences played an important role in defining and propagating a set of traffic management techniques, including a model "uniform vehicle code" designed to standardize local vehicle statutes.

Hoover also advocated a form of business progressivism called "associationalism," which emphasized using the power of the state to help business interests self-regulate.[19] Hoover's political philosophy made him a kindred ideological spirit with the NSC members, insurance executives, and auto club leaders to whom he delegated setting the conferences' agendas.[20] Given these men's taste for volunteerism, it is interesting that Hoover's conferences ended up recommending that state and local governments take a much greater role in disciplining drivers through vehicle registration, licensing, and the promulgation and enforcement of uniform rules of the road. Implementing those measures resulted in vast new state bureaucracies of unprecedented scale and power, including specialized traffic police forces (highway patrols), traffic courts, and departments or bureaus of motor vehicles staffed with inspectors and file clerks. Safety advocates and sympathetic legislators viewed this expansion of regulatory and police powers as justifiable because it fit well into much older legal frameworks. Just as with fast horses and dirty chimneys, the most likely victim of a reckless driver was a stranger. The ubiquity of automobile accident statistics also gave a new twist to the notion of the common good. The yearly toll of automobile deaths and injuries, not to mention the loss of property, carried the flavor of its own kind of conflagration.

Implementing the vision of well-regulated driving involved the coordination of state, local, and national governments. It also required citizens to accept and cooperate with the authority of the state. Therein lay the rub. Although the adoption of the automobile created an unprecedented flow of private and commercial vehicles between the states, legal consensus held that the federal government could advise states but could not dictate policy or enforcement.[21] The resulting piecemeal structures often gave off a whiff of arbitrariness—one more reason why people who would otherwise never dream of breaking the law routinely exceeded the speed limit or jaywalked.

The problem began with writing a motor vehicle code. State and local legislative bodies tended to build codes by accretion—often beginning with a framework translated from an older world of horses and streetcars or borrowed from another locality. Then they added layers of very specific statutes to address local concerns and noisy constituencies. In the mid-1920s, the consolidated traffic ordinances for Los Angeles, for instance, consisted of sixty-five pages of "highly technical" wording. The resulting collection of statutes presented an impenetrable and seemingly arbitrary mass of dos and don'ts to the average motorist.[22]

Substantial variations between different states and even cities further complicated matters. Regional differences in traffic laws posed a particular dilemma for travelers. When planning an interstate road trip, automobile associations advised motorists to write for information about local statutes in all the states they planned to visit. Most states provided booklets describing the most important laws. The ignorant risked not only getting a ticket but also having an accident. Rules about important issues such as who has the right of way at an intersection differed from state to state.[23]

To many reformers, nationwide adoption of a uniform traffic code seemed an obvious solution. Agreeing on content proved to be the sticking point. Local and state officials jealously guarded their prerogative to structure the vehicle code and its enforcement in whatever way they thought best. This effort to hold on to older laws was particularly remarkable given the fact that most cities (not to mention rural areas) had no traffic code to speak of before 1903. The statutes to which they clung had often been in use for less than twenty years. Many localities copied parts of the New York code based on William Phelps Eno's "rules for driving" and then added numerous layers of addendum in response to local interests.[24] In contrast, the culture of efficiency that swept through the country in the last hours of the Progressive Era enshrined standardization and uniformity as core principles, providing a counterweight. During World War I, the Highways Transport Committee of the U.S. Council of National Defense conducted an intensive study of traffic rules. In 1919, the committee asked both cities and states to adopt a code of "Uniform Highway Traffic Regulations." Perhaps foolishly, they called for suggestions from local officials, who flooded their offices with contradictory advice. The initiative failed.[25]

In 1924, the Committee on Uniformity of Laws and Regulations of the Hoover Commission and members of the Commission on Uniform State Laws drafted and published a model uniform vehicle code.[26] In theory, such a document should have not only laid out best practice but also suggested objective measures the police could use to draw a sharp line between legal and illegal behavior. The realities of lawmaking and an already-entrenched automobile culture hindered the kind of rationalizing revision some reformers envisioned. Instead, the Uniform Code epitomized the hybrid culture of risk around automobiles. For instance, in a section entitled "Rules of the Road," the authors defined "reckless driving" in language that echoed almost all of the most important keywords of preindustrial risk culture: "Any person who drives any vehicle upon a highway carelessly or heedlessly in willful or wanton disregard of the rights or safety of others, or without due caution and circumspection and at a speed or in a manner so as to endanger or be likely to endanger any person or property, shall be guilty of reckless driving."[27]

In contrast, the next section, entitled "Restrictions as to Speed," dictated numerically defined speed limits for specific scenarios: "fifteen miles an hour when approaching within fifty feet of a grade crossing" or "twenty miles an hour on any highway in a business district." It also set the (unrealistic) limit of "thirty-five miles an hour under all other conditions."[28] Twenty-three state legislatures rapidly voted to adopt the new uniform code.[29] More significantly, nearly as many decided not to, or continued to use a piecemeal approach. As anyone who has been ticketed for making a right turn against a red light in an unfamiliar city can attest, the resulting hodgepodge of regulations has, in fact, never reached complete uniformity.

If the rules of the road projected ambiguity, contestation, and variation on the page, enforcement added yet another layer of agendas and interpretations. The law empowered police officers to cite motorists for two different but related reasons: causing an accident by violating the motor vehicle code (e.g., hitting another car by running a stop sign) and acting in a way that might or might not cause an accident (such as speeding). Enforcing both required police to interpret the significance of driver behavior, but empowering the police to punish motorists for potentially dangerous behaviors caused much more resentment and resistance on the part of motorists because it was predicated on probabilistic reasoning. Outraged drivers pointed out that most criminal statutes penalized actions that had *already* caused a loss or that involved *criminal* intent. Traffic laws targeted behaviors that statistics showed as leading to accidents. The law did not make a distinction between skillful drivers who could avoid accidents while engaging in these behaviors and less skillful drivers who could not.[30] Experts recommended that police departments maintain "spot maps," marking each accident on a map in order to determine where enforcement of traffic laws was particularly needed. The strategy of "selective enforcement" maximized the use of police officers but also further undermined the authority of traffic law.[31] Locations read as potential accident sites by the police became "speed traps" in the eyes of motorists.

Others disputed whether violations had occurred at all. Until the invention of radar guns, Breathalyzers, and eventually video cameras, the evidence for most traffic violations was ephemeral—based on the word of the arresting officer that he had indeed witnessed a violation of the traffic code. The authority of this premise was particularly fragile in the 1920s, when the failures of prohibition undermined the already shaky reputation of municipal police and county sheriffs. It was also tinctured by race. Although rarely commented on in this era, the power to pull over and fine motorists was a particularly potent tool of harassment used by white sheriffs against black motorists.

Although judges and editorialists opined that violators should be sent to jail or the workhouse, courts mostly levied fines for moving violations. To dissuade motorists from contesting tickets, they set fees low enough that most motorists deemed it worthwhile to pay the fine, rather than risk losing in court. This practice led to the widespread belief that local governments used the power of issuing tickets primarily as a means of raising revenue, rather than to make the streets safer. Even a writer for the *National Safety News* perpetuated this idea. "Many a motorist has found to his sorrow that a dead town sometimes has a live cop," often hiding just behind a speed limit sign on the edge of town.[32]

Motorcycle-mounted traffic policemen became common on American roads in the 1920s. *Earl J. Reeder, "When New Paving Tempts the Speeder,"* National Safety News *(June 1926), University of Delaware Library.*

The perception was further reinforced by the not-unfounded belief that tickets could sometimes be "fixed" on the spot through a discreet cash offering to the arresting officer.

Whether or not legitimately exercised, enforcement of traffic codes also created a huge amount of work for police departments. Critics complained that handing out speeding tickets distracted from the more important business of catching real criminals. More affluent states with large numbers of automobiles, like New York and Connecticut, responded by creating special police forces, often called "highway patrols," whose officer corps could easily be distinguished from their more proletarian counterparts. Dressed in neat uniforms and equipped with riding boots, they were often portrayed standing heroically alongside their motorcycles. Accompanying publicity emphasized that their role included not only enforcing traffic laws but also helping motorists and catching criminals.[33]

Experts agreed: keeping potentially dangerous drivers from getting behind the wheel and removing those who had proven themselves unworthy complemented traffic law enforcement.[34] Giving state and local governments a monopoly on distributing and taking away licenses became the primary tool for achieving this goal. Licensing businesses and professions was a very old means of exercising police powers of the state. In the nineteenth century, state and local governments licensed a range of economic activities seen as potentially threatening to the common good, from distributing liquor to keeping stallions.[35] Very early in the history of automobility, a number of cities and states adopted licensing requirements. Many of these, however, followed the pattern of other kinds of licensing by targeting only people who drove for wages—chauffeurs, taxi drivers, and teamsters—on the theory that the public should know that they were entrusting their safety to someone who had been at least nominally credentialed. However, automobiles did not neatly fit earlier patterns. People who drove themselves proved just as dangerous to others as commercial drivers. By 1909, twelve states and the District of Columbia required all drivers to be licensed.[36] Over the next decade, another six states, almost all of them on the Eastern Seaboard, joined before the movement slowed as it met western resistance.[37]

By the late 1930s, twenty-three states still had no drivers' license requirement at all or issued licenses without a substantial screening.[38] Opposition came primarily from rural areas where residents viewed licensing as an unwarranted infringement on individual rights. Many strongly suspected that driver's license fees were actually a form of taxation, although the courts consistently ruled otherwise.[39] Less abstractly, licensing requirements seemed expensive, inconvenient, and unnecessary. Farm families, in particular, reserved the right to put their children (or themselves) behind the wheel as soon as it was convenient to do so. How, after all, was this fundamentally different from handing a child the reins, which did not require permission or payment to the government?

The push for licensing in the East came primarily from the Eastern Conference of Motor Vehicle Administrators. Its most activist members pushed for a very stringent examination system including a rigorous written exam, a physical, and a road test conducted by specially trained inspectors. In places like Massachusetts and Washington, DC, where such systems were actually put in place, the failure rate was very high. Frank Goodwin, the registrar of vehicles for Massachusetts, reported that his state rejected 21 percent of all applicants in

1922. Passing the exam once provided only a brief reprieve. Massachusetts drivers had to reapply each year, paying a four dollar fee for every renewal.[40]

Rigorous screening did not last for long. Not only did drivers (many of whom were also potential voters) resist, but state and local governments found the costs and bureaucratic challenges untenable. Just distributing licenses to the millions of new drivers was more than many states could manage. In practice, most utilized the initial licensing process primarily to prevent children and the obviously incompetent from driving. Almost any adult without obvious physical or mental impairments could obtain a license with a minimum of fuss. In small towns and rural areas, people also drove without licenses, out of either defiance, thriftiness, or ignorance, their choice ignored by local authorities as long as they avoided serious accidents.[41]

Instead, the license requirement functioned mostly as a form of leverage against drivers whose behavior violated prescribed norms and threatened pedestrians and other motorists. The threat of license suspension or revocation put real teeth in the vehicle code. Once committed to automobility, most drivers did not want to go back to walking or taking the streetcar. In 1925, New York State revoked 3,164 licenses. More than 1,500 were taken away for drunk driving.[42] This mirrors the pattern for other states (and says something about the culture of both drinking and driving at the height of Prohibition). "Reckless driving" provided traffic court judges with a second large, catchall category into which they could relegate drivers who they believed should not be behind the wheel.

Vehicle registration was the third and least controversial tool used by the state to discipline the use of automobiles (in contrast to conflicts over the registration of guns described in the next chapter). After a small amount of tacit resistance by automobile club members very early in the century, registration became the norm.[43] Authorities helped promote vehicle registration by touting numerical tags as a way of controlling crime. They claimed that centralized records allowed authorities to track down what slang of the day called "vampire cars" that had tried to slip away from the scene of accidents or robberies.[44]

As quickly as authorities produced measures to discipline drivers, drivers thought of ways around the system. Of course, we only know about the individuals who were caught, but their strategies suggest that a much larger group of people sometimes went to extraordinary lengths to drive the way they wanted to drive. In addition to the thousands of drunk drivers arrested each year in the 1920s and the Harry Oeschegers who fled the scene of accidents, a host of others put their children behind the wheel or sent a substitute to take a driving test or switched plates between vehicles.[45] A much larger number, perhaps even the majority of drivers, routinely exceeded speed limits either unthinkingly or with one eye peeled for the motorcycle patrolman. They slid through stop signs, shimmied around streetcars, passed on blind curves, failed to yield the right of way, and generally treated the vehicle code (in so far as they cared to understand it) as a set of suggestions they might or might not choose to follow, depending on the circumstances.[46]

Most state vehicle codes included regulations requiring certain kinds of basic equipment such as lights and horns (more on this later). But codes typically gave much more attention to defining the "rules of the road." By describing appropriate behavior, legislators hoped to create predictability and therefore safety in the highly disorganized system of roads and highways. However, as we have seen, compelling drivers to discipline themselves by creating statute law proved an uncertain means of controlling automobile risk, even with the threat of police enforcement. Having passed through the process of credentialing, most motorists found that the police rarely directly supervised their driving except at certain, largely predictable locations. Most of the time, drivers did not need to think about whether they were following the letter of the law. The actions of other drivers and the attitudes of friends and neighbors were more important in shaping their day-to-day behavior.

Remaking the roads themselves offered another means of tackling the problem of traffic safety. Like factories, streets and highways could be studied for risk and then safeguarded. Using a combination of signs and road markings, drivers could be instructed and disciplined on an ongoing basis and in relation to specific situations. The roads themselves could be rebuilt to protect drivers against the consequences of lapses in carefulness. Efficiency and safety would be the rewards. As with much else to do with automobiles, the gap between what should be done and what actually got done was slow to close.

Remaking the Roads for Safety

For a time traveler, one of the most striking features of most American streets and highways before the 1910s would surely be the absence of explanations, warnings, and admonitions about how to use the road. Signs cluttered big city streets and sometimes rural byways, but their function was primarily commercial.[47] The complex language of street markings dividing lanes, indicating flow of traffic, and setting aside space for pedestrians, buses, and bicycles had not yet been introduced. Moreover, motorists not infrequently ended up in a ditch (or worse) because the road itself offered no protection for someone who made a momentary bad decision: no curbs, few guardrails, and a scarcity of gently banked turns. Most early twentieth-century American roads were the physical embodiment of an older version of the vernacular culture in which information about how to preserve one's self and one's vehicle came from other travelers or from experience. Streets and highways functioned as fragmented, atomistic spaces in which individual drivers made decisions on a moment-to-moment basis, based on their own experiences. If the railroad was a tightly coupled system, the roads manifested virtually no system at all except in custom. In a slow-moving world, with little traffic and poor road surfaces, they worked well enough.

People familiar with the railroad operations would have recognized some of what was coming. The grade crossing sign, "railroad crossing—stop, look, and listen," was likely the first widely familiar, standardized traffic sign. Railroads already employed speed limit signs and red and green semaphores. They also tried to separate pedestrians and horses from trains long before such things were practiced on the public roads. Railroad engineers laid out tracks with the physical limitations of trains in mind—too tight a curve and derailment or uncoupling would be the likely consequence. Some technological methods of traffic control, such as the speedway separating Robert Bonner and his brother-drivers from slower traffic, also existed in big cities before the widespread adoption of automobiles. The dangers of streetcars, in particular, had already prompted the creation of more organized and segmented streets.[48] But adding automobiles into the mix, first in cities and then increasingly in suburbs and on the highways, created a very different kind of pressure to remake the roads.

At the beginning of the twentieth century, when most Americans considered driving an automobile a recreational activity, private organizations, including auto clubs and streetcar companies, took on responsibilities that would later fall to the state. It gradually became clear that a volunteerist approach was not viable. Consequently, Americans gradually gave themselves over to the idea that building, maintaining, and policing automobile-centric roads offered a public good, appropriately paid for out of the public purse and conducted by public agencies. That process received a huge push from federally funded road-building and improvement initiatives. The 1916 Federal-Aid Road Act established the practice of offering a 50 percent match of Federal Highway funds with money from the states. Legislators quickly realized that they could also attach a variety of conditions to the money, including implementation of practices to make the roads safer.

With or without federal money, building and maintaining a road system appropriate for motor vehicles proved an extraordinarily expensive undertaking. Consequently, officials took whatever voluntary help they could get. Automobile drivers venturing outside of city centers in the 1910s and 1920s were likely to encounter signs such as one erected on the Michigan border by the Detroit Auto Club: "Legal Speed Limit 25 miles per hour—15 miles in towns—POSITIVELY ENFORCED."[49] Auto club members also trimmed bushes, marked dangerous curves, and signposted narrow bridges. In big cities, streetcar companies paid for warning signs and, beginning in 1914, stoplights that could be powered from their preexisting power lines.[50]

Voluntary efforts had many drawbacks. Lack of standardization was among the most important. "At present there is a wide variety of signs for the same purpose, both those placed by authorities of different states and those placed officially by motor clubs or advertisers," a Hoover Commission report observed. The report further noted that the American Engineering Standards Committee had been developing a standard set of signs and symbols for signs, but these were not yet ready for adoption. In the interim, the commission recommended that "motor clubs and other responsible organizations should . . . be permitted to place standard signs, subject to the approval of the proper authorities, if not placed by the authorities themselves."[51]

Slowly, state agencies replaced volunteers. By the middle of the decade, the Pennsylvania Highway Department had installed nearly thirty thousand distance, direction, and warning signs on the state highway system. Taxpayer dollars paid for a variety of advisements, yet to be standardized: "Dangerous Hill. Go Into Second Gear" and "Narrow Bridge. Width for One Car Only."[52] As local and state agencies paved existing roads, they also took up the practice of painting instructions on the road surface itself to instruct and discipline drivers. The Pennsylvania department adopted the practice of painting a white line down the middle of two-lane roads to retrain rural drivers to stay to the right. "Formerly the custom was to drive in the middle of the road and to turn out for traffic passing in the other direction," the director explained.[53] This habit, functional on the dirt roads and with the slow speeds of an earlier era, led to head-on collisions with automobiles. In cities, officials used painted lines on the roads to train motorists out of another old custom—cutting the corner when making a left-hand turn.[54] In Los Angeles, the paint squad of the police department was recruited to paint "SLOW" and "STOP" on the approach to particularly dangerous intersections.[55]

In the newly emerging discipline of traffic engineering, experts categorized directives painted on the road or displayed on signs as "traffic regulation."

These instructions promoted “a safer and more orderly movement of vehicles” by helping individual drivers to make good decisions. Engineers distinguished traffic regulation from “traffic control,” which involved “absolute orders.” In nineteenth-century American cities, the traffic policeman situated in busy intersections to facilitate the flow of traffic acted as the primary instrument of traffic control. In the 1920s, human agents increasingly gave way to traffic lights wired into an automatic control system that coordinated the changing of dozens or hundreds of lights across a grid (or tangle) of streets.

Traffic signals represented an enormous conceptual departure from earlier methods of traffic regulation and control because, properly implemented, they turned city streets into a system that subordinated individual judgment to the smooth operation of the system as a whole. The mechanical, modern character of lights also gave them more authority than signs. Red and green traffic lights were originally used on railroads, and indeed traffic engineers borrowed liberally from the body of research on subjects such as color perception commissioned by railroad companies. In the early 1920s, engineers typically set signals so that all the lights would turn green at once along parallel streets, imitating the railroad practice of opening up a signaling block. After a few minutes, all the lights would turn red and traffic going on perpendicular streets would be allowed to proceed. The result was what one engineer called a “rabbit” effect, in which drivers would race through all the lights in a block at high speed. Accidents inevitably resulted.[56] By the mid-1920s, engineers began switching over to a modified “platoon” system in which timed lights rewarded drivers who maintained a steady rate of speed, often fifteen miles per hour.[57] Installing systems of lights worked best in cities built on a grid pattern. City officials in Syracuse and Pittsburgh, for instance, held off on adopting lights because they worried that the result would be additional confusion in a warren of streets characterized by nonparallel roads, erratic turns, and dead ends.[58]

In an increasing number of states, highway departments also rebuilt roads to prevent vehicles from going into ditches or down the sides of mountains. Although officials liked to brag about the scientific approach of their engineers, the process of figuring out what worked involved a certain amount of trial and error. “The picturesque white wooden highway . . . which has been painted into beautiful pictures of country scenes and recalled a thousand times in song and sentiment . . . must go,” the *National Safety News* proclaimed.[59] Rising highway mortality brought home the message. In 1927, the state of Wisconsin replaced guardrails consisting of posts sunk in the ground and topped with 2 × 6 boards. During the previous year, five people had been killed by being “impaled on the jagged ends of the guard rail as it splintered from the impact of the crashing automobile." The department also removed low walls erected next to culverts with the intended purpose of keeping vehicles out of ditches. Their records had revealed that too many nighttime drivers rammed into the ends of the walls while temporarily blinded by the headlights of oncoming vehicles.[60]

Greater attention to safety in highway engineering was one manifestation of a shift partially prompted by a larger percentage of fatalities involving people *in* vehicles. As long as pedestrians constituted the majority of motor vehicle accident victims, most safety experts gave little attention to this question.[61] When they envisioned a motor vehicle accident, they typically pictured a private automobile hitting a child playing in an urban street. By the mid-1930s, that image was being replaced in expert considerations and popular imagination with a mangled mass of metal, blood, and body parts strewn across a rural highway.[62] Reimagining automobile accidents led to efforts to introduce a growing number of safety features in the design of automobiles.

Remaking the Automobile

The push to adopt safety innovations in automobile design came from several quarters. Typically, automobile manufacturers first offered devices such as four-wheel brakes as optional equipment in luxury cars. If proven popular in the marketplace (or mandated by the state), they became available on less expensive vehicles. Automakers took the position that consumers should decide whether to pay a premium for the best available devices or use their own wits to compensate for inadequacies. Neither automakers nor consumers clamored for safety features. Instead, most viewed them as expensive luxuries that might or might not be worth the money. In the 1930s and again in the 1950s, automakers tried marketing specific car models on the basis of their safety. These consistently failed to attract large numbers of buyers.[63] Consequently, progressive legislators wrote them into vehicle codes as requirements, making state regulation the most important force behind the adoption of new safety features.

This shift in emphasis also reflected the slow emergence of safety as a distinctive category in automobile design, separate from general reliability and effective function. In the 1910s and into the 1920s, safety engineering meant improving basic technologies such as brakes and headlights that helped drivers avoid collisions, as well as sturdy bumpers that protected vehicles from damage. Most of these technologies also required skilled use and conscientious maintenance. In other words, they were active rather than passive safety devices. Car parts that contributed to safety were also only as good as the manufacturer was willing to make them.

Henry Ford was a hard man to work for. He was particularly notorious for clinging to certain design features despite the advice of his engineers, pleas from dealers, and complaints from car owners. Brakes were one of those subjects. As irate customers liked to point out, the brakes on the Model T often "shrieked like banshees" when applied. They tended to burn out quickly because Ford opted to economize by equipping brake shoes with cotton rather than more expensive asbestos liners. Experienced drivers knew the meaning of a telltale acrid smell and the sensation of pressing the brake pedal to the floor to no good effect. Fortunately for them, most states would not allow dealers to sell cars without two separate braking systems.[64] Their troubles did not necessarily end even if they successfully brought their vehicle to a stop. Changing brake shoes on early Model T's required removing part of the transmission assembly.[65] No wonder the 1919 Ford owner's manual, otherwise filled with copious technical detail, failed to describe the braking system or explain how to maintain it.[66]

Until the mid-1920s, the Model T (like most other cars) came equipped with brakes only on the rear wheels. If a driver slammed on the brakes, the car tended to pivot around the rear end. Improperly adjusted brakes amplified the effect. So did the decision of a thrifty owner (or mechanic) to replace one, but not both, brake shoes. Consequently, a sudden stop could turn a car into a ditch or oncoming traffic. In 1921, wealthy motorists could opt for four-wheel brakes for the first time if they purchased a luxurious Duesenberg. The innovation was so obviously superior to the two-wheel setup that by 1924 it was available on nearly twenty-five makes of automobiles.[67] Finally, in 1927, the newly designed Model A made the technology available to the mass market. Typically, Ford initially tried having both the regular and emergency brakes work off of the same hydraulic system to save costs and simplify the vehicle's design. Motor vehicle authorities in fifteen states told him they would not allow the vehicle to be sold with this arrangement, which defeated the purpose of requiring a backup in case of brake failure. Thanks to their intervention, the Model A that eventually took to the roads had two separate systems.[68]

Commentators prophesied that great improvements in safety would result from better braking systems. They did not count on the ability of drivers to adapt their risk-taking behavior to the potentialities of the new technology. As with antilock brakes in a later era, these braking systems gave drivers a greater sense of confidence in their ability to stop quickly and encouraged some to tailgate or take additional chances. In the 1920s, safety engineers carefully calculated the

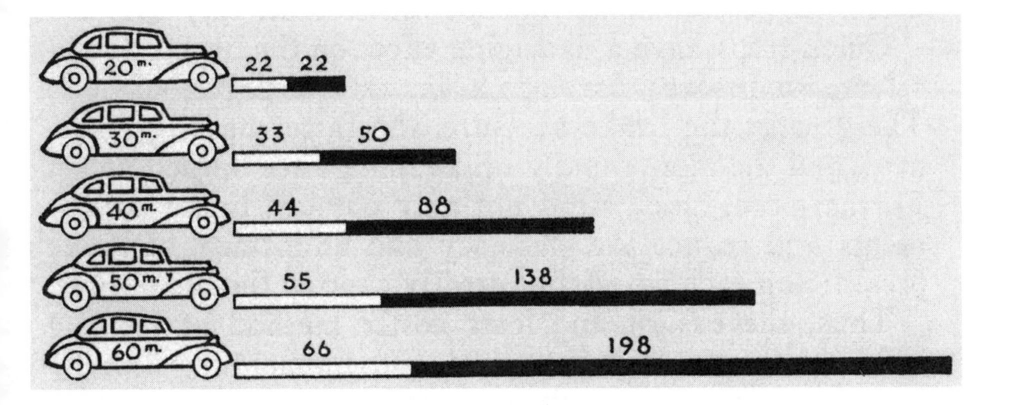

This little chart was supposed to teach new drivers about braking distances at various speeds. *Albert Whitney,* Man and the Motorcar *(1936). Hagley Museum and Library.*

distances needed to stop for a vehicle with properly adjusted brakes traveling at different velocities. These charts were translated into neat little graphics suitable for popular consumption.

"Most drivers are frankly skeptical about the stopping distances," Albert Whitney explained in his 1936 driver's education volume, *Man and the Motorcar*.[69] "They believe (without having made the tests) that they can stop their cars more quickly." In the minds of such drivers, braking was a matter of skill, not physics. Skill was undoubtedly involved, but it involved something more complex than quick feet on the pedals. In dense traffic, it required the ability to judge the intentions of other drivers—a particular challenge before brake lights became standard equipment. Countless drivers found themselves in a ditch after overzealously applying the brakes on an icy or slippery road. Sometimes, too much brake proved worse than too little.

Headlights, the other significant preventative technology, presented an even more complicated interplay between what manufacturers were willing or required to provide and driver behavior. Everyone agreed and statistics confirmed that nighttime driving was significantly more dangerous than taking to the roads in the daylight. Lack of visibility constituted the primary reason. Poor lighting reduced the pedestrian dashing across a dark, rainy street to a shadowy blur and hid the low stone wall or the Studebaker emerging from a side street until it was too late to take evasive action. The solution to the problem was less obvious. Brighter headlights illuminated more of the road and the objects in it but produced a blinding glare for oncoming drivers. Safety advocates pointed out that more illumination from street lights offered the best solution; however, this seemed expensive and unrealistic, particularly on rural roads. In the end, drivers were pulled back into the loop. In the 1930s, manufacturers introduced what was termed a "multi-beam lighting system" that allowed drivers to toggle back and forth between what were colloquially called high beams and low beams.[70]

Henry Ford did not resist all safety innovations, particularly if he could see their utility through the lens of his own experience. In the late 1920s, he took the lead in introducing safety glass into the mass market. Ford was motivated by guilt over terrible injuries suffered by Harold Hicks, Ford's chief aircraft engineer. Hicks had been brought onto the Model A design team to help increase the power of the new car's engine. Under orders from Ford, Hicks took the car out onto the public roads to test the prototype engine. Traveling at high speed, he rammed into the side of another car that unexpectedly turned in front of him, resulting in a wreck. Hicks hurtled through the windshield, suffering horrible lacerations. Ford consequently ordered that the new model be equipped with a windshield made of safety glass, despite the added cost.[71]

The glass came from the Triplex Safety Glass Company in Hoboken, New Jersey. The factory had been set up to exploit British inventor John C. Wood's patent for the process of laminating glass with cellulose citrate. The resulting substance broke into chunks, rather than slivers, under impact. Previously, vehicle manufacturers' only option was glass with wire embedded in it. "65 % of all injuries in Automobile Accidents are due to flying glass," a 1927 Triplex advertisement proclaimed. "What is the first thing that happens in an automobile accident? Windows and windshield fly into razor-edged, needle-pointed splinters of glass that cut, perhaps permanently disfigure, blind, or even worse."[72]

The image was compelling. Both the public and legislators rapidly grasped the significance of this particular innovation. By the mid-1930s, enough states had put laws on the books requiring that safety glass be installed in cars sold within their borders that it had become a standard feature, at least in windshields.[73] If the glass manufacturers, particularly Libbey-Owens-Ford, could be believed, consumers had also come to view safety glass as a necessity. "It's Modern to be Safe!" one of their ads trumpeted.[74]

Safety glass was the first widespread evidence of a new paradigm that accepted at least some accidents as inevitable and looked for engineering solutions useful for making crashes less lethal for vehicle occupants. The automobile engineering community initially produced piecemeal improvements: better locks on doors so that drivers would not be thrown from their vehicles, stiffer frames to prevent crumpling, some efforts at padding on the inside of vehicles, and the removal of "protruding metal gadgets."[75] By the mid-1930s, increasing numbers of engineers began to talk about what happened inside automobiles during a collision—a discussion that would eventually result in postwar efforts to render automobiles "crashworthy."[76]

In the meantime, automobile owners, public officials, lawyers, and insurance companies struggled with the unforeseen consequences of two other solutions to the inevitability of accidents: insurance and liability law. Here the complexity of automobile risk, the questions of causation and responsibility, and the hybrid culture of risk not only complicated the legitimate functioning of these ways of managing risk but opened the door to a new kind of moral hazard.

Insurance, Tort, and the Problem of Fraud

The advent of the automobile era marked the maturation and convergence of the insurance industry and tort law. As we have already seen, individual consumers had been purchasing insurance for centuries to protect against loss. Insurance against liability claims also had a longer history. While some nineteenth-century industrial employers purchased insurance to protect themselves from employee lawsuits stemming from industrial accidents, automobile insurance was the first widely marketed form of liability protection offered to users of a particularly dangerous kind of consumer technology.[77] Consumers' adoption of insurance and the explosion of personal liability suits around automobile accidents rapidly became symbiotic. Car owners bought insurance because they feared lawsuits. Accident victims sued because they knew (or believed) that insurance would guarantee compensation. By 1930, Americans collectively paid more for auto insurance premiums ($189 million) than for workmen's compensation ($150 million). The widespread acceptance of insurance also transformed the process of settling lawsuits. It became increasingly common for victims' lawyers to negotiate directly with insurance companies for settlements, in or out of court.[78]

In theory, marketing huge numbers of automobile insurance policies would seem like one more piece of evidence for the triumph of modern, quantitative ways of thinking about risk. But in the early years of automobility, the practice of selling insurance and settling claims depended as much on the ability to judge character as on the law of large numbers, offering an excellent example of the hybrid culture of automobile risk, catering to custom and desire, and mixing statistics and storytelling. These characteristics helped open up the possibility of manipulation not only by cagey lawyers and parsimonious adjusters but also by criminals and the merely desperate. The availability of insurance also provided a potential disincentive for safe behavior: an implicit acknowledgment that accidents did happen, but that the full price need not necessarily be paid. By commodifying risk protection, it allowed drivers essentially to buy their way out of some consequences of risk taking.[79]

The hybrid character of automobile insurance could first be glimpsed in the process of selling insurance, or "solicitation of risk" as it was evocatively called by insiders. Salesmen had two jobs: getting people to buy insurance and also making sure that the people who did buy were good risks. In an age before compulsory coverage, effective salesmen worked hard to convince people of its necessity. They recognized that probabilistic arguments carried little weight with potential customers because people do not like to see themselves as part of a risk group, particularly one that shared a common characteristic of carelessness or bad luck.[80] Instead, to make sales, insurance salesmen encouraged potential customers to imagine a set of accident scenarios resulting in dire consequences. Ambrose Ryder, the author of a 1924 auto insurance manual, offered advice to salesmen that combined these two functions. He described types of men who would make good "prospects" and then explained how to pitch insurance to them. "The business man," "the man of wealth," "the man who borrows money" all made good customers, but he particularly favored "the man with a conscience" because "he carries with him a mental picture of an accidental injury resulting in death to a breadwinner and he sees the wife and children in destitute straits, suffering privations for a number of years, all because of one unfortunate moment in the driving of a car."[81]

Insurance companies struggled with a problem economists call "adverse selection." In theory, the riskiest drivers should pay the highest premiums or be excluded entirely from the risk pool.[82] However, not enough aggregate statistical data existed to determine what kind of people in what kinds of circumstances were most likely to have accidents. Understandably, potential customers also resisted revealing if they had a prior history of accidents or other factors causing them to be a greater risk. Consequently, sales agents ended up winnowing out "careless and reckless operators" using their personal judgment.[83] Guided by the concept of "moral hazard," as it was then understood by the insurance industry, they also tried to identify people of weak character who were likely to succumb to the temptation of engaging in dangerous or immoral behavior if they thought themselves protected by insurance.[84] In much the same way that nineteenth-century employers informally screened potential employees, they also looked for obvious signs of mental or physical disability.

After hopefully eliminating poor risks from the pool, agents divided customers into rough actuarial categories that helped determine rates. Again, Ryder describes these in terms that imply a scenario of use. "If a man owns a one-ton truck used in the grocery business in the city of New York, he pays a certain public liability and property damage rate."[85] The subjectivity of the process was underlined by the fact that a person or situation considered a bad risk by one insurance company might be considered a good risk by another.

When underwriters drew up contracts, they included various exemptions. These described scenarios deemed excessively risky and thus threatening to the insurance company's financial stability. Some were relatively standardized. Policyholders could not collect, for example, if losses resulted from racing (formally or informally). A minority of potential customers approached insurance companies looking for policies to protect against unusual risks. In the 1920s, some drivers sought coverage against the possibility that their vehicle would be confiscated if they disobeyed liquor laws. Violating the law was an obvious form of moral hazard, so most insurance companies would not underwrite such policies.[86]

While agents and underwriters extrapolated from their experience and aggregate data to imagine potential accident scenarios, adjusters dealt with stories about real accidents provided by claimants. "It requires long experience and keen judgment to properly weight conflicting evidence and form a mental picture of what happened," Ryder explained.[87] Detecting misrepresentation and fraud was a critical part of the adjuster's job. But hard-nosed probing of claimants' stories had to be balanced with expressions of sympathy lest the insurance company end on the wrong end of a breach of contract suit. In the 1990s, Sam Black, who worked as an adjuster in this early era, dictated an extensive oral history, painting a fascinating picture of how adjusters balanced these conflicting demands. Black began working for the Philadelphia Indemnity Company in 1921 when he was little more than a car-crazy teenager, fresh out of high school. After a brief stint selling insurance, he was moved over to the claims department. Philadelphia Indemnity was one of the first insurance companies to sell auto policies as their principle line. The founder had tried to limit the company's risk by only insuring members of auto clubs—well-off people who cared about their cars. But, as Black quickly discovered, this strategy did not preclude the possibility of insurance fraud.[88]

Like other new claims agents, Black started out working on property damage (or "PD" as he called it) claims before moving into the more complicated business of bodily injury ("BI"). Into the early 1920s, collision policies were still written in a format borrowed from marine insurance. Policyholders could recover the entire purchase price even after several years of owning a car because policies did not take into account depreciation. Consequently, insured cars might be worth much more on paper than in the used car market. Black's supervisor sent him to Allentown to adjust such a claim. The car in question had been reported stolen, later turning up with extensive damage. "I interviewed the policyholder who just denied he had anything to do with the disappearance of the car or the damage to it," he remembered. But something seemed wrong about the man's story. "I leaned back mentally, rested, and told myself 'it's going to be a long session.'" Then Black went back to questioning the claimant about the circumstances surrounding the theft. Eventually, he extracted the true story. It proved far more interesting than the initial false claim. The claimant's wife had kicked him out of the house for chronic drinking. He cleaned up his act, and she had agreed to take him back. To celebrate, he consumed a few drinks in a local bar, consequently ramming his car into a tree. He feared telling his wife he had fallen off the wagon, so he said that the car had been stolen. To support his lies to his wife, he filed a false insurance claim.[89]

Despite the potential for misrepresentation, physical damage claims were relatively easy to settle because determining the cost of car repairs was more or less a straightforward process. Accidents involving injury or death offered more complications. As claims agents tried to negotiate out-of-court settlements, they kept in mind what courts might allow as damages if the settlement could not be worked out privately. Tort law allowed for cash settlements for not only medical costs but also loss of income, "pain and suffering," and even a wife's connubial services. Adjusters particularly dreaded dealing with the death of a child because putting a price on something priceless inevitably meant big money.

The possibility that a case might go to trial was made more real by availability of lawyers willing to work for contingency fees. Particularly in big cities, the smell of blood and money could set off an unscrupulous contest between adjusters hoping to make a quick settlement and lawyers looking for sad stories that would make good lawsuits. Accident attorneys employed "runners" who gathered potential cases by paying policemen and doctors to give the names and addresses of accident victims. Runners looked for good stories and sympathetic victims. They raced the insurance adjusters to the injured people's bedsides (or the home of the next of kin) and talked them into signing over power of attorney. Larger law firms used investigators to find out whether the cases were worth pursuing. The reward, of course, was a "tip" or a portion of the settlement.[90]

While the increasing number of automobiles led accident lawyers to rightly believe that the number of cases would continue to grow, in practice, lawyers sometimes felt there were not enough lucrative ones to go around. Some law firms resorted to manufacturing cases by staging accidents. It is difficult to know how common this practice was, since, if it were done well, nobody would know the difference. The rare examples of those that got caught illustrate how the falsification worked. In the mid-1920s, a young accident attorney named Morris Katz made a deal with two brothers named Daniel and Benjamin Laulicht who worked as laundry truck drivers. The Laulichts staged scripted accidents between laundry trucks and taxi cabs (both covered by fleet insurance). Katz represented the "victims." Between 1924 and 1927, Katz's law practice went from thirty cases a year to 1,193, and both he and the Laulichts profited handsomely (until they were finally arrested).[91]

Most of the time, the threat of a lawsuit was enough to force a settlement. But cases that went to trial added new opportunities for storytelling. Soliciting perjury was, of course, not only grounds for disbarment but also illegal. But as Henry Drinker, a prominent lawyer (and descendant of Elizabeth Drinker), pointed out as part of an investigation for the Philadelphia Bar, "the fallacy of human memory is such that two honest witnesses, without any prompting whatever, will often develop within a few days, diametrically opposed recollections of the same occurrence." He noted that in the months and years between an accident and a trial, an experienced lawyer can "with patience and ingenious suggestion, gradually mold the bona fide recollection of any witness of low mentality to the desired version of the facts." Done well, the results were virtually undetectable.[92] While Drinker referred to claimants' lawyers, insurance company adjusters and attorneys were not immune from the temptation. They also encouraged witnesses to package the facts to produce a desired outcome.

The obvious solution to this early tort crisis, besides tighter rules for lawyers, required instituting mandatory, no-fault insurance along the lines of workmen's compensation insurance, adopted in many states over the prior decade. Among its many virtues, workmen's compensation decreased the number of workplace-related tort cases in the courts and provided a more reliable and arguably fairer way of helping people injured in factories and on railroads. Drinker thought that the primary impediment to implementing such a system for automobile accidents was constitutional.[93] He was wrong.

In 1932, Columbia University sponsored a large-scale study on the problem of compensation for automobile accidents.[94] The blue-ribbon panel proposed a system of compensation modeled on the New York and Massachusetts workmen's compensation systems. Under the committee's plan, all automobile owners would be obliged to purchase compensation insurance. They could not register a car without proof of insurance. In the event of an accident, victims would be compensated for medical costs and loss of livelihood based on a schedule similar to that used by the New York workmen's compensation board. Death benefits would also be paid on a fixed schedule.[95]

The plan received serious consideration in the legislatures of New York, Virginia, Wisconsin, and Connecticut. However, in the end, it was rejected. Objections came primarily from insurance companies, insurance lawyers, and the personal injury bar. Insurance companies argued that no-fault insurance would provide an added incentive for fraud and moral hazard, would increase insurance costs for consumers, and was "socialistic." The real reasons probably had more to do with the fear, based on their experiences with workmen's compensation, that many states would either closely regulate insurance rates or would set up competing state-sponsored insurance funds. Lawyers argued that compensation schedules were unfair to people with high incomes (often described as "business executives") who could not recover their full losses.[96]

Insurance companies and their allies suggested an alternative. Since the late 1920s, they had pushed for the adoption of so-called "financial responsibility laws." Already in place in eighteen states when the Columbia report came out, they became the norm by the end of the 1930s. Financial responsibility laws required drivers to purchase insurance or provide some other type of evidence that they could pay for any accident they caused *after* they had been deemed negligent in an accident by the courts or had been convicted of a serious moving violation. On first glance, it is difficult to understand why insurance companies would prefer a system in which only the worst drivers were required to buy insurance. But financial responsibility laws left the decision as to whether to insure and, if so, how much to charge for premiums up to insurance companies. They could deny coverage or sell a policy at exorbitant rates, if an applicant seemed like a particularly bad risk. The financial responsibility approach also left the tort system untouched.[97]

Among the many themes running through the debate on reforming the insurance and tort system, several others stand out as important to our larger story. First, parties on either side had very different views about the function of insurance. Advocates of the workmen's compensation model viewed auto insurance as a form of social insurance that would ultimately protect the common good by providing a safety net against the (statistically very real) possibility of people being hurt or injured in a motor vehicle accident and consequently becoming a burden on society. They saw the courts as a risky and unfair way of redistributing the costs of those accidents. In contrast, those who preferred private insurance and the "no-liability without fault" system that made court cases necessary viewed automobile accidents as private events. They maintained that the ability to make decisions about how to manage their risks (e.g., whether or not to buy insurance or go to court) should remain an individual right. If individuals lost their houses because a family member had run over a child and they lacked the probity to buy insurance, so be it. This position says something important about the distinctive character of America's version of a risk society that, even in our own era of mandatory insurance, all efforts to implement so-called no-fault insurance policies have resulted in systems where victims still have some recourse to suing for damages.[98]

However, it should also be noted that, especially in the immediate postwar era, more and more people voluntarily bought automobile insurance. In the late 1920s, fewer than 30 percent of drivers carried insurance. By 1966, on the eve of another wave of reform ushering in mandatory insurance laws in a number of states, J. C. Bateman, president of the Insurance Information Institute, estimated that 85 percent of the cars on the road were insured.[99] The number of auto liability lawsuits also exploded, rising about 50 percent between 1955 and 1970.[100] Buying insurance and finding a lawyer after an accident had both become part of the culture. As we will see in the final chapter, both also became part of the larger consumer product safety movement of the postwar era.

IN THE FIRST HALF OF THE TWENTIETH CENTURY, automobile risk became the first large-scale problem tackled by risk experts and safety advocates in the context of an emerging consumer culture. To do so, various groups mobilized the full range of tools developed over the previous two centuries. The problem itself was revealed and analyzed through the use of statistics. Physical changes to vehicles and to the road itself informed and disciplined users and limited damages when accidents did occur. Implementation required unprecedented use of public resources and state power. The process of licensing, vehicle registration, and traffic law enforcement gradually became the most important way in which most adults in American society directly interacted with the government and its agents: police, traffic judges, and clerks at the Department of Motor Vehicles.

At the same time, automobiles offered ordinary people the unprecedented (and undesired) ability to unintentionally kill or injure strangers or to destroy thousands of dollars of property in a single moment of inattention. These scenarios, harnessed by lawyers and insurance agents who saw an enormous opportunity to commodify risk, set in motion another dimension of a modern American risk society: the spiraling relationship of insurance and tort. They created new temptations for the unscrupulous to succumb to the lure of moral hazard, intrinsic to insurance.

Automobility also revealed the durability and adaptability of the vernacular. In practice, vernacular notions of risk such as carelessness and common sense became incorporated in everything from the wording of traffic statutes to the design of automobiles. A sturdy body of beliefs and practices absorbed some expert prescriptions while generating rationalizations for resisting others, creating a kind of hybrid.

Over the course of the twentieth century, the number of automobile accidents steadily increased while the rate of accidents as measured by miles traveled decreased. Automobile accidents are now the single largest cause of mortality among young adults.[101] These statistics offer a complicated message about the conflicts and contradictions of a modern risk culture. The availability of information about driving risks has only partially changed behavior. Lawyers, insurance companies, and petty criminals all found ways to profit from the inevitability of accidents, while automakers found it exceedingly difficult to sell safety. And yet, the system as a whole seemed to work to most people's satisfaction, punctuated only occasionally by moments of crisis, often deemed to be so by specific actors—Ralph Nader's unsafe at any speed in the 1960s, conflicts over mandatory active and passive restraints in the 1970s and 1980s, Mothers Against Drunk Driving, and perhaps cell phone use in our own time. Like many other risks, motor vehicles gradually settled into a state of slow-motion negotiation. The same cannot be said for the next technology, firearms.

9

What's a Gun Good For?

"One of the reasons that firearms are of use is that they are dangerous," Bellmore H. Browne informed readers of his 1908 hunting manual, *Guns and Gunning*.[1] Most of Browne's contemporaries would have agreed but might have gone on to ask: Dangerous to whom or to what? Clay pigeons? Whitetail deer? Rabid dogs? Burglars? Curious children? Innocent bystanders? Would-be suicides? And dangerous under what circumstances? Without doubt, guns posed a threat in the hands of the wrong people, the ignorant, criminal, alienated, and the very young. However, many gun enthusiasts also argued that, in the right hands, guns offered not only an important means of self-defense and a deterrent to crime but also a source of health-giving outdoor recreation. And, of course, they were a symbol of a distinctive kind of American liberty and self-reliance.

Firearms and the risks accompanying them were hardly new in 1908. But public discussions about how to control the risks of this inherently dangerous technology acquired a new sense of urgency. Statistics and systematic observation indisputably showed that privately owned firearms caused increasing numbers of not only murders and suicides but also accidents. By 1920, shootings accounted for 6,739 annual deaths in the United States—about the same as the number of people killed in railroad accidents. Among consumer products, guns ranked second only to automobiles as a cause of death.[2]

At nearly the same moment that Bellmore Brown astutely summarized the distinctive nature of firearms, efforts to control their risks began to cleave into two parts. Motivated by worries about rising rates of mortality from gunshot wounds, public officials looked to regulation as a means of controlling the acquisition and use of firearms, particularly handguns. They quickly found themselves embroiled in a vociferous debate. The question of whether weapons designed primarily to kill human beings were an unacceptably dangerous technology for a modern society or a necessary and useful means of self-defense resided at its center. Initially, legislators, coroners, and other advocates of state-sponsored gun control tried to avoid the problem of defining appropriate versus inappropriate use by adopting a public health model. They aimed for a decrease in the number of weapons available to civilian users as a means to decrease the overall fatalities, whether accidents, suicides, or homicides. Resistance from organized gun enthusiasts eventually led to a different approach: regulation that equated criminality with the dangerous misuse of guns and reified "honest citizens'" right to armed self-defense.

Meanwhile, hunters, trapshooters, and other devotees of gun sports did not need to decide what a gun was good for, at least while engaging in sporting pursuits. Guns were for shooting animals and targets, not human beings. Consequently, little controversy followed the gradual adoption of safety education and safer gun-handling protocols in sporting circles. Manufacturers also cooperated in designing arms and ammunition that were safer. They did not want to be responsible for having their products explode in the hands of the shooter. Adrift in the middle, between sporting safety and curbing criminal use, the risk of homicides and suicides committed by people who had no history of criminality became largely invisible in public discussions about controlling the dangers of guns.

Guns for Nearly Everyone

Discussions about who should be allowed to own and use firearms took place within the context of a rapid expansion of gun ownership. A complicated mixture of motives, meanings, and emotions surrounded the acquisition and possession of these devices. Manufacturers and retailers of firearms and related products played an important role in telling consumers why they should need and desire to own a gun. Messages conveyed by marketers about citizenship, rights, and self-protection were readily absorbed into the vernacular. A host of different social groups, ranging from lynch mobs to women's clubs, also generated and exchanged ideas about who should possess guns, how they might be used, and what constituted an acceptable level of risk for people who found themselves at either end of the barrel.

Unlike many technologies discussed in this book, the invention and widespread adoption of firearms predated industrialization.[3] Scholars have argued at length about how common guns really were in early America, and when America became a "gun culture," without reaching any firm conclusions.[4] Everyone agrees, however, that in the second half of the nineteenth century, new technical innovations in design and manufacturing led to the introduction of guns that were simultaneously cheaper, easier to use, and more accurate and reliable.[5] New production techniques also made it possible to produce enormous numbers of firearms. By 1919, nongovernmental manufacturers reported producing more than a million firearms a year. Handguns represented a quarter of the total number.[6]

As manufacturing techniques gained in efficiency, makers confronted the great dilemma of mass production—they could potentially make many more firearms than the existing market could absorb. Moreover, demand was not driven by a continual need for replacements because firearms wore out slowly. Manufacturers responded to these challenges by diversifying their offerings, engaging in rapid technological innovation that rendered older models less desirable and advertising to create new markets and new uses. By the late nineteenth century, consumers could choose from thousands of different kinds of guns designed for a wide range of purposes and pocketbooks. A single-action revolver could be purchased for less than a dollar, while wealthy sportsmen might spend hundreds for a fowling piece custom fitted by a specialist. There were guns for children, notably the air rifle. There were guns for women, many of them small and with features that supposedly rendered them safer to use. Several manufacturers marketed versions of the "bicyclette"—a tiny handgun that lady cyclists could carry as they ventured out into the countryside.[7]

Changing strategies for marketing guns fit a larger pattern in which marketers increasingly tried to create demand by suggesting that their products would fulfill needs and desires previously unrecognized by consumers.[8] Beginning around 1910, the largest firms like Colt and Remington began pursuing this strategy in earnest. These businesses hired advertising firms to create sophisticated advertisements aimed at attracting new customers. The resulting advertising campaigns focused on the meaning and context of gun use (in other words, they explained the value of owning a gun). Risk played an important part in these explanations. Advertising firms placed ads in mass circulation periodicals such as *McClure's* and the *Saturday Evening Post*. These campaigns focused on expanding gun ownership in the middle class, emphasizing the themes of self-protection and recreational hunting. Because middle-class women were the great untapped market, a number of advertisements focused on explaining why women

Patented Oct. 4, 1887.
May 14, 1889.
April 2, 1896.
April 7, 1897.

H. & R. "Bicycle"
Double Action Revolver.

Description:

22 Caliber, 7 Shot. 2 Inch Barrel. Weight, 12 ounces. R. F., Long or Short Cartridge. 32 Caliber, 5 Shot. 2 Inch Barrel. Weight, 11 ounces. C. F., S. & W. Cartridge. Finish—Nickel or Blue.

We have given this revolver the name "Bicycle" because it is convenient for the cyclist to carry in the pocket when awheel.

Although designed for cyclists, this revolver is equally adapted to all cases where a small, light weight, effective and handy pocket weapon is desired. It has small frame and automatic ejector.

12

Patented Oct. 4, 1887.
May 14, 1889.
April 2, 1896.
April 7, 1897.

H. & R. "Bicycle Hammerless"
Revolver.

Description:

32 Caliber, 5 Shot. 2 Inch Barrel. Weight, 12 ounces. C. F., S. & W. Cartridge. Finish—Nickel or Blue.

This revolver is small of frame, compact, effective, reliable and safe. Has independent cylinder stop and automatic shell ejecting device. Cocks without catch or drag, and, as the name indicates, has no hammer to catch in the clothing when hurriedly withdrawn.

13

The "bicycle" revolver was marketed to women as a means of self-defense. *Harrington and Richardson Arms Company,* Catalog Number Six, *Warshaw Collection of Business Americana—Firearms, Archives Center, National Museum of American History, Smithsonian Institution.*

needed guns. In 1914, the N. W. Ayer & Son advertising company created a series of ads that marketed the Colt automatic pistol as a home protection. "House Robbed and Wife Scared in His Absence," one began. "It often happens. Men bent on crime wait until the husband's back is turned. Every woman living in the country ought to be protected by the safest and most effective firearm."[9]

It is difficult to determine how consumers responded to these advertisements, but they tellingly echo the themes of patriotism, self-protection, and self-sufficiency beginning to color the arguments of people who opposed gun control legislation. By the same token, their imagery also depicted a distinctive version of American history in which liberty and masculinity were tied to gun ownership. In 1919, Remington ran a series in the sporting periodical *Field and Stream* linking their products to various kinds of Americanism. One ad, emblazoned with the image of a mountain man and his trusty rifle, declared, "No poison-pollen of Old World Imperialism gone to seed can contaminate—nor any attempt of crowd-sickened collectivism undermine—the priceless individualism of the American who truly keeps his feet on the earth."[10] Other ads depicted cowboys and minutemen; they explained that guns could be used to ward off not only burglars but also dangerous ideologies such as anarchy and Bolshevism.

Gun manufacturers also zealously catered to a core market of recreational hunters and target shooters. In the two decades after the Civil War, a huge expansion occurred in gun sports.[11] Interest surged again in the 1910s as the well-off created gun clubs for weekend trapshooting and Northern businessmen bought up coastal plantations in South Carolina where they could retreat for a week of hunting in the company of their friends and business associates. Gunpowder and ammunition manufacturers saw particular potential in trap- and skeet shooting because it reliably used up much more ammunition than hunting. The DuPont Corporation even published a series of booklets promoting the sport and explaining how to set up a gun club.[12] Clubs and sporting magazines were critical to the formation of the community of sportsmen who played a central role in the gun control debates of the interwar years.

While the public images provided by gun manufacturers and sporting magazines portrayed one kind of consumer (and, implicitly in the self-protection ads, a second, illicit consumer, the dreaded bandit), the reality was actually much more complicated. There was not just one gun culture in early twentieth-century America; there were many. As with automobiles, ordinary Americans did not wait for the advice of experts or manufacturers to decide how to integrate firearms into their everyday lives.

According to Mr. H. J. Pinkett, the first act of an Omaha lynch mob he witnessed in the early 1920s was to arm themselves. "Some led the way to sporting goods houses, the pawn shops, and the wholesale hardware stores for guns and ammunition." The men used the guns to rally the crowd and fend off the police. "Now the firing began. Men hooting and yelling, wearing pistols and discharging them into the air," Pinkett remembered.[13] In her 1892 manifesto against lynching, *Southern Horrors*, Ida B. Wells wrote, "A Winchester rifle should have a place of honor in every black home, and it should be used for that protection which the law refuses to give." She cited the examples of African American men arming themselves against lynch mobs in Jacksonville, Florida, and Paducah, Kentucky, as evidence of the success of this strategy.[14]

In less sinister ways, firearms were a ubiquitous presence in rural life. The gun sitting next to the kitchen door was understood to be a necessary tool. One quick shot was the fastest way to put a horse or cow with a broken leg out of its misery or to drive a fox out of a chicken coop. Subsistence or "pot" hunting (as the sportsmen disdainfully called it) remained a source not only of recreation but also of much-needed protein for the rural poor, although much constrained by new conservation laws.[15]

An individual did not even need to own or use a gun to be part of a gun culture. Many Americans learned much of what they knew about guns from Wild West shows, popular novels, and, by the mid-1920s, movies. Public fascination with sharpshooting made a huge celebrity out of Annie Oakley until a train wreck in 1901 ended her career. In the 1920s and 1930s, a number of men earned at least a partial living by giving fancy shooting expositions. Ed McGivern, the most famous of sharpshooters, complained that too many of his students (mostly highway patrolmen and FBI agents) had gotten the wrong idea about how to shoot a gun from the movies.[16]

Guns had also become an increasingly visible part of city living. The pages of the *New York Times* paint a partial picture of the diversity of urban users and uses: telegraph boys taking shots at a sign while waiting for the next message to deliver; German immigrants carrying their rifles out to the countryside for weekend target shooting competitions; nervous commuters fingering hidden revolvers as they traversed dark streets and dicey neighborhoods. Fourth of July celebrations brought weapons out into the open, as did labor conflicts.[17] In 1904, the Brooklyn postmaster issued revolvers to postal employees who handled valuable mail.[18] One reporter imagined how a crowd on a New York street would look if subjected to an X-ray. It is, he wrote, "disquieting to know that the crowd that shoves and pushes you in the subway is more or less heavily armed. The conventional audience at the average New York theatre or even church carries a large number of pistols."[19]

Inevitably, members of this well-armed populace ended up shooting each other. A rapid increase in the number of homicides involving firearms did not escape the notice of public officials, many of whom believed that the easy availability of guns exacerbated urban violence. The problem of gun violence was particularly meaningful to the police chief who sent his men out to face a public carrying concealed weapons, the politicians who heard from their constituents about living in fear of being robbed at gunpoint or caught in the crossfire and sometimes feared for their own safety at the hands of armed assassins, and the coroners who counted and categorized bodies. As a consequence, big cities became the first sites of a series of battles about laws intended to control the risks attendant on the growing availability of cheap, effective, easily concealable firearms.

The Beginnings of Modern Gun Control

At first glance, twentieth-century efforts limiting who could own and carry firearms by statute might seem like a continuation of a very old practice. In a sense, it was. Since colonial times, Americans had created gun control laws that singled out particular social groups for disarmament. These statutes gave public officials the power to take firearms out of the hands of people they believed most likely to misuse them. The point of these laws was not to lower overall mortality from firearms injury. Mathematical probability was not part of legislators' calculations. Laws and lawmakers were also silent on the issue of preventing accidents and suicide. Instead, these statutes reflected the fears of dominant groups. Not surprisingly, African Americans, immigrants, and former confederate soldiers were all singled out for disarmament.[20]

A new generation of gun control advocates imagined using more scientific criteria for restricting access to firearms. They described mortality from guns as an epidemiological problem, akin to smallpox and automobile accidents. They imagined gun control legislation as functioning in a manner similar to traffic laws. Behaviors shown by systematic research as leading to injuries and deaths would be restricted, regardless of the social standing of the persons committing them.[21] They also aimed to decrease the overall prevalence of firearms in civilian hands. No one really needed a gun in an urban, industrial environment, they argued, except as a weapon against other people or a form of self-defense. As a slogan of the time put it, "If nobody had a gun, nobody would need a gun."[22] From their perspective, private ownership of firearms should be treated as a privilege rather than a right.

When it came to firearms, this very modern way of thinking about risk faced stiff opposition from gun enthusiasts who argued that gun ownership was the right of "honest citizens," who were entitled to arm themselves for self-protection. This perspective implied a causal link between gun risk and "criminality," which was difficult to document in statistical terms but seemed commonsensical to its advocates.[23] The position of gun enthusiasts also incorporated a very old idea that social standing was a good indicator of competence in managing risk. At the same time, gun enthusiasts as represented by sportsmen's organizations and publications also promoted the idea that people who were skilled in the use of guns had particular authority to talk about how to control their risks.

These two ways of thinking about restricting gun risk first came into collision in New York as legislators struggled to find a way to address public concerns about the role of firearms in crime and social disorder. In 1905, the New York State Legislature passed Penal Law 1897, a gun control statute that prohibited "foreigners" from carrying firearms unless specifically licensed. It also contained a ban on selling to minors that was, by most accounts, virtually unenforceable.[24] The Armstrong Law, as it was popularly known, seems to have passed into use with little public comment because its focus on particular social groups fit comfortably with the precedents of nineteenth-century gun statutes. The state of New York's gun policy might not have extended any further except for the reckless actions of a former New York City employee named James Gallagher five years later. Gallagher shot William J. Gaynor, the mayor of New York, in the neck as Gaynor stood on the deck of the steamship *Kaiser Wilhelm der Grosse* saying goodbye to well-wishers before embarking on a vacation in Europe.[25] The shot was not fatal, but it did wound the mayor.

Assassinations and assassination attempts were to gun control as train wrecks were to train regulation—they shifted the balance in public perceptions of risk and made possible regulation that, under normal circumstances, would have never been passed.[26] Newspaper coverage, including photographs of the shooting, provoked public fear and outrage and propelled politicians (some of whom must have worried about their own safety) into action. The next day, Assemblyman Goldberg introduced a bill requiring all prospective gun owners to obtain a permit from the police before buying a gun. The bill died in committee, prompting Goldberg to write an irate letter to the *New York Times* claiming that gun manufacturers had inappropriately used their influence to kill it.[27]

As we have already seen in the response to automobile risk, progressive reformers viewed licensing and the accompanying process of registration as modern ways to control public use of dangerous consumer technologies. At the same time, individual consumers resented these measures and questioned the state's motives and authority to judge their competency and intentions. Goldberg's missive prompted a letter to the editor entitled "Right to Bear Arms" from a writer who signed himself "88 Colt." Colt's letter is one of the earliest published statements of the twentieth-century anti–gun control argument later promulgated by the National Rifle Association (NRA). All the pieces are there, from the

Second Amendment argument alluded to in its title, to the "if you make guns illegal, only criminals will have guns" position.[28] Perhaps Goldberg's suspicions were correct and the letter was penned by a representative of gun manufacturers. More likely, however, the letter represents an emerging discourse that would become increasingly common as the so-called pistol issue heated up.

This legislative project required a craftier politician than Goldberg. No one better fit that description than "Big Tim" Sullivan, a longtime ally of Tammany Hall and representative of the district that encompassed the Bowery—that dense warren of sweatshops and tenements crowded with impoverished immigrants.[29] Only newspaper accounts survive to help us imagine how Sullivan maneuvered this bill through the legislature. Apparently, statistics about homicides and suicides helped carry the day. The result was a series of amendments to the 1905 penal code that collectively have come to be known as the Sullivan Law.[30]

This statute was partly an old-fashioned, anti-immigrant measure and partly the first modern gun control law. One clause upped the punishment for immigrants carrying unlicensed weapons from a misdemeanor to a felony. But it also required all civilians who wished to carry a concealed weapon to first obtain a permit. It also mandated a permit simply to possess a concealable weapon even if the firearm theoretically never left an individual's dresser drawer or the shelf under the cash register. Local magistrates, justices of the peace, or, in New York City, the police had discretion in approving permits. The law also obligated dealers to keep a record of whom they sold guns to—not a very effective form of registration.[31] For the historian, characterizing the law is further complicated because what Sullivan intended, how the law actually worked, and what it came to mean in the debate over gun risk were three different things.

The scant surviving evidence suggests that Sullivan saw it as a solution to a local problem: excessive gun violence in New York City. Asked immediately afterward why he sponsored the bill, the politician described it as a means to protect his constituents from predation by armed gangs in their own neighborhoods.[32] His statement, if taken at face value, offers a strikingly rare public acknowledgment that, although middle-class people made the most public noise about being targets for criminals, the poor were more likely to be the victims of violence and least economically able to arm themselves for self-defense. Additional evidence suggests that Sullivan meant what he said because he also tried to have patrolmen disarmed to curb police brutality.[33]

Docket books listing arraignments for Sullivan's district show that arrests for violating the law were relatively rare, especially in the earliest years. Courts arraigned far more people for public intoxication or cruelty to animals than for carrying concealed weapons. The charge was also unfamiliar enough that the court recorder sometimes resorted to writing "vis. Sec. 1897" under "nature of complaint" in reference to the part of the penal code, something that was not done for any other charge. Moreover, the vast majority of those arrested under the law seem never to have gone to trial. They were either fined a small amount or released with a warning (and presumably without their guns).[34] Police appear to have applied the law haphazardly, mostly as a way of temporarily getting violent people off the streets.

Significantly, court docket books also show that the police did not single out the foreign-born for arrest under the law. On the contrary, the majority were listed as "native born" and "white." Most, like John Savage, a thirty-year-old living on West Twenty-Sixth Street, had English surnames.[35] The fate of Joseph Russo and Adolf Benigno, rare examples of Italian immigrants who were charged, in this instance as codefendants, is indicative of what happened to cases that were sent forward to the district attorney. Even though they were initially charged with a felony, they ended up pleading guilty to a misdemeanor and paying a fine of twenty-five dollars each.[36] When defendants faced multiple charges such as assault and possession of a deadly weapon under the Sullivan Law, they were ultimately sent to Sing Sing on the assault charge.[37] Despite the inflammatory stories in the newspapers about innocent gun owners sentenced to hard labor, grand juries avoided letting cases go forward, juries hesitated to convict, and the district attorney regularly offered plea bargains to reduce felony weapons charges down to a misdemeanor.[38]

Did the law help lower the overall crime rate? No one was sure, despite efforts to compile relevant data. Statistics on homicides and suicides showed little change in the first year. Defenders of the law argued for more time. They also pointed out that to make the statute truly effective, adoption would have to be nationwide (or, minimally, by the state of New Jersey) in order to staunch the flow of guns over bridges and through tunnels.[39]

Critics of the law pointed out that innocent people were being arrested for unintentionally violating the law. The most sensational stories made not only the New York newspapers but also the wire services and monthly magazines, helping to turn a local issue into a national one. The *Charlotte Daily Observer* told its North Carolina readers about Mrs. Fannie Emery, who was arrested for fending off two burglars with an unregistered weapon. "You did a very wise thing in firing your revolver," the judge reportedly said. "If you had not fired the shot these men would have escaped. I do not think your action comes under the Sullivan Law."[40]

The process of permitting, however, brought the Sullivan Act to the attention of the wider world of gun enthusiasts. As it turned out, many magistrates were unwilling to issue permits to suburban commuters who felt uncomfortable walking home in the dark without a pistol in their pocket. Judges apparently felt that privilege should be reserved for people who routinely confronted the risk of violence in their jobs. Letters and editorials from people who felt entitled to bear arms began appearing in the media nationwide. Most agreed that criminals should not be allowed to have guns, but they did not see why "honest citizens" should be inconvenienced and humiliated by being required to justify having a gun. Moreover, they questioned the authority of judges to deprive legitimate gun users of their weapons.[41]

Despite the outcry and dubious evidence of success, the Sullivan Law remained on the books. In the interwar era, it became the most important point of reference in national debates over firearms regulation. Gun controllers looked to it as a model for a national gun control law. Gun enthusiasts viewed it as a frightening warning about excessive regulation. Into the 1930s, mentioning the Sullivan Law could be used to inflammatory effect in places as distant from New York as Lexington, Kentucky.[42]

Defending Gun Ownership

Karl T. Frederick was no ordinary New York pistol owner. His abilities as a marksman had earned him two gold medals in the 1920 Antwerp Olympics. Over the next two decades, the champion marksman emerged as the most ubiquitous and influential representative of middle- and upper-class gun owners who deeply resented the Sullivan Law and its legislative legacy. By arguing for gun control laws based on a distinction between "honest citizens" and "criminals," Frederick successfully constructed a durable alternative to the public health model of controlling gun risk. This distinction appealed to both gun enthusiasts and legislators because it catered to the way these two overlapping social groups imagined themselves using firearms.

Frederick wielded a notable set of credentials in carrying out his mission. A law degree from Harvard complimented his skills with a pistol. He supported himself for more than fifty years by practicing mostly corporate law as a partner in the New York firm of Kobbe, Thatcher, and Frederick.[43] In testimony before various legislative bodies, he repeatedly asserted that he was not a lobbyist for the gun industry. Money was never the object of his efforts. His Olympic medals and role as an officer in the U.S. Revolver Association and later the NRA reinforced this assertion. His reputation as a skilled marksman also clearly dazzled many of the state and federal legislators he encountered in the twenty years he worked to suppress strict gun control.

Sometime soon after his Olympic victories, Frederick seems to have concluded that the spread of gun regulation was inevitable in the face of public anxiety about crime.[44] Less temperate gun enthusiasts responded to efforts to promote gun control legislation either with outbursts about how Americans were abandoning their traditional values or with suggestions such as arming civilians to combat crime.[45] Frederick, however, was a political and legal pragmatist. Rather than see the nation go the way of New York, he concocted an ingenious strategy. He would participate in the legislative process, offering legislators alternatives that would protect the interests of gun owners like himself, while satisfying the public's demands for protection against crime.

Frederick began his campaign in the early 1920s by drafting a model law that could be adopted by state legislatures.[46] In all its various incarnations, Frederick's statute seemed on the face of it not that different from other proposals being floated at the time. For instance, it included a provision for a twenty-four-hour waiting period before purchasers could pick up a new gun; it forbade dealers from selling to minors, felons, and aliens; it even required a special license to carry a concealed weapon. Several critical differences, however, separated it from the Sullivan Law, laws proposed by crime prevention organizations, and European laws being introduced at the same time. Gun owners were not required to obtain a license to purchase a weapon or to register firearms already in their possession. They did not need a license at all for rifles or shotguns or for handguns kept in the home or in a place of business. Rather than let judges have free rein in deciding who could have a gun, some versions provided a list of people who had lost that privilege.[47]

Wearing his title as vice president of the U.S. Revolver Association, Frederick approached the National Conference of Commissioners on Uniform State Laws with a draft of his firearms law in 1922. The commission was a group of lawyers appointed by their respective states to draft model laws, mostly to smooth the path of interstate commerce by eliminating discrepancies between the laws of different states. Firearms legislation was far more controversial than most of the issues the commission took on, but Frederick somehow convinced them to appoint a committee to study his proposal.

If other kinds of experts no longer needed to have direct, personal experience or skill to speak authoritatively about risk and safety, discussions about gun control in the 1930s proved a powerful exception. Frederick's experience and skill with firearms served as important tools in convincing policy makers that he knew what he was talking about. On the first of many such occasions, Frederick proved to be a compelling advocate for his cause. He was careful to speak to both sides—the need of some kind of legislation to control criminal misuse of guns and the importance of protecting the rights of legitimate gun users. Charles Imlay, the firearms committee chair, was not only won over by Frederick's carefully crafted arguments but also dazzled by his displays of technical expertise about guns (not to mention his sporting accomplishments).[48] Imlay would later join Frederick in his efforts.

Strikingly, the commission member with the most experience of civilian gun violence was least convinced by Frederick's claims that guns were useful for self-protection. "I hail from a state, where up to a few years ago, the carrying of a pistol was more nearly unanimous than was the wearing of neckties," A. W. Shands, the representative from Cleveland, Mississippi, told his fellow commissioners. "I have been personally connected in the practice of law with trial of something over 100 homicide cases . . . and in that number of 110 cases to be exact about it, there were only three cases in which the dead man was unarmed." If the point of the law was, as Shand perceived it, "to permit a man to carry a weapon in self defense," he thought the underlying assumption was naïve. "I daresay there is not a man in this assemblage . . . in whose defense a pistol would be of any value whatever in a combat with a professional bandit, a highwayman or aught else." Moreover, he thought that the presence of a gun unnecessarily escalated confrontations that would otherwise end in nothing worse than a bloody nose. Shand's comments were ignored. The discussion moved on.[49]

In 1927, the Uniform Firearms Law committee thought that it finally had a viable draft. They sent it out to the American Bar Association (ABA), which had already endorsed the general idea, and to various other interested groups for comment. Along with the law, they published a statement of principles underlying the act. The first item in the list drew a compelling distinction between crime control and self-protection, as well as between legitimate and illegitimate users. "Without making it difficult for a law-abiding citizen to secure arms for the protection of his home, the Act seeks by strict regulations of dealers, identification of purchasers, and strict licensing of those who carry concealed firearms to keep such weapons out of the hands of criminals."[50] The committee had reason to be confident that their approach would soon come to define regulation of firearms at the state level. The law had already been adopted by a half-dozen states and was being considered by the legislatures of several more.[51] They were, however, in for a nasty surprise.

G. V. McLaughlin, police commissioner for the City of New York, wrote a scathing letter to the president of the ABA in which he concluded that the "whole bill seems to be a compromise affair gotten up for the benefit of the manufacturers of firearms."[52] McLaughlin was not alone in his sentiments. Members of the Bar Association Committee on Law Enforcement claimed that the draft law was "directly opposed" to their stated position that existing gun laws were not working and that the manufacture and sale of handguns for civilian use should be prohibited by law.[53] Worse yet, the National Crime Commission had been busy drafting their own model law that included several measures disdained by Frederick and his allies, including requiring prospective purchasers to obtain a permit to purchase a gun as well as to carry one.[54]

The National Crime Commission presented their evidence at the annual meeting of the Commissioners on Uniform State Laws. In the wake of these revelations, Imlay and Frederick considered the options presented to them, including redrafting the law or abandoning the process, as too controversial. In the end, they decided to stick with what they had, dismissing McLaughlin and the Crime Commission as "expressing a decidedly minority view."[55] In 1930, a final, definitive model statute was published.

The District of Columbia adopted a version of the law in 1932 after a vote of Congress.[56] Mrs. Gifford Pinchot, the Pennsylvania governor's wife and a formidable social reformer in her own right, was the first person granted a license to carry a concealed weapon after her state implemented the law in 1931.[57] Frederick also engineered a nearly successful effort to replace the Sullivan Law in New York State. The Hanley-Fake Bill made it through the legislature, only to be vetoed by the governor, Franklin Delano Roosevelt.[58]

There things might have rested except for the election of FDR to the presidency and the well-publicized exploits of bank robbers like John Dillinger, Bonnie Parker, and Clyde Barrow (who reportedly favored a sawed-off shotgun hidden in a special pocket in the waistline of his pants).[59] As part of his New Deal agenda, Roosevelt declared a war on crime and authorized his attorney general, Homer Cummings, to draft appropriate federal legislation, including gun control laws using the federal government's power to tax and regulate interstate commerce. In the spring of 1934, committees in both the House and the Senate considered bills based on Cummings's recommendations. Concerned about potential constitutional challenges, the authors of both bills had built their proposed legislation around the federal government's power to tax and regulate interstate commerce. Crucially, "firearms" covered under the bills were defined as including "pistol, revolver, shotgun having a barrel less than sixteen inches in length, or any other firearm capable of being concealed on the person, a muffler or silencer thereof." Neither bill attempted to regulate in any way long guns such as rifles and shotguns, which, as hunting weapons, were considered sacrosanct. Both bills included clauses requiring dealers to be licensed and obliging sellers to file a form with the IRS every time a gun was transferred from one owner to another. The bill of sale with revenue stamps functioned as a de facto license and registration. Law enforcement officers could presume that people discovered with undocumented guns had obtained them illicitly (what legislators called the "bank robber clause").[60]

Meanwhile, opposition forces were gearing up for a fight. Organized gun enthusiasts could accept some government regulation of machine guns—weapons primarily associated with criminal activity. But they opposed both the inclusion of handguns and the scope of licensing and registration requirements. The National Rifle Association led the opposition. Over the previous decade, a membership drive that harnessed the power of local gun clubs had transformed the NRA from a small organization mostly known for teaching marksmanship into a substantial political force.[61] In order to apply constituent pressure on Congress members, the organization sent out a press release about the bill and published a number of articles arguing against handgun control. Its readers were asked to write to their representatives in opposition to the bill. Since they did not actually have the text of the bill, they described it in vague and inflammatory terms as being worse than the Sullivan Law.[62]

As the congressional hearings began, Karl Frederick reappeared, this time wearing the title of president of the NRA. He was partnered in his testimony by Milton Reckord, adjutant general of the Maryland National Guard and executive vice president of the NRA. Representatives of the major gun manufacturers, especially Colt, also accepted invitations to testify but largely restricted themselves to technical questions because of the fear of appearing self-interested. As a result, Frederick and Reckord (with a guest appearance from Charles Imlay) made the case against including handguns.

The ensuing discussions revealed the complexity of interest groups involved, as well as Frederick's skill in convincing legislators to weaken legislation. However, it took a while for his abilities to become apparent. In the first set of hearings, Frederick and Reckord were unprepared and caught off guard by the hostile tone of the inquiry. Members of the Ways and Means Sub-Committee who questioned them were clearly irritated by the NRA's press release about the bill and the deluge of letters from angry constituents that resulted. Frederick responded by deploying his honest citizen argument. He told the hearing that the proposed law would "deprive the rural inhabitant, the inhabitant of the small town, of any opportunity to secure a weapon which he perhaps more than anyone needs for his self-defense and protection" by making it too difficult to buy a handgun. The chair of the committee leapt on Frederick's claim that handguns were essential to the self-defense of farmers. "Why is not a rifle or a shotgun, the possession of which would not be prohibited under this act, sufficient for the self-defense of an individual or an individual's home?" In response, Frederick changed the subject to states' rights.[63]

The first set of hearings adjourned without a vote. They were set to convene in a month's time. In the meantime, Frederick and Reckford went to see the attorney general's deputy with a proposal that the words "pistol and revolver" be removed from the definition of "firearm" so that the law would apply only to machine guns and sawed-off shotguns. Frederick claimed that this provision, along with other anticrime legislation moving through Congress, would be enough to counter any "Dillinger" threatening the public welfare. The deputy attorney general countered with an offer to write in an exemption for pistol club members, but "we did not want it," Reckord testified, "because we did not want members of our association to be exempted as such over and above any other honest citizen."[64]

Rebuffed, Reckord now brought the same proposal to the committee as the hearings began again. His questioners were still dubious about his reasoning for removing handguns from the bill. Charles Imlay, the former chair of the Uniform Firearms Law committee, came to testify in support of his longtime collaborator. He became embroiled in a telling discussion with Representatives Treadway and McCormack comparing the risks of automobiles and handguns. Imlay did not see why it should be legal to loan a car to a neighbor, but not a handgun (since permits, under the proposed law, were not transferable without applying to the federal government). One of the congressmen probed Imlay's reasoning. "Suppose I asked you to borrow a gun; would you loan it to me with the same state of mind that you would loan an automobile?" Imlay replied, "If I knew you." "You are a remarkable man," McCormack rejoindered. "I would not loan a gun to my best friend without an explanation from him as to what he wanted it for."[65]

A few weeks later, H.R. 9741 was forwarded to the floor of the House for a vote with a critical change. Sometime between the hearing and the recommendation, the words "pistol and revolver" had been removed from the definition of "firearm," leaving machine guns and sawed-off shotguns as the most important types of guns being regulated. Apparently, Frederick and his compatriots had successfully applied pressure to key committee members. Representative Carl Mapes from Michigan wanted to know what had happened since he had been receiving protests from "prominent women and women's organizations in my district" about the changes. Robert Doughton, who was presenting the bill, explained: "Protests came to the committee from some ladies' organizations throughout the country objecting to the elimination of pistols and revolvers. The majority of the committee were of the opinion however, that the ordinary, law-abiding citizen who feels that a pistol or a revolver is essential in his home for the protection of himself and his family should not be classed with criminals, racketeers, and gangsters; should not be compelled to register his firearms and have his fingerprints taken and be placed in the same class with gangsters, racketeers, and known criminals."[66]

Another congressman wanted to know, "Have we the gentleman's assurance that the sportsmen's organizations have withdrawn the opposition they formerly expressed to the measure?" Doughton responded, "They have; and they heartily support the pending bill." He added, "The Department of Justice has agreed to an amendment which makes the bill acceptable to sportsmen and sportsmen's organizations."[67] After a few more questions, the bill was voted on and passed. Thirteen days later, the president signed it into law as the National Firearms Act.

Frederick and Reckord's tireless efforts on behalf of gun owners helped ensure that, by the end of the 1930s, gun ownership and gun regulation in the United States diverged significantly from much of Western Europe, where an individual needed a permit to purchase or possess any kind of gun, long or short.[68] Attorney General Cummings made a number of public pleas for universal registration of firearms but was unsuccessful in convincing anyone in Congress to pursue his agenda. In the next few years, Congress voted for several minor changes to the act and in 1938 passed the Federal Firearms Act, which further regulated firearms dealers and repairers. Reckord and, more rarely, Frederick made return appearances, but there were no more probing questions about the honest citizen's right to have a handgun for self-defense. A wave of assassinations and urban violence would have to occur in the 1960s for Congress to take another look at this issue.[69]

Not all gun enthusiasts were appreciative of the strategy of an elite eastern lawyer and retired army general. Especially in the West, there was a widespread suspicion that any regulation would eventually lead to confiscation. A substantial subculture shared the sentiments of Mr. L. L. Donmeyer of Turlock, California, who, as the gun control debate heated up in the 1920s, had written in to the NRA's journal the *American Rifleman*, "It is distressing to know that we have in our midst a few weak-kneed imitations who are attempting the passage of laws to prohibit the American from purchasing and possessing firearms." He thought that "if these spineless imitation Americans are so deathly afraid of firearms, they have our permission to step out of the good old U.S.A. any time they feel so inclined."[70] Jack Miller and Frank Layton, who were arrested for transporting an unlicensed 12-guage sawed-off shotgun across state borders from Claremore, Oklahoma, to Siloam Springs, Arkansas, took their case to the Supreme Court. They challenged the National Firearms Act as violating their Second Amendment rights. The court disagreed. They reiterated a narrower interpretation of the Constitution consistent with more than a century of rulings.[71]

In contrast, gun ownership was developing an increasing stigma among segments of the middle class. Like other forms of regulation, the public was also in favor of stricter laws as long as they were not personally inconvenienced. Seventy-nine percent of respondents to a 1938 Gallup Poll agreed that "all owners of pistols and revolvers should be required to register with the government."[72] As the century wore on, the gap would only grow wider.

Sportsmen and Gun Safety

When the midwinter weather grew bleak in eastern centers of commerce, many executives liked to retreat south for a week of hunting at private clubs carved out of old plantations or empty tracts of land. William A. Law, president of Penn Mutual Life Insurance of Philadelphia, made his last such pilgrimage in the winter of 1936. In the company of S. Clay Williams, chairman of the R. J. Reynolds Tobacco Company, and two other men, he was walking down a narrow, winding trail at the Brush Creek Hunting Club in Colbert, Georgia, when Williams slipped, discharging his gun into the back of Law's leg, hitting an artery. Law's companions applied a tourniquet to the wound, carried him two miles to their automobile, and then drove him to the local hospital, where he died. One of them later reported that the wounded man had admonished Williams to "refrain from self-reproach," quoting Law as saying, "It was just one of those things that happen."[73]

The state had little to say about Law's death. Other than requiring a hunting license, which provided revenue for enforcing laws about hunting out of season, most localities did not normally get involved in regulating the possession and use of hunting rifles. As in most such shootings, the coroner ruled Law's death an accident. In the rare cases in which unintentional shootings went to court, juries almost always returned a verdict of not guilty. Hunting accidents were considered largely a private matter—one of the risks of an inherently dangerous sport. However, shooting one's hunting companion was not exactly like falling off one's horse. Sportsmen were acutely aware that gun sports were unique in the risks posed to bystanders and other participants.

By the 1930s, accidents such as this one were not supposed to happen. For decades, the sporting community had made a concerted effort to inculcate a simple set of rules to prevent hunters from shooting each other (or themselves) accidentally (most of which would have been familiar to hunters of Benjamin Franklin's era). These included never pointing a loaded gun at something one did not intend to shoot, never shoot at something moving at the bushes, and—the rule that might have saved Law's life—never walk along with your gun pointed at your hunting companion.[74] At the start of each hunting season, sporting periodicals reminded readers of the previous season's toll of accidental shootings and rehearsed the rules again. In an effort to bring the message to the broader public, sportsmen's associations offered press releases to newspapers. Formal training programs, such as those offered by the army, the NRA, the Boy Scouts, and schools, also emphasized these rules.

In target shooting, the effort to make gun sports safer had been extraordinarily successful. Gone were the days when a Sunday afternoon's trapshooting could turn tragic because someone fired wildly at escaping pigeons. Gun clubs carefully laid out sight lines and policed the way participants handled guns. Competition and club rules generally specified that guns could only be loaded once participants were "at the score" and must immediately be unloaded of any remaining cartridges before leaving the firing box. In competition, marksmen could only shoot when called upon by officials and then only within a specifically marked area. Pointing a gun, loaded or unloaded, at another person was grounds for discipline or disqualification.[75]

Hunting was a different matter, however. Hunters walked around with loaded guns because it would take too much time to insert ammunition once game actually appeared. The target itself was uncertain—where would it come from? What would it look like? Was the rustling in the bushes a deer or another hunter? Excitement also blurred judgment. Since hunting, particularly in elite game camps, was also a bonding exercise, it was sometimes socially difficult to walk away from a companion engaging in dangerous behavior. Few sources were as blunt on the matter as the *American Shooter's Manual*: "I would advise any gentleman, who shall be so unfortunate as to be in the field with a companion, who will persist in such improper conduct, to leave him without ceremony, return home, or shoot alone, in another direction. This may savor of incivility; but a man's politeness should never jeopardize his life or limbs."[76]

How could hunting be made safer? Before the gun control debates of the 1920s, many writers recommended criminalizing careless shooting. Bellmore Browne, with whom this chapter began, thought hunters who accidentally shot someone should be forbidden by law from ever shouldering a gun again.[77] A 1906 editorial in *Forest and Stream* drew a parallel with the growing movement to test and license automobile drivers for competency. "May it not be that some action on the same general lines will have to be taken to protect the shooting public from itself?" the author asked.[78] Such suggestions about state intervention grew rare as sporting organizations went on the offensive against new gun laws.

The safety-first movement offered an alternative model of public education and awareness. In 1938, the Sporting Arms and Ammunition Manufacturer's Institute organized a meeting of major manufacturers to discuss the problem of gun safety. They declared that members of the institute desired to "make shooting one of the SAFEST sports for the youth of this nation."[79] Members found it easy to create a list of things manufacturers could do to make guns and ammunition safer. For instance, they recommended testing cartridges to ensure they would not blow apart the chambers of poorly made guns. Dealing with factors they labeled "beyond the Industry's control" was harder. In the end they offered the "Ten Commandments of Safety," which they proposed to distribute as package inserts in guns and ammunition. They also intended to publish posters and distribute information through sporting periodicals and magazines.[80]

Not until after World War II did the NSC break its silence on firearms. In 1947, they published a study that revealed that the death rate from firearms accidents had increased 11 percent that year, lagging right behind railroad fatalities in the tally (although far fewer than automobile accidents). Safety experts argued that war souvenirs, including the prized German Luger brought home by many returning servicemen, were the culprits.[81] It would take much longer for state and federal legislators to pass regulation specifically aimed at decreasing the number of accidental shootings.[82] By the 1960s, the conspicuous presence of firearms in the hands of more radical participants in the civil rights movement reopened the question of guns as a legitimate means of self-defense.

Rethinking Self-Defense Arguments in the 1960s

Soon after the formation of the Black Panther Party in 1966, Huey Newton decided that guns were essential to the Panther's mission of protecting African Americans against police brutality.[83] As his cofounder Bobby Seale explained in his 1970 memoir *Seize the Time*, "There was a law service section up in the poverty program office, and Huey studied those law books, backwards, forwards, sideways, and cattycorners; everything on gun laws."[84] He discovered that, under California and federal firearms laws, the Panthers could legally carry rifles, shotguns, or handguns in public or in their cars as long as they did not conceal them or point them directly at someone (which could be considered assault with a deadly weapon). At the time, Newton was on parole. Consequently, he was told by his parole officer that he was allowed to carry a long gun but not a handgun. So Newton armed himself with an M1, a semiautomatic rifle extensively used by the military, which he put in plain sight between the seats of his VW bug. Seale carried a 9mm pistol.[85]

To raise money to buy more guns, Newton hatched a plan to resell cheaply bought copies of Mao's *Little Red Book* to would-be radical Berkeley students. With some of the proceeds, Newton, Seale, and several others visited the sporting goods section of the B.B.B. department store, where they bought a shotgun. Guns were not just a means to an end, though. Even Newton and Seale were not immune to their peculiar emotional appeal. "We were drunk admiring those pistols, shotguns, rifles, what have you," Seale remembered. After a minor exchange with a saleswoman, they walked out with the weapon they wanted.[86]

Wednesday evenings, Newton required members to attend political education classes. At the first meeting, he announced, "Before you go through political education you have to go through use and safety of weapons for one hour, and at Saturday meetings you have to go through use and safety of weapons for one hour."[87] He also decided that when Junior Black Panthers turned sixteen, they could take classes on "how to use a gun and defend themselves and the people of the black community," if they brought a note from their mothers—a plan that in its own way reflected the linking of boyhood and guns common in white America of the time.[88]

Despite Newton's efforts to inculcate safety practices, the strategy of armed self-defense proved very dangerous to the Panthers themselves. By 1970, an estimated thirty members had died in violent confrontations with the police. Newton was wounded in a shoot-out with police officers who pulled over his car (as they had done many times before). He was convicted of voluntary manslaughter for the death of one of the officers but later released on appeal. Seale was also tried for murder but acquitted by a hung jury. Eventually, what remained of the Black Panther Party disavowed armed violence as a tool of social change.

Karl Fredericks probably did not have Huey Newton in mind as a gun owner when he laid out the blueprint for gun laws like California's in the 1920s. Although race was never directly mentioned in public discussions about gun control in the interwar years, the testimonies of Frederick and Reckord and others implied that the state had an obligation to protect white male privilege by not making it too "inconvenient" to purchase and use guns, especially for hunting and self-defense. They assumed that the police and other more informal mechanisms of racial hegemony would prevent "negroes with guns" from becoming a serious threat. By the late 1950s, more radical civil rights leaders had begun to openly challenge these assumptions by suggesting that blacks arm themselves.[89] In the aftermath of Newton's shoot-out with the police, legislators hastened to tighten up California's gun laws.[90]

THE PROGRESSIVES (IN THEIR VARIOUS GUISES) LAID MUCH of the groundwork for a modern American risk society. In some ways, their efforts on the gun issue are consistent with the way they tackled the problem that came with the widespread availability of inherently dangerous technologies. They recognized that legislation offered a crucial tool in limiting the ways people could injure themselves and others. Despite resistance from gun enthusiasts, reformers did indeed achieve more stringent controls in some states. By the end of the 1930s, federal regulations began requiring gun dealers to be licensed and put some limitations on transportation over state lines and sales through the U.S. mail.

But in other ways, the gun story did not follow the paradigmatic path of other parts of this developing risk society. By the 1920s, most policy makers rejected a public health approach that would limit the overall number of injuries and deaths from firearms by whatever means necessary. Instead, a sharp distinction was made between intentional and unintentional misuse of guns based on the notion of criminality. In policy debates, a small group of elite users were chosen repeatedly to represent the broader public. The settlement these men negotiated reflected their own sense of privilege and beliefs about what a gun might be good for but neither solved the gun risk problem nor satisfied the extraordinarily broad range of people whose lives were affected by guns. Consequently, the debate became increasingly polarized between different groups who could not agree about the nature of the problem or indeed whether guns were a problem at all.

As a result, at a time when the United States was beginning to export the techniques of the safety-first movement to the rest of the world, American efforts to regulate guns remained far less stringent than those in many European countries, particularly Britain. The long-term consequence is a striking kind of American exceptionalism. Today, approximately 40 percent of American households contain at least one gun. Out of the estimated 200 to 250 million privately owned firearms in the United States, 65 million are handguns. Sixty-six percent of the 15,517 homicides committed in 2000 used guns.[91] Moreover, as a society, we are still arguing about gun risk in virtually the same terms as people did almost one hundred years ago.

As this and preceding chapters suggest, potentially dangerous technologies are not without their pleasures—a fact not lost on industrialists looking to expand their markets. The trick was to provide pleasure without having your customers end up dead. No one understood this better than the entrepreneurs who created the first mechanized amusement parks at nearly the same moment as the emergence of the safety movement, the introduction of the automobile, and the proliferation of mass-produced firearms—some of which ended up in the shooting galleries of America's amusement parks.

10

Risk as Entertainment

Amusement Parks

Just outside Pittsburgh, Kennywood amusement park sits on a bluff across the Monongahela River from U.S. Steel's Edgar Thompson Works. From the top of the park's Racer coaster, generations of riders could glimpse a vast gray expanse of rolling sheds, smoke-belching smelters, and, on some days, carloads of molten metal rumbling across the river to the rail yards situated just below the park.[1] With a south wind, the sound of the factory whistles and the distinctive odor of burning coke and molten metal wafted through the picnic grove and across the artificial lake. Nobody noticed. In the heyday of American industrialization, mills were everywhere in Pittsburgh. But their dangers and tensions belonged to a parallel universe, seemingly unconnected to the pleasures of a day in the park.

Kennywood depended primarily on the patronage of urban, working-class people. These men and women had plenty of experience with accidents involving industrial technologies. As we have already seen, they also belonged to the first generation subject to the broadening ambitions of safety experts. Despite the ever-present anxiety of risk in their everyday lives, they sought out and willingly paid for risk as entertainment. However, they made a strong distinction between the risks they were forced to tolerate as part of wage earning or urban life and the pleasant sensation of risk purchased as a commodity. For admission-paying visitors, Kennywood was a site of consumption, a place that reordered the meaning and distribution of risk endemic to the landscape of factories, dense urban neighborhoods, roads, and streetcar lines across the river. Once inside the park gates, ticket holders believed that the price of admission entitled them to experience the excitement of elective risk taking, without any of the negative consequences.

The Old Racer Coaster at Pittsburgh's Kennywood Park overlooked the rail yards and railroad bridge leading to the Edgar Thompson Works steel mill. *Courtesy, Charles and Betty Jacques Amusement Park Collection, Penn State University Archives, The Pennsylvania State University Libraries.*

The people who helped to create and run Kennywood and the hundreds of other parks built across America in the early twentieth century did their best to comply with public expectations.[2] Protecting the public from actual accidental injury and death, while maintaining the illusion of teetering on the razor's edge of disaster, required an enormous effort of engineering, maintenance, performance, and marketing. It did not always succeed. A fickle public ignored attractions they perceived as either boringly safe or truly dangerous. They also sometimes chose to add their own quotient of danger, particularly on thrill rides. When accidents did happen, lawsuits and bad publicity often followed.

Behind the scenes, more progressive members of the industry struggled to adapt the tool kit of modern risk management to the peculiarities of their industry. Andrew Stephen McSwigan, Kennywood's original manager, was also the founder and first president of the industry's trade association, the National Association of Amusement Parks (NAAP). His son, Brady, succeeded him as both manager of Kennywood and a leader in the association. In the 1920s, the NAAP priorities were the creation of a code of safety standards and a mutual liability insurance pool. At the same time, members reserved the right to reject input from outsiders who claimed abstract knowledge but lacked intimate experience with roller coasters, merry-go-rounds, tilt-a-whirls, or any of the hundreds of other kinds of rides and attractions. They were particularly dismissive of government inspectors. In all but a few places, the industry managed successfully to fend off state-sponsored regulation until the 1970s. Until then, self-policing and occasional insurance company inspections remained the principle means for containing risk.

The Many Forms of Risk as Entertainment

Journalists and other observers often used the term *pleasure seekers* to describe the crowds that wandered the grounds of amusement parks.[3] Each season, park managers looked for new ways to cater to every imaginable source of pleasure seeking (within the boundaries of legality and conventional public morality): sweet, salty, fatty food for the hungry or merely gluttonous; cool water beckoning swimmers and boaters on hot days; dancing and dark rides for those who craved sexual contact; and strategically positioned seating for those that just liked to watch.[4] And then there were the thrill rides, the disaster scenarios, and the dangerous acts, daredevils, and aerialists—each catering in a different way to a deep, human craving for controlled risk taking. These three different approaches to the commodification of risk allowed consumers to decide whether to watch or participate, whether to see someone take real chances or simply enjoy a theatrical representation.

For most people, enjoyment of risk-centric entertainments depends on what psychologists call a "protective frame."[5] This psychic boundary separates pleasurable thrills from anxiety-inducing fear by indicating whether or not real harm might occur. In an earlier world, pleasure from risk taking came mostly from exploiting the potential of the everyday. A sense of skill or mastery constituted an important part of the frame. Some people, particularly young men, made a habit of playing with the uncertainties of fire, animals, tall masts, and deep water. Others preferred to pay to watch someone more skilled and daring take chances. The nineteenth-century audiences who paid to see John Rarey face down a seemingly unbreakable stallion or gathered to watch Dan Patch jump from the roof of a mill building into the Pawtucket River stood outside the frame, looking in.[6]

In commodifying risk, amusement park entrepreneurs continued the long tradition of catering to the public's desire to watch someone else face danger. Their strategy, after all, required supplying entertainment for a wide range of appetites. Each summer, Brady McSwigan hired dozens of performers to entertain Kennywood's visitors. In 1929, the Dutton Circus provided a series of aerialists—Vivian DeVere, "a thrilling high aerial trapeze act presented by a chic and accomplished little lady"—and animal acts, mostly involving horses.[7] In the continual struggle to find a novel combination of heights, water, and animals—the conventional elements of many acts—some performers strayed into the realm of the ridiculous. Mills and Mills featured two high-wire aerialists teetering above the crowd in an elephant suit. No record survives to tell us how "Jumbo Jr.—The Human Elephant" was received in Pittsburgh, but the Mills brothers' attempt at originality spawned no imitators.[8]

As the twentieth century unfolded, new technologies—bicycles, automobiles, motorcycles, and airplanes—played an increasingly important role in demonstrations of skill and daring. The pleasure of watching someone ride a bicycle in a cage or walk along the wing of an airplane had multiple dimensions. Just as audiences familiar with horses enjoyed horse tamers, a generation increasingly familiar with machines appreciated performers doing something imaginative and dangerous with them. Crashes and calamities could expand the audience for daredevil acts.[9] Promoters made much of the element of danger, using words like "death defying" to generate public interest. Like horse breakers and railroad brakemen, performers who risked their lives for enter-

High-wire and animal acts were both popular at amusement parks. Two men in an elephant suit creatively combined the two. *Courtesy, Charles and Betty Jacques Amusement Park Collection, Penn State University Archives, The Pennsylvania State University Libraries.*

tainment were proletarianized (and largely poorly paid) workers who gambled their lives and bodies in exchange for wages. Consequently, the public generally expressed little concern about whether those risks were excessive and unnecessary. Most commentators on the riskiness of amusement park employment followed the line of Edwin Slosson, who worried most about the moral effect on audiences: "Our feelings for the dangers and suffering of others is easily jaded, so the feats must each year be more and more dangerous to excite the interest of spectators."[10] A more astute observer pointed out that members of "the public," in their position as consumers of the spectacle of risk taking, "don't comprehend that their hunger for shudders forces the management to gratify it, or that it is they who have put another's life in jeopardy." He added that "they overlook the necessity . . . that has driven him [the performer] to adopt such an atrocious calling."[11]

Relying on fallible and fragile human talent also had a downside for park managers. They preferred reliable attractions in their factories for fun. Performers too often failed to deliver. In the early 1930s, Kennywood manager Brady McSwigan received a letter from a fellow park manager, Rex D. Billings, describing an act Billings had recently employed. Peejay Ringens "rides a bicycle down an incline at terrific speed," Billings told McSwigan. "Upon going into the air, the machine drops from under him and he goes on to dive in about three feet of water." Peejay's outstanding characteristic, as it turned out, was not the quality of his performance, but rather the fact that he was willing to perform every day and in adverse weather conditions.[12] Season after season, McSwigan struggled with performers who begged off or did not show up at all if injured, ill, or fearful of the conditions under which they were being asked to perform.[13]

Many parks (though not Kennywood) also offered the opportunity to watch paid performers endanger themselves in a form of entertainment sometimes called "disaster spectacles." Of the three types of commodified risk presented by parks, this second form most directly prefigured what would become the twentieth century's dominant form of risk as entertainment: cinematic and televised stories of ordinary people confronting mortal dangers.[14] In these shows, actors performed a series of dramatic scenarios against a painted backdrop of natural or man-made disasters such as the Galveston flood of 1900 or the San Francisco earthquake of 1906. Real disasters are, by definition, unpredictable and dangerous to witness. Reenactments gave spectators the opportunity to see an event they had read or heard about at their convenience. They also gave meaning (often redemptive) to events that were, in real life, mostly chaotic and destructive, through plots emphasizing heroic rescues and happy endings.

The Coney Island parks originated a popular variation on the theme—a simulation not of a natural disaster, but rather a familiar, man-made one—the tenement fire. Several times a day, crowds filled the bleachers in front of an enormous stage to watch as a building burst into flames and terrified women screamed from the windows and the rooftop. Onto the scene galloped horses pulling fire engines. Gallant firemen turned the hoses on the building, mounted tall ladders to rescue the helpless women, and held trampolines while male tenement dwellers jumped to safety. In the end, the fire was extinguished and the victims rescued as the crowd cheered.[15]

For most park managers, daredevils and disaster spectacles were of secondary interest. Mechanical rides, the third form of commodified risk, defined amusement parks, setting them apart from other kinds of public amusements, such as circuses. Unlike performances, they offered the public the opportunity to be inside the protective frame, to engage all the senses in the pleasurable exercise of being just a little bit scared. In the first decades of the century, the category of "amusement devices," as they were sometimes called, included slides, trick stairs, and other spinning barrels that literally tossed patrons to the ground. Edward F. Tilyou, proprietor of Coney Island's Steeplechase Park, explained his rationale for investing in these types of entertainments. "The average person likes to have a share in making his own fun, instead of having attendants or mechanical contrivances do it all for him." These kinds of amusements pleased bystanders as well as participants. As Tilyou noted, "Folks also take great glee in seeing other folks in embarrassing positions."[16] However, most parks gradually abandoned these devices. They were too dangerous and too prone to human error on the part of both patrons and attendants.

Instead, parks increasingly adopted rides in which patrons were confined to their seats while the design of the ride itself determined how much of their own fun they could actually make. Like railroads, these small socio-technological systems could be rendered predictable, and therefore safe, by limiting human input from both patrons and employees. At the same time, the potential for thrills could be literally built in. In effect, mechanized rides deskilled risk taking as entertainment. And the public loved them. By the mid-1920s, Kennywood patrons could enjoy three different roller coasters, as well as a variety of other exciting novelty rides, including the "Whip," "Circle Swing," and "Caterpillar." End-of-season receipts showed more than four hundred thousand tickets sold, making them the biggest moneymakers in the park. The merry-go-round and the two funhouse attractions, the "Mysterious Knockout" and the "Bug House," ran a close second.[17]

Besides being popular, the devices had another advantage: they lacked many of the uncertainties of human and animal performers. As long as rides did not break down (hardly a given), park owners could know with certainty how often and for how long they would run and what their capacity would be. As the technology improved, less and less skill was required to operate them. The large capital investment needed to build rides constituted their principle disadvantage. Contracting with concessionaires to bring in smaller, temporary rides helped spread the burden. A few parks, such as Chicago's Riverview, went so far as to lease six of their eight coasters to concessionaires. "We prefer to pay dividends rather than keep on re-investing in steel and lumber," a manager explained.[18]

Circle swings, merry-go-rounds, and endless variations on spinning cars all had their enthusiasts. But giant wooden roller coasters gradually became the centerpieces of parks and the most prestigious area of design.

Designed and Operated for Thrills and Safety

Roller coasters provide a significant, if not altogether typical, example of how amusement parks utilized mechanical devices to reconcile consumer demands for thrills, safety, and novelty, while also turning a profit. Through a series of ingenious innovations, designers improved the safety of coasters while simultaneously making them more exciting to ride. To a certain extent, the systemic logic of railroads applied: limiting human agency during operations made roller coasters more predictable and therefore safer. But successful designers also tapped into an intuitive and empirical understanding of risk perception to manipulate the expectations and emotions of park patrons. Attention to the technological side of controlling risk did not end when rides were put into operation. Like railroads, roller coasters required careful maintenance and a skillful and judicious operation to perform reliably.

Various kinds of rides based on the idea of a railroad began appearing in the late nineteenth century. The earliest "scenic railroads" sought to entertain riders by taking them on an imaginative journey, often aided by elaborate lights and sets. The Edward C. Boyce Company described their version as "2,500 feet of bewilderment . . . passengers find themselves being dashed along down sharp inclines and up again . . . through caves of darkness and tunnels containing beautiful scenery, and lighted by many colored electric lights.[19] Other companies built versions featuring scenes from the Bible or a trip around Pike's Peak or some other tourist locale. After the turn of the century, the term "roller coaster" gradually emerged to describe a different kind of ride.[20] Although roller coasters often had names and sometimes themes, designers mostly abandoned the use of scenery evoking a journey in favor of a more visceral experience. These rides entertained patrons through rapid motion and a series of psychological and physiological tricks balancing anticipation with stimulation, a feeling of imminent danger with the exhilaration of surviving intact.

Travel on early scenic railroads and roller coasters proceeded at a leisurely pace.[21] But slowness did not equate with safety. On a June morning in 1910, two cars left the track of the Rough Rider scenic railway on Coney Island, throwing seventeen people onto the pavement beneath the ride, killing three. Derailments and collisions were commonplace.[22] Designers counted on gravity to keep cars on the tracks, occasionally supplemented by a second set of wheels mounted on the side of the cars and/or the use of guardrails. Consequently, heavily loaded cars, which generated more centrifugal force as they traveled around curves, sometimes flew off of the tracks. Differing loads also meant that some trains generated more momentum than others, catching up with cars ahead if platform operators or drivers employed on some rides failed to assiduously apply the brakes. Loading areas proved particularly dangerous. Sometimes trains traveling up the first hill broke loose, rolling backward into the station, ramming into stationary trains being loaded or unloaded. Other accidents happened because trains failed to slow down at the end of a ride.

Better braking systems offered one obvious solution. In the 1890s and early 1900s, inventors experimented with brakes inside the cars operated by a paid driver or more occasionally by the passengers themselves. Braking from inside cars allowed quick responses to uncertainty caused by excess loads, but it also created other problems. Inside the protective frame, some passengers felt the lure of speed while others, less willing to suspend disbelief, argued for a sedate pace, unappealing to their fellow riders. Neither a railroad nor a roller coaster could operate as a democracy. It also rapidly became apparent that even if drivers remained immune from the requests and exhortations of passengers, they could not necessarily be trusted to maintain safe speeds.

Brooklyn city officials suspected that the derailment of the Rough Rider resulted from "reckless operation." Many operators on Coney Island were, as they told the press, "inclined to speed them beyond the limits of safety."[23] Ten-hour shifts of riding cars in the hot summer sun taxed the judgment of even the most conscientious. W. Earl Austen, a former driver interviewed in the 1970s, remembered, "I ran the thing until I would actually find myself falling asleep. Dozens of times I've eaten sandwiches in them. They'd hand us coffee or something on the way up." Austen also noted that brakes on the trains he drove only worked during certain parts of the ride.[24] Braking from outside the car offered an alternative. Brakemen sat next to the turns applying a brake as the cars came around. However, this strategy was still very much tied to human judgment.[25] Careful design to eliminate the need for braking except at the end of the ride offered the best solution, but this could only be achieved if a way could be found to ensure that cars would stay on the tracks in the turns.

More than any other person, roller coaster designer John Miller played a key role in figuring out how to design roller coasters that felt very dangerous but were, in fact, more predictable and safer than their predecessors. Miller stood out from his competitors in having a particularly astute sense of both changing public perceptions of what constituted a thrilling ride and the ways the public behaved during the course of a ride (including the ways they might endanger themselves). In the course of a forty-year career, Miller filed more than one hundred patents, including a number of widely used safety devices. He also designed some of the fastest, tallest, and most exciting coasters of the 1910s and 1920s.[26] Kennywood had three: a rebuilt Racer that featured two sets of cars, rumbling side by side on parallel sets of tracks; the Jack Rabbit, a high-speed coaster that utilized a natural ravine in its design; and the Pippin, which also featured a long drop, this time ninety feet into a ravine.[27]

Surprisingly, Miller had no formal academic training in engineering. In the early years of the industry, design, construction, and operation of these enormous, complicated devices emerged out of the craft tradition—the ambitious expression of a slowly disappearing part of the vernacular. Miller was the son of a German cabinet maker who, like Henry Dorn a generation earlier, had immigrated to Ohio in search of economic opportunities. Miller left school at fourteen to help his father build houses, barns, and bridges. In 1896, he traveled east to work for G. E. Dentzel, a prominent manufacturer of merry-go-rounds and other amusement devices. After Dentzel learned that Miller had experience supervising work crews as part of his father's bridge-building business, he sent the young man to act as foreman for a scenic railroad the company contracted to build in Torresdale Park, an early amusement park in northeast Philadelphia. The magnitude of the assignment did not frighten Miller because, as he later recalled, "I did not know what a scenic railroad was."[28]

Once on the site, Miller "did not have any plan at all, only [a] grade chart of the ride." As he recalled, "we worked things out fairly well." He and his fellow ride builders later relied on more elaborate plans and specifications. But designing and building a wooden coaster was not like assembling an automobile or even a ship. Each coaster was site specific. Trial and error constituted an essential part of the process. Teams of carpenters assembled thousands of pieces of lumber, improvising as they went. Routine building practices included tearing out and rebuilding sections that did not seem to be working. Designers gradually learned to include features such as moveable mounts under the tracks that allowed for small adjustments to obtain the desired ride or to compensate for settling in the structure as a whole.[29]

Although experience gave Miller and his team a good idea of how a particular coaster would feel to riders, in reality, they needed to rely on the on-site trials. The final step of the process required builders to run a train over the newly constructed tracks. Empty cars loaded with sandbags made the first transit. If all went well, human passengers followed to see how the ride would feel to park visitors. On the Torresdale job, Miller sat in the rear four cars of a train. He later described it as "the experience of my life." Hearing something give way, he looked down to see a part called a "dog lever," which helped keep the train on the tracks, falling. At that point, he knew enough about scenic railroads to scramble up to the train's whistle, warning the engineer to stop all the other cars so as to prevent a collision. When Miller's train finally came to a halt, it was "at all angles but this on [the] structure." Miller had found his calling. He remained at Torresdale for the rest of the season as a ride operator. The following year, Dentzel sent the now-seasoned veteran to the Buffalo World's Fair, where he built a variety of attractions, including a "mill stream"—a water flume ride.[30]

As Miller gained experience, he began experimenting with technological devices that might make rides safer. In 1910, he patented a device called a "chain dog," a safety ratchet for keeping cars from slipping backward into the loading area if the chain broke. As an added benefit, the slow, rhythmic "clunk, clunk, clunk" of the car moving over the teeth of the dog added to the psychological suspense of waiting for the fast part of the ride to begin. Secondly, he locked the cars down to the track, using a third wheel mounted under the track rather than on the side of the car.[31] This innovation prevented derailments and eliminated the need for braking in the corners, making high-speed coasters like Kennywoods' Racer, Jack Rabbit, and Pippin possible.

Once they gave up the idea of using drivers or brakes inside the cars, Miller and his fellow designers did not have to contend with passengers on roller coasters having a great deal of control over how the coaster operated. However, the problem with an ersatz risk is that repetition teaches riders that the coaster is not dangerous when operated as intended. As one observer put it, "If tickets are bought for sensation by the strip, I honestly believe anyone can become callous to any shock."[32] In a sense, the protective frame eroded from the inside, taking the pleasurable excitement of risk taking with it. Those who craved a little extra tingle or who wanted to show off rapidly discovered that they could add extra thrill by standing up or hanging out of cars. To control impulsive or foolish passengers, Miller patented the lap bar.[33] This device remains an elegant solution to many of the design constraints of coasters. Unlike seatbelts (a World War II invention), it is cheap, durable, and easily adjustable (to aid in the rapid loading of cars). One size fits most passengers and does not interfere with clothing. Employees can bring down the bar without actually touching passengers, avoiding the risk of offending delicate sensibilities.

While Miller's innovations made it possible to build rides incorporating sharper turns, steeper hills, and faster speeds, roller coaster designers now found themselves confronted by the limitations of the human body. In an age before ergonomic studies, designers relied on experience, guesswork, and experimentation to know how bodies would react. Sometimes the first test ride provided an unwelcome surprise. Violent, high-speed turns left patrons more terrified or nauseous than thrilled. Some rides also had the potential to break bones and snap necks.[34] Park owners recognized that roller coasters with a reputation for being "wicked" attracted particularly avid thrill seekers, but they also needed rides that appealed to a broader public and were good "repeaters" (meaning that ordinary patrons would want to ride them again and again). Moreover, complaints and lawsuits made owners painfully aware of the expense that could occur when they transgressed popular expectations about the balance between thrills and safety.

Slower scenic railway–type rides equipped with new safety devices offered one solution. However, increasingly sophisticated audiences became gradually less willing to suspend the disbelief required to imagine cheaply painted backdrops as scenic vistas. Moreover, the public seemed to crave the sensation of speed. Interviewed about why some rides proved unpopular, well-known designer William F. Mangels analyzed one of his failed creations. "It was paced too slowly. I was ten years late." Mangels went on to explain: "The public wants speed . . . the automobile has completely revolutionized riding devices. People used to be willing to poke along, but now when they go out for a good time they want it quick and fast."[35]

Confronted with the bodily limitations of sheer speed and violence, ride designers, including Miller, turned to a series of psychological tricks to make the rides feel more dangerous without being more dangerous. First and foremost, roller coasters depended on the tension between anticipation and acceleration. The slow ascent up the first hill, accompanied by the ominous clunk of the chain dog, set the stage. In a well-designed ride, passengers glimpsed the first drop only at the moment they reached the top of the hill. Passengers at the rear of the train could also enjoy the disconcerting experience of watching the front of the train disappear. To increase the sensation of speed, designers positioned beams and overhangs close to the track. "Easily the best sensation on the Island is the scenic railway with a wooden beam that looks as if it was going to hit you in the head," one journalist exclaimed. "It's great!"[36] In the 1920s, tunnels were also very popular as designers played with the disorienting impact of darkness. Many wooden roller coasters were also intentionally designed to sway and creak slightly to heighten tension in riders. As one roller coaster restorer explained it, "This structure is more like a piece of furniture than a building because it is a moving object."[37]

Roller coasters required experienced and trustworthy operators. At the end of the 1917 season, George A. Dodge from Boston's Paragon Park lost his principle coaster operator and supervisor, Herbert Schmeck, to the army. Like Miller, Schmeck helped to build the ride as an employee of the Philadelphia Toboggan

Company (PTC) and then stayed on to operate it. On the horns of the dilemma about how to replace him, Morse sought the advice from a manager at the PTC. He considered promoting Schmeck's assistant, a man named Dunn, who had many virtues. "He is a good mechanic, does not drink, knows the structure, as he worked on it all through the job, helped assemble the cars, and knows the cars," Dodge explained. The key phrase was "does not drink." The person Dodge really wanted had far more experience and a special feel for coasters (at least when he was sober). "Baucher," he wrote, "is a first class all around Coaster Man, ready for any emergency, as well as keeping the structure and cars in proper shape at all times." In other words, Baucher might do a better job of managing risk. Yet, Dodge added, "Liquor is his only fault and if he did not let liquor entirely alone, he would be worthless for the job."[38] Neither man had formal engineering credentials. Intimate knowledge of this particular device constituted their most important qualification. Both understood its quirks and weaknesses, how it should feel and look and sound, and how to make adjustments when something seemed wrong.

Like most, if not all, early twentieth-century technologies, roller coasters required continual maintenance. The constant pounding of loaded cars and changes in temperature loosened fastenings and shifted tracks. Bearings, couplings, and other essential components of cars also gave way under stress. Exposure to extreme weather also took its toll on the machinery. Rain, snow, and salt spray at seaside parks rapidly rotted timbers and tore bracing apart. Rides taken out of operation cost parks or concessionaires money. But accidents cost more. Operators cajoled, tinkered, and patched to keep rides going, relying on their experience and judgment to know when real disaster might be imminent. By looking, listening, and feeling for changes, they continually measured the actual status of the coaster against a mental picture of how it should be. In well-run parks an employee walked the tracks each morning looking for broken timbers, missing bolts, and loose track. He also removed "baby dolls, canes, caps, or fur pieces that might have fallen out of the hands of some person riding the night previously," which might get caught between the wheels of the cars. Employees then ran an empty train around the track. If it returned without incident, one or more of the crew would take the ride. A man with an oil can sometimes also rode along, lubricating on the fly.[39]

Even in well-run parks like Kennywood, these measures were not always enough. Accidents involving patrons invariably happened. Moreover, just because amusement parks provided patrons with a respite from the outside world did not mean that the industry was immune to changing attitudes about risk and safety taking hold in the rest of American society.

Bringing in the Risk Professionals

As the industry regrouped in the aftermath of World War I, people such as the McSwigans began to think in earnest about adding other measures to control for risk. They worried that a "good enough" approach increasingly threatened not only the livelihood of individual park owners but also the reputation of the industry as a whole. As ambulance-chasing lawyers and auto insurance adjusters already knew, the public was increasingly less willing to silently bear the costs of accidents. Accident victims wanted compensation from capitalists who had accepted their money and unexpectedly delivered pain and suffering in return. One insurance adjuster claimed that judgments on public accidents were "300 to 500 percent higher in 1923 than a decade earlier.[40] As purveyors of a commodity that was, by definition, supposed to be carefree, amusement park owners found themselves vulnerable to liability even when injured patrons had, by legal standards, also shown negligence. "It doesn't appear to make much difference, gentlemen, what sort of evidence you can accumulate from eye witnesses" about carelessness on the part of an injured patron, the adjustor told an industry gathering. "The jury is going to stick you in ninety-nine cases out of a hundred."[41]

Faced with the specter of skyrocketing costs and a potential public relations nightmare, activist members of the industry's trade association began to spend funds investigating ways of controlling risk and liability. Like many of the progressive business people we have already met in this book, they aspired to the adoption of more modern, efficient, scientific ways of doing business. Aware of Herbert Hoover's ideology of associationalism, they looked to reform from within by providing better information and expert advice. Their efforts borrowed from practices already in place in other industries, including safety inspections, safety campaigns and first aid, and liability insurance. They also anticipated measures to control liability such as warning labels and safety standards that would become commonplace in the battles over consumer product liability in the postwar era.

J. P. Hartley occupied a unique position for identifying what could go wrong in an amusement park. In 1920, his employer, the United States Fidelity and Guarantee Company, became the industry's largest provider of public liability insurance through a mutual insurance scheme arranged by the NAAP.[42] Parks and concessionaires had long purchased fire insurance. Since 1910, new laws also obliged them to pay for workmen's compensation in some states. But coverage against patrons' personal injury suits remained unusual. A few parks had long-standing relationships with insurers willing to underwrite such policies. Most found insurance companies unwilling to take the risk. If park managers did succeed in finding an insurer, the policy was likely to be extraordinarily expensive because insurers did not have enough actuarial data to calculate premiums realistically reflecting potential losses.[43]

When it first offered coverage, United States Fidelity also lacked that kind of information. The company initially gambled that their risk pool would be large enough to cover any unexpected losses while they figured out a more scientific way to do business. Two years later, Hartley and his colleagues had collected enough data to offer a snapshot to the NAAP's membership at their annual meeting.[44] Hartley's conclusions about which rides carried the most risk ran counter to popular perceptions. Of the 1,013 claims filed, "the Fun House is a spot that causes us a lot of trouble and costs money," he explained. Serious accidents, such as one in which a young boy "came in contact with a pair of gears . . . and had his leg just chewed right off," were rare, but bumps, breaks, and contusions seemed all too common. Merry-go-rounds took second place as a source of injury. Hartley speculated about the reasons. "These fellows take the girls out and they fill them full of steak, and then they sit on one of these horses and they get dizzy, and the first thing you know, off they go."[45] Others suspected that the dizziness might be from another cause. During a similar discussion two years prior, another insurance man claimed that "as many as 50 to 60 percent of the accidents which occur in amusement parks are due to intoxication." He thought that Prohibition had significantly worsened the situation.[46] Accidents befalling children gave insurance adjustors and park owners particular anxieties. Hartley claimed that many children were hurt "with their mothers sitting right close by." Even if the park were not legally negligent, he thought that juries would be inclined to make awards. "I don't know how that can be guarded against," he admitted.[47]

Disciplining the public, without ruining their fun, proved challenging. Amusement park owners could hardly bombard patrons with the kinds of reminders about safety first that the NSC recommended to industry. Making people truly aware of potential dangers threatened to break the protective frame. Not doing so not only opened the door to risky behavior but exposed park owners to legal claims of negligence.

Technological restraints offered a costly but appealing alternative. These devices required minimal cooperation from patrons but left riders little latitude for self-endangerment. Once mechanical restraints became a normal part of the amusement park experience, most people gave them little thought. The click of the bar locking into place provided one more aural prompt stoking happy anticipation. Ride manufacturers already offered devices such as Miller's lap bar as an option to park owners. As liability became a concern, more park owners began to view them as a necessity. After seeing a demonstration of a kit allowing parks to retrofit older rides with locking lap bars at the 1937 NAAP convention, Brady McSwigan contacted the manufacturer. At twenty-five dollars a set, the investment probably seemed worthwhile. In return, he received a list of operators who "will not operate without them," including some of the biggest names in the business.[48]

Other devices such as flat rides, swings, and merry-go-rounds proved much more difficult to safeguard. Trade publications advised a policy of stopping devices such as the Whip and Aeroplane swing if patrons refuse to sit down.[49] Some parks also installed warning signs inside the loading area or in the cars themselves. "We sometimes furnish our coasters with a sign reading 'Don't Stand Up' or 'Hold Your Hats,'" Henry Traver of Traver Engineering told Brady McSwigan. "I have had considerable experience testifying on roller coaster accidents and it is always worth something to be able to say that there is a sign over the loading platform."[50] Like warning labels on postwar consumer products, these functioned not just (or even primarily) to instruct the public, but rather to fend off legal accusations that the ride owner had "failed to warn" users of potential dangers. The growing prevalence of such warnings testified to a diminishing expectation that common sense should suffice in encounters with technological novelty.

When accidents did happen, time was of the essence in fending off lawsuits. Insurance companies recommended that park owners instruct employees to take every accident seriously, make a report immediately, and offer the injured patron medical aid. By the mid-1920s, many parks had established first aid stations staffed by nurses and sometimes doctors. The timely application of solicitousness and free medical aid often convinced patrons not to pursue further action.[51] When insurance coverage became visible to accident victims, the problem of moral hazard followed. A lawyer, who worked for some of the parks, told his clients they should allow their employees to make small settlements of five or ten dollars. "Don't tell him about the insurance company," he advised, "because that puts it in his mind at once that here is a chance to get some money."[52] Adjusters could also offer their share of stories about slip-and-fall artists hoping to extract large amounts of money from small amusement park accidents.[53]

In comparison to the steady stream of minor mishaps from patrons tripping, falling, and sticking their limbs in the machinery, catastrophic mechanical failures proved relatively uncommon. Park owners and insurance companies

Posting safety rules and warnings where amusement park patrons could see them served a dual purpose: preventing accidents and protecting parks and ride operators from liability. *Courtesy, Charles and Betty Jacques Amusement Park Collection, Penn State University Archives, The Pennsylvania State University Libraries.*

dreaded them, however, because these accidents created terrible publicity and resulted in much larger settlements than the average of twenty-five dollars per person doled out for bumps and falls. J. P Hartley worked as the adjuster for a series of increasingly dramatic accidents on a ride called the Aerostat at Chicago's Riverview Park. The Aerostat consisted of airplane-shaped cars that swung around a central pole. During the 1924 season, a cog broke in the gear at the top of the pole, causing the cars to swing wildly and collide with each other. One woman was thrown over the fence. Hartley estimated she landed at least sixty feet away. Thirty people filed for compensation. The insurance company quickly settled with all but one: a young man who had previously worked in the insurer's Chicago office. "I suppose he is wise to the game," Hartley remarked with chagrin. "He is hanging out for a nice little sum, but I think we can get that compromised in time."[54]

As public liability insurance became more common, inspection of rides by insurance company representatives emerged as a standard industry practice. Like state-employed factory and building inspectors, ride inspectors carefully went over machinery and made recommendations. However, they had several advantages over their publicly employed colleagues. Insurance companies could and did cancel coverage if their policy owners refused to comply with recommendations or failed to reduce the number of claims.[55] Inspectors did not have to worry about political pressure to look the other way. On the other hand, something of a revolving door existed between the industry and the insurance companies. Former ride operators made the best inspectors and ride inspectors might also be drawn back into the industry by high bidding. In a few places like Chicago, city-employed inspectors worked directly with insurance companies, but this relationship was unusual.

Even in the 1920s, reliance on internal and insurance inspection did not seem modern or systematic enough for some NAAP leaders. They looked at the flurry of standard-setting and scientific testing happening elsewhere and thought that their industry should do likewise. A series of "safety codes," compiled in cooperation with the American Society of Mechanical Engineers, the insurance companies, and the Department of Labor, constituted their chosen method.[56] The idea seems to have come from Albert W. Whitney, from the National Bureau of Casualty and Surety Underwriters, whom we met in chapter 8 as the author of *Man and the Motor Car*, the bureau's contribution to driver education.[57] In 1922, Whitney was appointed chair of the American Engineering Standards Committee. By the time he appeared before the amusement park convention, he had helped to shepherd through the creation of fourteen industrial safety codes providing guidelines for subjects such as the use of grinding wheels in industry.[58]

The amusement park men were not particularly interested in turning amusement parks into safer workplaces. Instead, they imagined that the codes would serve to inform and, to a certain extent, discipline the industry—particularly smaller players—with regard to consumer safety. More idealistically, some of them also seemed to have believed that it might be possible to put down in writing what would later be called "best practices," marrying expert and vernacular ideas into a seamless whole.

As a prelude, one of the committee members attended an industrial accident prevention conference in Washington, DC. Standing in the hallway, he chatted with Ethelbert Stewart, the U.S. commissioner of labor, about writing codes for the amusement park industry. With a mixture of pride and chagrin, he later reconstructed the conversation. "You and your confreres have the d—st proposition on your hands that we know of," Stewart reportedly declared. The Department of Labor had been writing "codes" (by which he meant statutes and guidelines) "to educate and compel leaders in all industries to safeguard machinery, put their plants in good sanitary condition, and use every precaution known to science in order to prevent fatal and non-fatal injuries" for years. Stewart saw the proposed amusement park code as something else entirely—a measure intended to protect consumers rather than workers. "Now you come along with a proposition and an admission that you are erecting and furnishing amusement devices which will give the patronizing public all the thrills and feeling of great danger and at the same time will not injure or harm them in the least." Nevertheless, he was willing to give his blessing, acknowledging the growing importance of giving consumers what they desired while ensuring their safety. "The present generation demands and will have these thrills and whilst we are somewhat stumped, we are heartily in accord with your movement to have a code that will make these devices safe."[59]

Over the next decade, the committee managed to turn out a series of codes for attractions as diverse as swimming pools and shooting galleries.[60] How much impact they actually had was harder to say since compliance remained completely voluntary.

The National Safety Council and the Amusement Park Industry

For almost four decades, the amusement park industry and the NSC seemed to coexist without making any public acknowledgement of each other's existence. It is hard to imagine that a lack of awareness was the cause. Cities such as Pittsburgh and Rochester hosted both parks and extensive safety-first campaigns. Like other steel mills, U.S. Steel's Edgar Thomson works, across the river from Kennywood, helped pioneer the safety movement in industry. The NAAP had its headquarters in Chicago, as did the NSC. Experts such as Whitney shared their expertise with both groups. Business organizations such as the Chamber of Commerce provided other points of connection. More likely, neither side saw the public relations value in making amusement parks an exemplar for the value of thinking safety first. Once again, the tricky business of maintaining the protective frame and amusement parks' status as a sight of consumption as well as work put the industry into a special category.

In 1949, an article in the *National Safety News* entitled "Thrills and Chills without Spills" broke the silence. The subject was Riverview Park in Chicago. The article had been brokered by H. A. Dever, the park's new director of picnics and outings. Mr. Dever formerly worked as an insurance claims adjuster. "Like you," he told Brady McSwigan, "we are continuously asked if we have many accidents on the rides." Dever thought that evading the question resulted in "wild speculation as to how many we do have." He also felt compelled to contradict what he called "accident propaganda"—sensationalist articles in the popular press. "Thrills and Chills" constituted one part of that effort.[61]

The article informed readers that "here the adventure-loving spirit finds expression and lets off steam in daring exploits." Making an appeal to the *National Safety News*'s safety-minded audience, the author indulged in a bit of unsubstantiated but appealing speculation. "Though no psychologist has told us so," he wrote, "it's possible that this form of spine-tingling may provide an outlet for that inner compulsion to take chances in such a way that we are less prone to take dangerous chances in more hazardous situations."[62] In other words, amusement parks might be making a previously unacknowledged contribution to the larger mission of the safety movement.

Like Kennywood, Riverview Park was a family-run company. George Schmidt, the father, graduated from the University of Chicago with a law degree. He had presided over "the plant" at Riverview since 1904. He was also a stalwart member of the NAAP. "Safety had always been a prime consideration in Riverview's operation," the reporter explained, "but accident prevention has become more intensive and more scientific since George's son, William B. Schmidt, came home from college with degrees in electrical and mechanical engineering." In a gesture toward modernizing the business, William Schmidt, now suitably credentialed, was installed as vice president in charge of engineering and safety.[63]

Fitted out with the tool kit of modern safety expertise, Schmidt set to work. He counted and studied. He calculated that accident frequency in the park was 4.83 per 100,000 patrons admitted in 1935. By 1948, the rate had dropped to .77. According to the article, Schmidt "visualized a direct application for all the electrical and mechanical things he had learned in school. Many of the safety inventions on the rides are patented in his name." Borrowing directly from the railroad industry, Schmidt installed several of the key railroad safety innovations on the "Bobs," the park's signature roller coaster. It had Westinghouse air brakes and a block signaling system, complete with traffic lights. Schmidt erected a control tower next to the ride from which the operator, with his hand on a dead man's switch, controlled the cars.

Schmidt's expertise was complemented by oversight from a number of outside experts. In early spring, as the park readied the rides for the season, Chicago's commissioner of buildings made a site visit. In a bit of safety (and probably political) theater, the commissioner and an "inspection party" walked the tracks of one of the roller coasters. Riverview's insurance carrier sent a safety engineer to be available full-time until the opening. As required by city ordinance, an independent consulting engineer also inspected the rides.[64]

Like the steel mills showcased by the safety movement decades earlier, Riverview did not really reflect industry norms. In 1949, it claimed to be the largest amusement park in the world. Operated by people vested in being leaders in their industry, it was situated in a city at the epicenter of the industrial safety movement. Inspection of amusement park rides by the building commissioner dated to the Columbian Exposition of 1893.[65] A small incident, regarding a smaller park on the other side of the country, helps put Riverview in context. In 1955, a Mrs. Clark who owned a small amusement park in San Diego commissioned a new ride to be designed by John Allen, a well-known coaster designer, and built by Zambreno & Illions Amusement Devices. The city of San Diego required that an independent structural engineer assess the plans with an eye to various safety issues, including earthquake resistance. R. C. Illions advised Mrs. Clark to find a structural engineer and "make a deal with him" in order to get around the requirement. As far as Illions and Allen were concerned, they knew how to build a safe ride. Bringing in an outside expert just slowed down the process.[66] In much of the industry, self-policing to control risk remained the norm and the way industry members preferred to conduct their businesses.

IN THE FIRST DECADES OF THE TWENTIETH CENTURY, a new industry took on an old task: selling risk as entertainment. Some of what they offered would already have been familiar to the crowds flowing through the gates of Coney Island, Kennywood, Riverview Park, and hundreds of similar places across America. Daredevil and circus acts featured skilled performers pushing the limits of human physical ability and daring. Disaster spectacles, like print accounts of nineteenth-century railroad wrecks, urban conflagrations, and natural disasters, provided audiences with the opportunity to participate vicariously in narratives of endangerment and rescue. In contrast, mechanized rides offered something new: the opportunity to experience both thrills and safety not just as a viewer, but as a participant.

In order to run their businesses successfully, amusement park managers and ride designers relied on both their knowledge of vernacular culture and their ability to tap into expertise when needed. They proved themselves deft manipulators of risk perception, inserting patrons into a "protective frame" where pleasure rather than fear resulted from encounters with risk. They also recognized that the public would not tolerate accidents in this context. Bad publicity drove away customers, and, by the 1920s, shifts in popular expectations about safety made amusement parks the target of an increasing number of lawsuits. Consequently, park owners banded together to obtain insurance coverage, to secure expert advice, and to attempt to set safety standards for the industry as a whole.

In contrast to many examples in preceding chapters, however, Progressive Era reformers largely failed to institutionalize regulation of amusement parks. In most places, other than Chicago and New York, there was no state-sponsored oversight at all. Only in 1984 did the state legislature of Pennsylvania pass its first amusement park ride inspection law.[67] In the view of many industry insiders, insurance company oversight and self-policing protected the public adequately and perhaps more effectively than state regulation. Preventing accidents was in park operators' best interests, they argued. Rejecting the more radical claims of the expert culture that knowledge could be separated from experience, they argued that too many government employees lacked the necessary familiarity with their business to realistically determine what constituted an acceptable level of risk. As the 1950s drew to a close, this issue became central to a national debate about consumer product safety.

11

Consumer Product Safety

The story was both tragic and familiar. Eleven-year-old Carole Hackes accidently set fire to her blouse while playing with matches. She sustained serious burns on her neck and chest before her mother managed to put out the flames.[1] Both Benjamin Franklin and Elizabeth Drinker would have recognized the nature of the accident, but they might have been astonished by where Carole's story was retold and the lesson drawn from it. In their eighteenth-century world, even eleven-year-old girls supposedly knew the risks of mixing fire and clothing. Those who ignored that bit of common knowledge were labeled careless or unlucky or perhaps both. The law offered neither a means to prevent such incidents nor a way for victims to obtain compensation by holding someone else responsible. As far as we know, Franklin never turned his fertile imagination to the problem of making everyday clothing less flammable.

But Carole Hackes's accident took place in the mid-1960s, by which time a growing number of activists, safety experts, and ordinary citizens had come to believe that consumers, particularly children, could not be expected to anticipate or ward against the hidden dangers of everyday technologies. In their formulation, accidents happened to people who were "unsuspecting," or "victims," implying that these events were neither random nor the fault of the injured person. Increasingly, courts and legislators agreed. In this new safety paradigm, manufacturers should be held responsible for anticipating risks and should be willing to redesign, install guards on, or at least warn consumers about the potential dangers of their products. If they refused to do so, the state should step in with appropriate regulations. Strict liability in tort cases would provide an additional incentive, while simultaneously compensating victims. Testifying before a U.S. Senate hearing on proposed new standards for cloth and clothing manufacturers, Carole's father, CBS news commentator Peter Hackes, voiced this twentieth-century set of expectations: "Consumers like me think we are buying safe products. I assumed that under the law I was protected against any harmful item, be it an explosive toy, a poisoned food, or a hazardous electrical fixture."[2] In Hackes's view, responsibility for his daughter's injury rested with fabric and clothing manufactures. They had a moral if not a legal obligation "to produce a safer clothing fabric."[3]

The most influential postwar advocates of consumer product safety also departed from their earlier counterparts by rejecting consumer safety education as a viable means of controlling risk. Some went so far as to describe the safety-first approach as "notoriously ineffective."[4] Instead, they put technology at the center of the frame as both the cause and potential cure for accidents. This approach acknowledged the extraordinary difficulty of compelling changes of behavior, especially from people engaging in elective behaviors in the privacy of their own homes. It also reflected mid-century Americans' complicated relationship with scientific and technological progress. Over and over again, advocates argued that the technologies of everyday life had grown too complex and were changing too rapidly to be understood, let alone managed, by ordinary users. The subtext, voiced most loudly by Ralph Nader, alleged that manufacturers had the scientific and engineering know-how to make consumer technologies safer. They were just too greedy to do so without government compulsion. As Mike Pertschuk, a leading consumer advocate as a congressional staffer in the 1960s and later head of the Federal Trade Commission, reflected, "We never thought to question the capacity of business to meet any standard imposed, efficiently, and at a minimal cost."[5] To set standards, Pertschuk and others worked to create a government agency staffed with experts who could dictate the terms under which products would be designated safe enough to be made available to consumers. Thus, adherents of what came to be called the consumer products safety movement set about to breach the last strongholds of the vernacular risk culture—the house, the yard, and the nursery.

Save the Children

In the summer of 1955, Texas representative Martin Dies shared a story of childhood trauma with his congressional colleagues. "I was locked in an icebox myself once, and it scared me half to death, and I stayed there until I was just about frozen stiff. It was one of those great big things."[6] Psycho-historians could probably have a field day analyzing the impact of this event on a man who had made his political name chairing the first House Un-American Activities Committee (HUAC) and waxing apoplectic about the evils of the New Deal. But, in fact, Dies was just using a personal example to support his colleague Kenneth A. Roberts's proposed bill requiring that refrigerator manufacturers redesign their products so that children could free themselves if trapped inside. "I'm glad that you got out" was the only response his colleague Ertyl Carlyle could muster.[7]

Both the iceboxes of Dies's childhood and the electric refrigerators of 1950s America closed with a mechanical latch. A handle on the outside released the latch. There was no way to open the door from inside because manufacturers had not imagined why anyone would intentionally climb into a refrigerator, let alone somehow pull the door shut. And yet, every year a few children somehow managed to lock themselves in empty appliances forgotten in basements, garages, and dumps in empty lots.

Children suffocating in refrigerators hardly constituted the most pressing issue confronting Congress that year. But heartbreaking newspaper accounts and campaigns by women's groups portrayed an epidemic in the making and propelled lawmakers into action (and perhaps gave these Southern Congressmen a less divisive issue than the fallout from the *Brown* decision the year before). The bill itself required refrigerator makers to find a way to prevent such accidents but did not propose a specific technical solution. Roberts and his House colleagues were vague about whether one actually existed. Roberts defended his position by arguing that a nation that had recently blown up a Pacific atoll with a hydrogen bomb could surely develop a small, inexpensive device for securing refrigerator doors.[8]

Congress's power to regulate interstate commerce had never before been employed to order manufacturers to change the *design* of a consumer product to make it safer. On the face of it, the odds seemed against the bill. The representative of the National Electrical Manufacturers Association testified that it was not technically possible to manufacture an acceptable device at a reasonable price. He also pointed out that modifications to new appliances would not solve the problem of an estimated fifty million iceboxes, refrigerators, freezers, and airtight cabinets already in use.[9] Thirty-four states had already passed laws requiring their citizens to padlock or dismantle such appliances before discarding or storing them. Industry lawyers also expressed concern that accidents involving a less-than-perfect latch, advertised as child-safe, would increase manufacturers' liability.[10]

This National Safety Council safety poster was included in the 1955 congressional report arguing for safety devices on household refrigerators. *Courtesy, National Safety Council.*

The manufacturers had already begun an alternate approach: make users more responsible through an educational campaign conducted in cooperation with the NSC.[11] Boy Scouts prowled junkyards and empty lots looking for abandoned iceboxes and refrigerators. A California chain store offered a free turkey to anyone who brought in an icebox door.[12] The NSC printed up posters and pamphlets and arranged for special programs in the public schools.

While supporters of the bill portrayed a generation lost to abandoned refrigerators, the statistics told a different story. No one was sure about the exact number of victims, but even the bill's advocates admitted that the count had hovered between ten and twenty individuals per year since the late 1940s when the Public Health Service began collecting such statistics.[13] One witness put this figure in perspective: in 1952, 14,851 children died in accidents. Cars were the big killer, followed by burns, drowning, accidents with firearms, and poisoning, mostly by drugs found in the family medicine cabinet.[14] Attendees at the hearings could easily have drawn the conclusion that there was barely a problem and that whatever problem did exist could be handled by procedures already in place.

Despite these objections, the bill requiring a safer form of closure on refrigerators passed, becoming Public Law 84-930. Even Roberts, the measure's architect, was surprised. He had intended only to use the threat of regulation to pressure manufacturers to hurry the process of creating an appropriate closing device and voluntarily regulating themselves on its installation. Perhaps more surprising to observers, passage of the refrigerator safety-closing bill proved not to be a legislative anomaly. Instead, it set an important regulatory precedent. For the first time, Congress mandated design changes in a consumer product for the sole purpose of protecting users from potential accidents and, more particularly, from injuring themselves through misuse.

The key to regulatory success could be summed up in one word: children. Over the course of the 1950s and early 1960s, child safety became an important wedge opening the door to federal regulation of consumer products. In the increasingly child-centric culture of mid-century America, even the most hard-boiled industry lawyer or statistician knew better than to argue with the often-repeated sentiment that even if only one child's life is saved, it will be worth the cost. Moreover, while children did not vote, their parents did. Martin Dies, never the most subtle of politicians, made this explicit by repeatedly suggesting that fixing the refrigerator door problem was about putting mothers' minds at ease.[15] Children also provided an opening wedge because it was difficult for opponents of change to argue that they really ought to know better and that it was a parental, not a societal, obligation to make sure they did. Manufacturers had their own reasons for considering cooperation. Besides the opportunity to garner public good will, they could glimpse the specter of increasing liability floating among the carcasses of their abandoned products. They hoped a good-faith effort to exorcise the possibility of accidents would ward off future lawsuits.

As we have seen in previous chapters, reformers and legislators found it much easier to make the case for regulatory protection of children than adults, particularly adult men. Lacking life experience and adult judgment, children could not be counted on to make appropriate judgments about risk and safety. They are vulnerable to the self-serving or uninformed decisions of adults on their behalf. Fear of liability also provided a strong incentive when it came to remedying technological risks that might kill or injure children. Protecting adults was trickier.

Legislators need to be able to explain why grown men and women could not protect themselves within the existing legal and social framework. The intended beneficiaries themselves did not always appreciate measures put in place supposedly for their own good. Using accident statistics to make the case that the accumulation of injuries threatened the common good was one tried and true way around these objections.

Protecting Adults

Since the Progressive Era, public officials and private organizations had been compiling improved statistics on accidental injuries and deaths. By the 1950s, these efforts had coalesced into a new field: injury epidemiology. Consequently, various organizations amassed increasingly fine-grained data on nonfatal injuries. Through the Public Health Service and other government agencies, the federal government also funded research on the number, causes, and prevention of accidents. Practitioners and adherents argued that the United States was in the midst of a public health crisis: an epidemic of accidents. The numbers showed that accidents had indisputably become the most important killer and disabler of young people. Massive intervention was required to stem the resulting human and economic losses.[16]

Some of epidemiologists' findings came as no surprise. Thanks to the workmen's compensation insurance system, in place in many states since the 1920s, as well as data compiled by the NSC, the nature and extent of workplace injuries were well established. Daniel Patrick Moynihan's influential 1959 article "Epidemic on the Highways" helped popularize a new way of looking at the all-too-familiar problem of automobile wrecks.[17] But data collected from the first large-scale surveys also revealed that almost half of all nonfatal injuries happened in the home.[18] Falls accounted for a significant proportion of injuries, particularly to the very young and the very old. But a significant number of cuts, burns, broken bones, and minor amputations resulted from using tools, machines, and other consumer goods.[19]

If domestic accidents really constituted a public health crisis and if consumers really did have a right to safety, legislators could not legitimately draw the line to exclude such risks, just because the victims were adults acting on their own volition outside the wage nexus, particularly if, as many prominent advocates argued, constant technological innovation meant that consumers could not be expected to understand, let alone manage, such risks on their own. And so, a growing number of products, old and new, were held up to a new kind of scrutiny by legislators, government agencies, private consumer groups, and the courts. The results of that process seemed to satisfy no one, but they proved extraordinarily effective at expanding the reach of safety professionals and the state into what had previously been the domain of private enterprise and personal choice. The saga of efforts to remake a quintessential suburban machine, the power lawn mower, illustrates the complex and vexing nature of this process.[20]

Power rotary mowers, as they were known in the industry, entered the mass market in 1946. From their introduction, they were extremely popular. By the early 1960s, manufacturers produced about four million a year. When sales finally started to level off in the mid-1970s, about seven million mowers were being sold annually.[21] Many users of these devices were first-time homeowners (or their wives or children), newly well-off settlers of postwar America's "crabgrass frontier."[22] For them, the power mower embodied the bounty of small machines that promised to make everyday life easier and more pleasant. In an era in which the middle class could no longer count on hiring working-class people to clean the floors or tend the garden, the mower offered a compromise to the do-it-yourselfer—with the added thrill, at least for some users (mostly male), of getting to play with a machine.

Manufacturers could offer these devices to consumers at affordable prices because they were very simple to design and manufacture.[23] Most manufacturers bought simple two-cycle engines, usually from Briggs and Stratton, purchased or made their own frames and blades, and assembled the mowers. As long as the blade turned fast enough to overcome the inertia of grass, pretty much any variation on this design theme would hack (or, technically, tear) a path through a lawn.

However, that fast-moving blade was the invisible snake in this crabgrass paradise. If it came into contact with an object with more inertia than a dandelion—say, a foot or a hand—the results were often a swift and bloody amputation. If the impediment was made of solider stuff—brick, rock, or wire—the momentum of the blade got transferred to the object, creating what was called, in the lawn mower safety world, a "missile."[24]

As it turns out, a lot of people stuck their hands under their mowers (usually to clear grass clogging the exit chute), ran over their own or their children's feet, or inadvertently launched rocks and pieces of wire toward unsuspecting passersby. Others were injured by handling the gasoline used to power the mowers or by mowers that came apart with use. By the late 1950s, power mowers caused about eighty thousand serious injuries a year.[25] A great deal of user agency was involved in mower accidents, including consumers using mowers in ways unintended, unimagined, and sometimes expressly warned against by manufacturers.

No one, not even manufacturers, ever claimed that power mowers were harmless. Instead, initial public debate centered on whether manufacturers should make any effort to protect consumers from what they called "careless" behavior. Underlying the discussion about what, if anything, to do about mower accidents was an unstable distinction between voluntary and involuntary risk—between the kinds of accidents users could anticipate and avoid and those they could not.

The law provided one context. In the late 1940s (the years when the first mowers entered the market), if a mower flew apart because it was defectively assembled, the manufacturer had a legal responsibility for damages to users and their property. But people who put their hands and feet under mowers paid their own doctor bills, and people who got hit by missiles could only sue the person pushing the mower.[26] While the state regulated some products such as drugs, where risks were invisible to the consumer, because the risks of mowers were supposedly obvious, regulatory law was silent about how manufacturers of this kind of consumer product should take safety into consideration.

Since the legal status of mower risk seemed clear, the marketplace became the principal site of negotiation about mower risk. The stakes were the hearts, minds, and dollars of consumers. On the consumers' side, the players included consumer advocacy groups that independently tested and recommended products. The Consumers Union, publisher of *Consumer Reports*, gradually took the lead in the lawn mower story. It is worth noting that the organization was under scrutiny from HUAC for its class-based politics in the 1930s, so this firmly middle-class technology was a safe choice for advocacy.[27] Manufacturers were represented by a trade association, initially called the Lawnmower Institute, and later rechristened the Outdoor Power Equipment Institute or OPEI.

Consumer advocates were the first to publicly address the issue of mower risk. In 1947, the Consumer's Research Institute tested the "Mow-Master"—the first rotary mower on the market—and found it unacceptably dangerous.[28] A year later, the Consumers Union tested nine brands of rotary mowers and labeled them all unacceptable.[29] Consumer's Research folded its tent soon thereafter, but the Consumers Union went on to set a pattern that continued through the 1960s of product reviews condemning the majority of rotary mowers tested on the grounds of user safety. It is not clear how much influence these reports had on consumer behavior. They did, however, give the Consumers Union a reputation early on for independent technical expertise on lawn mower safety among policy makers.

Meanwhile, some of the leading mower manufacturers also concluded that the industry had an issue with product safety—a problem they blamed on small-scale manufacturers churning out shoddy goods, as well as on those careless users.[30] In 1955, the Lawnmower Institute (OPEI) commissioned the American Standards Association (ASA, later the ANSI—American National Standards Institute) to write a safety standard for power mowers. The OPEI's likely motivations included a genuine desire to make a safer product, a response to critics like the Consumers Union, and a tool for pushing the most fly-by-night manufacturers out of the market. Nearly five years later, the ASA delivered a very peculiar document. Unlike most ASA standards of the period, it had two parts: safety specifications for manufacturers, and an appendix entitled "Suggestions for Users."[31]

Consumer advocates would later critique the ASA standards as pandering to the industry, but it is striking how many of the constituent elements eventually became common practice among product design engineers—including what became the holy trinity of guarding, warning, and redesign. More problematically, the standard was voluntary and manufacturers tested and certified their own products. The reward for complying was permission to put an ASA sticker on the company's product that would supposedly attract customers. The first power mowers bearing the ASA stickers came on the market in 1962 in a fanfare of publicity.[32] Even *Consumer Reports* paused its unrelenting critiques to tell readers that the ASA standards made rotary mowers safer.[33] Unfortunately for the mower manufacturers, the moment for self-congratulation was brief. Dramatic changes in liability law and the regulatory climate would soon mean that the marketplace was no longer the primary battleground for acceptable standards of lawn mower safety.

Strict Liability, Regulation, and Design

Manufacturers (or more precisely, their trade associations and lawyers) were already keeping a sharp eye on the civil courts, where judges and juries increasingly took the side of injured plaintiffs in product liability cases. As had happened with streetcars, amusement parks, and workplace injuries earlier in the century, juries showed a tendency to ignore capitalist claims of contributory negligence, if they thought plaintiffs deserved compensation. The inclusion of "instructions to users" and warning labels in the ASA standards was, in part, a strategy for fending off lawsuits based on the premise of "failure to warn" or implied warranties.[34]

The real watershed in product liability came in 1963. That year, the Supreme

Court heard a case involving a man who had been severely injured while using a lathe designed for hobbyist woodworkers. The justices' ruling in *Greenman v. Yuba Power Products, Inc.*, established the principle of "strict liability" in consumer product safety. Under strict liability, plaintiffs no longer needed to show evidence such as documents or testimony demonstrating that a producer had behaved in a negligent manner. If the product caused injury, that was generally proof enough that manufacturers had not given enough attention to warning, safeguarding, or redesign. The following year, a set of guidelines issued to the legal profession, the *Second Restatement on Torts,* sealed this change into legal doctrine. Strict liability forced manufacturers to imagine the user not as the proverbial "reasonable man" but as a person who was likely to mow a hedge with a lawn mower or leave the house with an iron turned on.

At the same time, the regulatory climate was also changing. In 1962, President Kennedy had sent a message to Congress calling for an expansion of federal regulatory oversight of consumer products.[35] He followed up with a four-point Consumer Bill of Rights. In the 1950s, protecting children from unsafe products was a safe issue for Dixiecrats and anti–New Dealers. In the 1960s, it attracted more liberal politicians because, in practice, it was a middle-class issue—a piece of the Great Society for white suburbanites. As such, it helped offset extensive government efforts on behalf of the urban poor. Senator Warren Magnuson, who chaired the flammable fabrics hearing and would also play a role in regulating mowers, adopted consumer protection as a signature issue after nearly losing a close election. Magnuson learned of the issue from his aid Jerry Goldstein, who had read about the subject in a *New Yorker* article.[36] "Maggie liked the idea," Goldstein later said. Not only did it offer something to the "ordinary citizen," but it also helped offset Magnuson's growing reputation for being in the pocket of business interests.[37]

As the door to regulation opened, Congress began considering how far the mandate might extend. In a 1966 hearing on a law regulating children's toys, the committee chair, Senator Magnuson, asked Dr. James L. Goddard, the Food and Drug Administration commissioner, whether the government set safety standards for power mowers.[38] Magnuson, of course, already knew that the answer was no, but the exchange gave him the opportunity to state for the record, "We have hundreds of letters, of course about safety in power mowers. We even had a couple of Senators who had their toes cut off while using the mowers themselves. We ought to have a law to protect them against their own folly in being too good around the house on the weekend."[39]

The beginning of the end for voluntary safety standards for mowers came in 1968 when Lyndon Johnson signed a bill authorizing the creation of a fact-finding commission called the National Commission on Product Safety. In the foreword to its final report, the commissioners stated, "When it authorized the Commission, Congress recognized that modern technology poses a threat to the physical security of consumers. We find the threat to be bona fide and menacing. Moreover, we believe that, without effective governmental intervention, the abundance and variety of unreasonable hazards associated with consumer products cannot be reduced to a level befitting a just and civilized society."[40] The authors of the report featured power lawn mowers as one of sixteen "unreasonable hazards" exemplifying the nature of the product safety problem.[41] The report offered a blistering indictment of the voluntary ASA standard. Federal investigators found that "one-quarter of the 216 models we examined did not comply with the industry's own consensus safety standards," even though many of those mowers bore the ASA sticker. The commission also thought that the standard itself was too lax.[42] After the commission's initial study in 1969, a panicky OPEI did a self-study that confirmed that the existing standard was inadequate and that they had, in effect, been handing out the safety seal without asking for proof of compliance. They launched a "crash program" to revise the standard and to force all manufacturers to obtain independent laboratory certification of compliance.[43] The Commission Report voiced doubts that the OPEI could actually make a voluntary standard work.[44]

As a result of the commission's efforts, the Consumer Product Safety Commission opened its doors for business in 1972. Soon thereafter, lawyers from OPEI submitted a petition requesting the development of a consumer product safety rule for power lawn mowers—in other words, they wanted to be regulated. They were explicit about at least part of their strategy: a federal regulation would bring noncomplying manufacturers into line with industry standards or force them out of business.[45] The unstated motivation was that a federal regulation would help protect them against strict liability by stating explicitly how far they had to go in protecting and warning consumers. The OPEI deftly suggested that the easiest way to write a regulation would be simply to adopt the voluntary standard. The strategy backfired. The Consumer Product Safety Commission's new director of engineering science responded with a scathing memorandum in which he called the voluntary standard "window dressing." He thought that a mandatory standard was necessary but that the ASA, now the ANSI, should not write it.[46]

Instead, the Consumer Product Safety Commission put out an open call for bids. And, perhaps not surprisingly, the Consumers Union won. The terms of the contract specified that the Consumers Union had 150 days to write the standard and had to invite extensive public participation.[47] Making everyone happy turned out to be a big order. The Consumers Union took three years to come up with a final draft standard. They were then required to advertise for public comments. As the letters began to roll in, the other shoe dropped.

While the vernacular culture had shifted to accommodate the expectation of safety as right, injury epidemiologists and consumer product safety advocates had not succeeded in shifting the popular perception that experience, skill, and carefulness were more important than statistics in predicting accidents. A very vocal minority of Americans made it clear that they opposed the new mower standard. They were all for safety in the abstract or in circumstances where they had no sense of personal control over risk. But they were furious about what they saw as unnecessary government interference in their rights as citizens and consumers. They also resented paying for or being inconvenienced by what they saw as other people's carelessness. Many correspondents singled out the recommendation that manufacturers include some means with which users could quickly stop the mower blade. The standard was only supposed to specify performance, not design, but it seems to have been understood that the regulation was calling for a so-called deadman's switch, which would stop the engine every time the user let go of the handle.[48]

One self-identified conscientious consumer, Susan Wendeln, a schoolteacher from Indianapolis, told the commissioners that she always wore her seatbelt and believed in using safety devices. However, she resented paying extra for a deadman's switch and other safety devices on a lawn mower. Like many of the correspondents, she believed that accidents only happened to careless people—a category that obviously did not include her. She went on to tell the commissioners, "I have taught school for ten years and believe that people should be allowed to remain ignorant. Even with free safety lessons some people will refuse to heed warnings until their own mistake teaches them."[49] Donald Roberts worded his letter more succinctly. "GET OFF OUR BACKS! The lawnmower I have now is an abomination compared to the old one. No more of your do-gooder 'improvements.'"[50]

Responses such as these must have stung some members of the commission. To distinguish their efforts from the previous voluntary industry system and to put into practice what many of them truly believed, they had bent over backward to ensure that the public had a voice in the standard-setting process. They had also picked the Consumers Union to write the standard, in part because it had the reputation of being an advocate for users. The problem seemed to lie in the Consumers Union testing process, which ironically did not include a good way to imagine or observe how ordinary people used products. In their thirty years of testing mowers, Consumers Union engineers gained credibility by developing a set of rigorous "scientific" testing techniques that satisfied the Consumer Product Safety Commission's need for a seemingly objective basis for the standard.[51] Using the epidemiological information about mower accidents the Consumer Product Safety Commission had collected, they constructed a picture of a user who was unself-consciously bound for self-destruction.

The commission staff did take into account all those irate letters. They rewrote the standard, retaining the requirement for stopping the blade, but adding a clause requiring manufacturers to include means through which users could easily restart their mowers (what was known as the "easy restart" clause). This, in turn, infuriated the manufacturers because it presented them with a technical problem for which there was no easy, low-cost solution.[52] Over this and other issues, they sued the Consumer Product Safety Commission, eventually losing in federal court.[53]

In 1979, the new mower standard was published in the *Code of Federal Regulations.* And then the political and cultural climate changed again. In 1980, Ronald Reagan took office and set about fulfilling his promise to restore American economic competitiveness through deregulation. The Consumer Product Safety Commission ranked high on conservatives' list of federal agencies needing to be dismantled. To survive, it agreed to a series of compromises. Somehow, the OPEI made the repeal of the "easy restart" mandate one of the conditions of the agency's survival. In 1981, that clause was rewritten. The rest of the standard remains in effect.[54]

In terms of accident prevention, the mower safety problem was not really ever solved. Instead, it reached a new level of negotiated stasis. In the 1979 standard, the Consumer Product Safety Commission justified the new regulation by arguing that the statute would reduce "blade contact injuries" by 77 percent, to about eight thousand per year. Yet, by the year 2000, the College of Hand Surgeons estimated that, once again, almost eighty thousand people suffered amputations or severe laceration in lawn mower accidents.[55] A trip around a suburban neighborhood on a warm spring weekend would reveal one of the reasons why: users routinely tie the deadman's lever in the engaged position so that they do not have to pull the cord to restart the mower every time they pull a grass clog out of the chute.

Deregulation and the Liability Crisis

In 1981, the new Reagan administration brought to Washington the message that a quarter century of intense federal regulatory activity had crippled the American economy. They cited studies claiming that the annual cost to the economy of federal regulations might have been as high as sixty-six billion dollars by the mid-1970s and that regulation had also contributed substantially to the decline in American productivity.[56] Whatever the most radical free market believer might have advised, realistically it was politically impossible to disengage the federal government from the business of ensuring the public's safety. Once managing the risks of everyday objects such as cotton blouses, refrigerators, and lawn mowers moved from the realm of private to public, and once a significant segment of the public came to believe they were entitled to protection from the unforeseen consequences of their own actions, there was no going back. Instead, a familiar set of elements—liability, regulation, engineering, public education—was reconfigured in the 1980s into a form that is more or less what we live with today.

At the center of this reconfiguration lurked a very old set of questions: What is the value of safety? How should it be weighed against other values—moral, economic, social? Who should bear the cost of accidents if they did occur? For at least some observers and influential policy makers, the answer had to be different in the grim economic landscape of an America scoured by deindustrialization and massive inflation from what it had been during the high point of the postwar economic boom. As one Reagan administration economic advisor stated, "The primary problem of current federal health and safety regulations is that in many areas Congress has set a standard of lowest feasible risk, regardless of the incremental benefits and costs of meeting this standard"—a generalization that did not reflect the complex negotiation that had gone into creating regulations, but did capture the philosophical stance of consumer product safety advocates of the 1960s and 1970s.[57] In contrast, supply-side economists argued for calculating something they called "maximum net benefit." This approach theoretically required that regulators balance the economic value of lives (or fingers and toes) saved against the costs to business of affixing warning labels and implementing design changes. Viewed from a maximum net benefit perspective, the regulations implemented in the 1950s to force refrigerator manufacturers to change the design of their products to save the lives of a half-dozen children per year would not have made sense in economic terms, if indeed anyone had dared do the calculation. More importantly, making such a Malthusian statement in public would have spelled political suicide for most elected representatives.

"Stockman Moves to Kill Consumer Panel," a 1981 *New York Times* headline read. The Consumer Product Safety Commission was near the top of Reagan budget director David Stockman's list of regulatory agencies that should be disbanded. He had informed the Senate subcommittee overseeing the commission that he believed that the panel's benefit to the public did not exceed its cost. The congressional members in charge were not willing to follow his recommendation.[58] Instead, the administration appointed a faithful member of the Republican Party, Nancy Harvey Stoerts, who had served Nixon and Ford in various consumer positions, as chair and, for good measure, a conservative Democrat, Terrence M. Scanlon, as part of the five-member commission. This move underestimated the power of the gospel of safety to trump party loyalty, particularly when the safety of children was at stake. Hence, a series of dramatic disagreements between Stoerts and Scanlon began to emerge within the commission and were eventually leaked to the press. These disagreements revolved around one familiar subject, amusement parks, and one new one: the exportation of risk to other nations. Stoerts took the position that one dead child was too many. Scanlon sided with Stockman and the supply-side economists, asserting that the costs of regulating risk had to be weighed against other goals and values.

Early in 1984, the commission met to discuss whether manufacturers should be allowed to export consumer products that had been recalled in the United States. They had already voted 4 to 1 to specifically allow the exportation of products banned in the United States under the Flammable Fabrics Act but were hesitating to broaden the exception. The attraction for manufacturers was obvious. As one commission member stated, "If manufacturers felt they could recoup their losses without damages to their reputations, they might try to sell things in the third world." Among the products he imagined might be exported were "pajamas treated with the fire-retardant Tris, which has been shown to cause cancer," and "probably an awful lot of paint containing lead, spackling compounds with asbestos, a miter saw in which the blades went flying off." In his view, the deciding factors were moral and epidemiological. "Things that cause asbestosis here are going to cause it anywhere." Riskiness was a physical fact that did not fundamentally change if the victim was in Kuala Lumpur rather than Boise. R. K. Morris, spokesman for the National Association of Manufacturers, disagreed. "The standards about complying or not complying are cultural issues."[59] If foreign governments had lower standards (or none at all) for acceptable risk, the responsibility for protecting foreign consumers should not devolve onto the United States government at the expense of beleaguered manufacturers. Eventually, the commission voted to prohibit the export of banned products. The dissenting vote came from Scanlon.[60]

In August, Scanlon even more visibly broke ranks with his fellow commissioners. The subject was amusement parks, which the Consumer Product Safety Commission had begun regulating a few years earlier. It had been an unacceptably dangerous summer for park patrons, as far as Commission Chair Nancy Stoerts was concerned. By August, twelve people had died. Congress had stripped the commission of its short-lived jurisdiction over rides in permanent parks in 1981, leaving them only the authority to regulate temporary fairs and carnivals. Stoerts wanted the authority returned. "The consumer assumes that these rides are safe and that they are inspected by someone other than the operator and indeed, they are not," she argued, echoing Peter Hackes's statement that consumers were entitled to assume that the government was looking out for them. Scanlon shot back that Stoerts was creating a "regulatory mirage" that the federal government could ensure that rides were safe. The federal government had neither the resources nor the expertise to perform such inspections. Scanlon was not actually opposed to state-sponsored oversight. Instead, he argued that "it can be done for far less money by the states and local jurisdictions," leaving it to others to point out that only twenty-five states actually had such laws on their books. An industry representative added, "It is critical to our industry that our rides be safe. There is no reason to believe that a Federal program would reduce the odds of injury."[61]

This time Stoerts lost not only the argument but also her job.[62] She was replaced by Scanlon, albeit not until a Government Accounting Office investigation cleared him of abusing his position on the commission to stymie regulation of all-terrain vehicles—a charge, not coincidentally, launched by Joan Claybrook, the president of Ralph Nader's watchdog group, Public Citizen."[63]

Like many other aspects of Reagan-era efforts at deregulation, the free-market rhetoric was not matched by an accompanying substantial decline in regulatory activity. The commission ceased to create explicit design standards like the one established for lawn mowers, but in Scanlon's tenure, it kept up a steady stream of recalls and warnings. Commission members were particularly interested in controlling any risk associated with children, especially if the hazard was connected to toys. The focus on recalls was symptomatic of the new approach to regulation. The CSPC identified specific risks but did not specify design remedies. Recalls made voters happy by giving them the sense that government was watching out for their well-being without overtly mandating the kinds of design changes that had annoyed mower users. The recalls also put the onus on consumers to bring defective goods back to retailers. As it turned out, many did not respond. For instance, a recall of electric fans mandated by the CSPC resulted in the return of only 1 percent of the devices.[64] Advertising or honoring a recall cost manufacturers money but also provided some potential protection from liability.

The Liability Crisis

Increasingly, it was lawsuits, not mandates from the government, that worried manufacturers. The 1970s witnessed an explosion of consumer product liability cases. It is impossible to count the number of cases in the lower courts, but one historian has estimated that federal cases quadrupled in the decade after 1976.[65] Many of the largest and most publicized involved the innovative use of class-action suits, resulting in enormous settlements. Novel medical products including the Dalkon shield, breast implants, and the antinausea drug Bendectin topped the list. A whopping 44 percent of federal cases during the decade involved a single material: asbestos. These cases typically involved exposure not to the raw material but to asbestos incorporated into some manufactured product, thus spreading liability to a variety of producers.[66]

For manufacturers and their allies, the overwhelming sense was of a tort system run amok. The sheer number of cases was not, in and of itself, the problem. Twenty years after the *Second Restatement on Torts*, most corporations were accustomed to taking measures to deal with liability. The problem (or at least the feeling) was that manufacturers could not predict how safe was safe enough, as far as the courts were concerned. In 1986 this uncertainty resulted in the so-called liability crisis. Despite what the term might imply, the crisis concerned insurance companies denying manufacturers coverage against consumer liability suits, not the direct economic impact of awards from successful suits. Introduced in the 1940s, this type of coverage had become a standard and seemingly necessary risk management tool for businesses. Underwriting such policies had always been difficult, but the combination of double-digit inflation rates and the specter of multimillion-dollar settlements for injuries from products sold decades earlier led insurance firms to question whether selling such coverage was worth doing at all. Manufacturers had functioned for centuries without product liability insurance. Many of the larger ones had enough assets to self-insure, but chief executive officers, like everyone else, were part of a distinctive postwar risk culture. Many could not imagine operating without insurance.[67]

Whether or not the number of tort cases involving more familiar products such as automobiles, power tools, and children's toys increased at the same exponential rate has been a matter of debate among scholars.[68] For contemporaries in the business community, the actual number did not matter. They drew two conclusions: the government needed to implement "tort reform," and they needed to increase their internal efforts to ensure that consumers did not injure themselves with their products, even through abuse or misuse.

Tort reform proved a hard sell for a number of reasons. In a consumer society, everyone, including judges, juries, and legislators, is a potential victim of a consumer product tort. No one really wants to give up the right to sue. The radical solution was something like the workmen's compensation system utilizing a schedule of fixed payments for different kinds of physical losses. But neither the increasingly powerful personal injury bar, which worked for a percentage of settlements, nor many plaintiffs would have been willing to entertain the idea of limited settlements. The modern liability system allows recovery for so-called pain and suffering, which includes loss of livelihood. In theory and sometimes in practice, this system meant that the already well-off could receive higher awards because they had more to lose. Finally, insurance companies could benefit from a rise in the number of lawsuits as long as they did their underwriting in a way that accurately predicts how much they might have to pay out. The threat of liability motivated manufacturers to buy insurance. If the insurance companies did not like the odds, as happened in 1986, they simply refused to write the policies.

Meanwhile, attention to both safety and liability had become part of the culture of design and marketing. The threat of liability and the perception that consumers and their advocates would heap bad publicity on companies that dumped dangerous products on the market prompted manufacturers to incorporate consumer product safety into their design and testing procedures, joining a long-standing tradition in the electrical appliance industry and a newer one in automobile production. This change was evidenced by a growing number of specialized product-testing firms offering individualized services to small-scale producers (and expert witnesses for hire for product liability suits). Sensing a need, university-based engineering programs also began offering product safety as part of their curriculum. Tellingly, one of the early reference books for product safety engineers offers a tutorial on product liability law.[69] To their surprise, manufacturers of certain kinds of products, especially automobiles, also found that consumers were beginning to view safety features and specialized safety devices as actually adding value to products.

IN THE LATE 1940S, CONSUMER PRODUCT SAFETY stood as the last redoubt of private, voluntary risk taking still governed by the protocols of a vernacular risk culture. Over the course of the next quarter century, a loose coalition of consumer advocates, public health officials, legislators, judges, lawyers, and legal theorists created new legal tools designed to protect users of consumer goods not only from the machinations of manufacturers but also from their own unwitting self-endangerment. A set of modernist beliefs underpinned these efforts. Everyday technology was too complex, too opaque, and too lethal to be successfully managed through the informal mechanisms of a vernacular risk culture. However, because every accident was analyzable and predictable in the probabilistic sense, every accident was theoretically preventable if experts studied it closely enough. Contrary to the beliefs of an earlier generation, the best solutions involved engineering rather than user education. And finally, in a consumer's republic, all citizens were entitled to be protected against risks and to compensation should they fall victim to someone else's actions or decisions.

By the 1980s, the most ardent safety advocates busied themselves lamenting what had not come to pass. Their dreams of a powerful regulatory state that put safety first had been sacrificed to a declining, deindustrializing American economy and a new political ideology that suggested that safety at any cost might be a luxury American society could not afford. Others argued that greedy lawyers were at fault; they had taken advantage of the liberalization of tort designed to provide compensation for people who had been injured by no fault of their own. Because of their avarice, they had used regulations and lawsuits to ruin businesses and drain the fun out of life through too many safety devices and warning stickers.

The reality is, of course, more complicated. Epidemiological research suggests that efforts at deregulation have not had a significant effect on the accident rate. They point, in particular, to the steady decline in the motor vehicle death rate as measured by "vehicle miles traveled."[70] A variety of evidence suggests that since the 1980s, public expectations have continued to rise. Most notably, middle-class parents have voluntarily adopted a series of risk-limiting behaviors that would not have crossed the minds of the generation who worried about abandoned refrigerators in empty lots. Gone are siblings squabbling about who would sit in the front seat. Both are in the back now. Allowing them to ride unbuckled in the front seat would be not only dangerous and illegal but immoral. Bicycle riding requires helmets. Children no longer walk to school alone, if they walk at all. Any piece of equipment associated with babies will have prominent warning stickers, perhaps more than one.

On the other hand, the parent in the driver's seat is likely to be talking on a cell phone. Sixteen-year-olds, new to the road, are more likely to text. Automobile accidents and firearms remain the most common causes of what epidemiologists call "injury deaths."[71] Open the newspaper on any morning to see that negotiations about what constitutes an acceptable level of risk and how to manage it are ongoing.

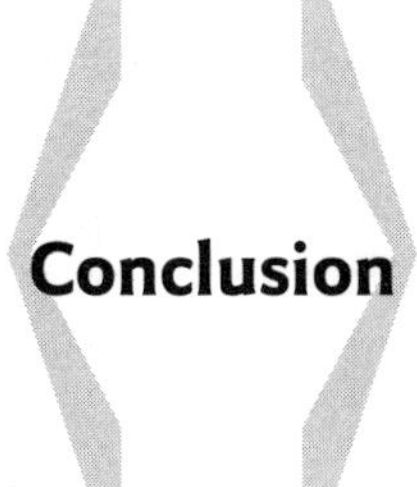

Conclusion

Here we are in the twenty-first century, living with a long history of thinking, negotiating, and managing risk. Vestiges of the past are most apparent in the material things we still use: lightning rods poking up along rooflines, a little more subtle in design now that they no longer signal the enlightened attitudes of building owners; stop signs and crosswalks, the legacy of early twentieth-century traffic engineers' efforts to discipline pedestrians and drivers; and the warning labels and safety features of many products. Yes, you can remove the warning label from that pillow, but consider the fate of poor Carole Hackes as you do so. Open the daily paper or a web browser: neither accidents nor voluntary risk taking have gone away. Wealthy men no longer race fast horses on the streets of Manhattan. Instead, they drive fast cars or fly airplanes in the Nevada desert. Expert and vernacular ways of thinking about risk still compete or, more often, intertwine. Consider the way information about risk is presented in the media: a combination of morality tales, sensationalism, and statistics Benjamin Franklin might have found familiar. The list could go on: insurance contracts and liability suits, traffic cops and Occupational Safety and Health Administration inspectors, not to mention the seemingly endless shouting about what to do about gun violence.

Although we live with the legacy of the past, surviving devices and practices do not necessarily carry the same meaning or serve the same function as they did for our predecessors. When Elizabeth Drinker looked at smoke curling out of a chimney or William Lloyd Garrison heard the rumble of a train, their thoughts and feelings were very much products of particular experiences, cultures, and historical moments. To an extent, risk perception is conditioned by context. The importance of context helps explain why people in different eras have focused on (and sometimes obsessed about) different kinds of risks. Before machines became a common part of everyday life, risks from nature provided the focus. In the tumult of industrialization, railroads became the most conspicuous symbol of the dilemmas of controlling risk. In the twentieth century, consumer products, particularly automobiles, have taken center stage. In each era, these risks were both worrisome and good to think with.

On the other hand, some aspects of risk perception have remained remarkably consistent over time. Eighteenth-century spectators at fires, such as Elizabeth Drinker's servant, Polly Nugent, and twentieth-century roller coaster riders share a taste for the pleasurable excitement of risk as entertainment. Despite more than a century of concerted expert efforts to align the perception and behavior of the public with probabilistic prescriptions about risk, individuals continue to insist that their skill and luck will allow them to engage in risky behavior without serious consequences.

Among all this complexity, several patterns in the process of historical change stand out. Over time, Americans have significantly increased the resources dedicated to controlling risk. Whether experts or members of the public, most people expect that something can and should be done to control risk at a level unimaginable to our ancestors. The state, in its many manifestations, has played an increasingly important role in defining acceptable levels of risk, disciplining the risk-taking behavior of individuals and organizations, and redistributing the costs of accidents when they do happen. Once-private sorrows gradually became the subject of public policy making. At the center of the gradual expansion of state oversight are ideas about the common good and social responsibility. Who decides what is safe enough? And who should pay when something goes wrong?

Regulatory activities are one means through which the state mediates risk. The court system is equally important. Since the mid-nineteenth century, tort cases have been the most significant means through which Americans sorted through the question of who should be held responsible when accidents happen. The fear of liability has likewise motivated a whole range of activities, from railroad managers' attempts to pay off railroad accident victims rather than take a chance in court, to the redesign of the power lawn mower, to the proliferation of various kinds of insurance in the twentieth century. Private organizations ranging from the National Safety Council to the American National Standards Institute have also made protection of large social groups—children, workers, amusement park patrons, and suburban homeowners—their business, spreading a distinctively American way of managing risk across that nation and the world.

The transition from a society dominated by a vernacular risk culture to one influenced by experts was marked by the emergence of safety as a special category of knowledge. Until the end of the nineteenth century, cultural authority to tell others what to do about risk resided primarily with people who either had specific experiential and skill-based knowledge or held positions of social status. Henry Drinker might direct a fire brigade not because he was a trained expert in fires, but because he was known as a person of wealth and standing in Philadelphia. Safety professionals drew authority from their ability to predict accidents and prescribe preventative measures in a wide variety of contexts.

Though often described as a calling, selling safety also provided employment for many people. The business of safety is one piece of evidence for another overarching trend: the commodification of risk. From the beginning of this story, a wide variety of actors have figured out a way to make money from uncertainty. Eighteenth-century fire insurance companies helped pioneer methods for creating risk pools, sometimes gathering in tidy profits in the process. Lightning rod salesmen and insurance brokers also preyed on the public's willingness to pay for protection from accidents that might or might not ever befall them. A host of other innovations from the Westinghouse brake to Henry Dorn's fire escapes to magnetic releases on refrigerator doors profited their inventors. Taming risk for entertainment has also provided a livelihood for many, including the long-forgotten grooms and horse trainers who helped Robert Bonner survive his trips down Harlem Lane and John Miller, purveyor of thrills and safety in the form of early twentieth-century roller coasters.

Together, these changes reflect many of the transformations that created the modern world. While their effects can be seen in many aspects of everyday life, the conquest has never been absolute. Vernacular ways of understanding and managing risk remain and, in fact, have formed around new technologies like the automobile. The risk society, as we know it, turns out to be the temporary manifestation of an ongoing process—messy, dialectical, and ultimately resulting not in safety in some absolute sense, but rather in a series of agreements about what is safe enough.

NOTES

Introduction

1. The circumstances surrounding the creation of Salignac's photograph of the bridge painters are documented in Michael Lorenzini and Kevin Moore, *New York Rises: Photographs by Eugene de Salignac from the Collections of the New York City Department of Records / Municipal Archives* (New York: Aperture Foundation Books, 2007), 34.

2. F. J. H. Kracke to B. de N. Cruger, Aug. 27, 1918, Box 12, New York City Municipal Archives (hereafter NYCMA).

3. The term *vernacular* is borrowed from linguists and architectural historians, who use it to describe practices that emerge from local knowledge and everyday practices and are "native or peculiar to a particular country or locality." *OED Online*, 2nd ed., 1989. *Culture*, in this context, refers to a combination of beliefs and practices held in common by a group of people.

4. Department of Bridges, City of New York, *Report for the Year Ending December 31, 1914*, 60–61; "Memo for the Commissioner," Office of the Mayor, John P. Mitchell Administration, Mar. 30, 1915, NYCMA. The following year the mayor's office began asking for accident reports. F. J. H. Kracke to John Purroy Mitchell, Aug. 30, 1915, Box 12, Office of the Mayor, John P. Mitchell Papers, NYCMA.

5. Federal census data first began generating comparable statistics on accidents in 1910. See "Death Rates for Selected Causes, 1900–1970," Series B, 149–66, *Historical Statistics of the United States, Colonial Times to 1970*, Part I, 58. In 1921, the National Safety Council began compiling and publishing accident statistics from the census and other sources for a broad audience as *Accident Facts*.

6. Glynis M. Breakwell, *The Psychology of Risk* (Cambridge: Cambridge University Press, 2007), 82. For instance, extensive contemporary research shows that men often view particular risks as less likely and less consequential than do women. This phenomenon has earned the label the "white male effect" among risk psychologists. See Dan M. Kahan et al., "Culture and Identity-Protective Cognition: Explaining the White-Male Effect in Risk Perception," *Journal of Empirical Legal Studies* 4 (Nov. 2007): 465–505; J. P. Byrnes, D. C. Miller, and W. D. Shafer, "Gender Differences in Risk Taking: A Meta-Analysis," *Psychological Bulletin* 125 (Fall 1999): 367–83; Paul Slovic, "Trust, Emotion, Sex, Politics, and Science: Surveying the Risk-Assessment Battlefield," in *Environment, Ethics and Behavior*, ed. M. H. Bazerman et al. (San Francisco: New Lexington, 1997): 277–313.

7. The *Oxford English Dictionary* traces the term *risk* to the seventeenth century, when it

meant "hazard, danger; exposure to mischance or peril." *OED*, online version, 1989 edition. Judith Greene's *Risk and Misfortune: A Social Construction of Accidents* (London: UCL Press, 1997) provides a history of the concepts related to accidents and risk. For two very different examples of histories of financial risk taking, see Peter L. Bernstein, *Against the Gods: The Remarkable Story of Risk* (New York: John Wiley and Sons, 1996); and T. J. Jackson Lears, *Something for Nothing: Luck in America* (New York: Viking, 2003).

8. Claude Lévi-Strauss, *Totemism* (Boston: Beacon Press, 1963), 89.

9. For more on the concept of social construction, see Nina E. Lerman, Arwen Palmer Mohun, and Ruth Oldenziel, "Versatile Tools: Gender Analysis and the History of Technology," *Technology and Culture* 38 (Jan. 1997): 3. See also Langdon Winner, "Do Artifacts Have Politics?," *Daedalus* 109 (1980): 121–31; Donald A. MacKenzie and Judy Wajcman, *The Social Shaping of Technology: How the Refrigerator Got Its Hum* (Philadelphia: Open University Press, 1999); Wiebe E. Bijker and Trevor J. Pinch, eds., *Shaping Technology / Building Societies: Studies in Socio-Technical Change* (Cambridge, MA: MIT Press, 1994).

10. On national styles in constructing railroads, see John Habakkuk, *American and British Technology in the Nineteenth Century: The Search for Labour-Saving Inventions* (Cambridge: Cambridge University Press, 1962).

11. For a useful overview on the theory-oriented sociological literature on risk, see Deborah Lupton, "Risk and Sociocultural Theory," in Deborah Lupton, ed., *Risk and Sociocultural Theory: New Directions and Perspectives* (Cambridge: Cambridge University Press), 1–11. Surprisingly, the anthropological literature has not proved very useful for my purposes. Much of it responds to Mary Douglas and Aaron Wildavsky, *Risk and Culture: An Essay on the Selection of Technological and Environmental Dangers* (Berkeley: University of California Press, 1984).

12. Works closest in themes and intention to this include the pioneering collection of essays Roger Cooter and Bill Luckin, eds., *Accidents in History: Injuries, Fatalities, and Social Relations* (Amsterdam: Rodopi, 1997), as well as more recent works, notably John C. Burnham, *Accident Prone: A History of Technology, Psychology, and Misfits of the Machine Age* (Chicago: University of Chicago Press, 2009); and Kevin Rozario, *The Culture of Calamity: Disaster and the Making of Modern America* (Chicago: University of Chicago Press, 2007). The vast majority of the safety-oriented literature deals with railroads, automobiles, and industry. Particularly significant works include Mark Aldrich, *Safety First: Technology, Labor, and Business in the Building of American Work Safety, 1870–1939* (Baltimore: Johns Hopkins University Press, 1997); Aldrich, *Death Rode the Rails: American Railroad Accidents and Safety, 1828–1965* (Baltimore: Johns Hopkins University Press, 2006); James W. Ely, *Railroads and American Law* (Lawrence: University Press of Kansas, 2002); David Rosner and Gerald E. Markowitz, *Dying for Work: Workers' Safety and Health in Twentieth-Century America* (Bloomington: Indiana University Press, 1987); David Blanke, *Hell on Wheels: The Promise and Peril of America's Car Culture, 1900–1940* (Lawrence: University Press of Kansas, 2007); and Peter D. Norton, *Fighting Traffic: The Dawn of the Motor Age in the American City* (Cambridge, MA: MIT Press, 2008).

13. On the sociology of accidents, see Green, *Risk and Misfortune*.

14. Ulrich Beck, *Risikogesellschaft: Auf dem Weg in eine andere Moderne*, translated as *Risk Society*, trans. M. Ritter (London: Sage, 1992). For an overview of Beck's ideas and their impact, see Sheldon Krimsky and Dominic Golding, eds., *Social Theories of Risk* (Westport, CT, and London: Praeger, 1992); and Deborah Lupton, *Risk (Key Ideas)* (New York: Routledge, 1999).

15. Ulrich Beck, *World Risk Society* (Cambridge: Polity Press, 1999), 50. The term *fatalism* is not consistently defined in the risk society literature. For a somewhat different definition that treats fatalism as a present-day as well as historical phenomenon, see Anthony Giddens, *Modernity and Self Identity: Self and Society in the Late-Modern Age* (Cambridge: Polity Press, 1991), 110–12. Giddens sees fatalistic attitudes as characteristic of "a refusal of modernity." He contrasts modern fatalism with premodern stoicism and belief in "fortuna."

16. Comparative and non-American studies of risk in history include Jamie L. Bronstein, *Caught in the Machinery: Workplace Accidents and Injured Workers in Nineteenth-Century Britain* (Stanford, CA: Stanford University Press, 2008); Richard H. Steckel and Roderick Floud, *Health and Welfare during Industrialization* (Chicago: University of Chicago Press, 1997); César Batiz, *La desgracia de ayer: los primeros accidentes en la historia del automovilismo en Venezuela* (Caracas: Fundación Empresas Polar, 2007); Denis Varaschin, *Risques et prises de risques dans les sociétés industrielles* (Bruxelles: PIE Lang, 2007); François Ewald, *Histoire de l'Etat-providence: Les Origines de la solidarité* (Paris: Grasset, 1996).

17. Some historians have already questioned the periodization and supposed "radical" nature of the shift. See, for instance, Soraya Boudia and Nathalie Jas, "Introduction: Risk and 'Risk Society' in Historical Perspective," *History and Technology* 23 (Dec. 2007): 317–18; and Alan Scott, "Risk Society or Angst Society? Two Views of Risk, Consciousness, and Community," in *The Risk Society and Beyond: Critical Issues for Social Theory*, ed. Barbara Adam, Ulrich Beck, and Joost Van Loon (London: Sage, 2000), 34.

Chapter 1: Fire Is Everybody's Problem

1. Stephen J. Pyne, *Fire: A Brief History* (Seattle: University of Washington Press, 2001), 106–13.

2. William J. Novak, *The People's Welfare: Law and Regulation in Nineteenth-Century America* (Chapel Hill: University of North Carolina Press, 1996), points out that fire prevention and control are among the most important uses of the law to protect public safety in early America (54).

3. On efforts to improve firefighting in early modern Europe, see Daniel H. Winer, "The Development and Meaning of Firefighting, 1650–1850" (PhD diss., University of Delaware, 2009).

4. The Romans, for example, used pumps to move large quantities of water, and Roman emperors employed slaves to fight fires. John Peter Oleson, *Greek and Roman Mechanical Water-Lifting Devices: The History of a Technology* (Toronto: University of Toronto Press, 1984), 51, 303–9.

5. Elaine Forman Crane, ed., "Introduction," *The Diary of Elizabeth Drinker* (Boston: Northeastern University Press, 1991), 1:vi (cited hereafter as *Drinker Diary*). Crane describes Drinker as having a "tendency towards timidity," xvi.

6. Ibid., 15.

7. Arthur D. Pierce, *Iron in the Pines: The Story of New Jersey's Ghost Towns and Bog Iron* (New Brunswick, NJ: Rutgers University Press, 1957), 33.

8. "Philadelphia, March 7," *Pennsylvania Gazette*, Mar. 7, 1738.

9. Alice Morse Earle, *Colonial Dames and Good Wives* (Boston and New York: Houghton Mifflin, 1895), 277, 279.

10. Henry J. Kauffman, *The American Fireplace: Chimneys, Mantelpieces, Fireplaces, and Accessories* (Nashville, TN: Thomas Nelson, 1972), 299.

11. *Drinker Diary*, Dec. 5, 1794, 623.

12. The use of punishments to teach children about fire has also been observed in developing countries by anthropologists. Norman Schwartz, personal communication, Oct. 23, 2010. Idioms are from http://idioms.thefreedictionary.com, accessed Oct. 23, 2010.

13. Only a few of the thousands of nineteenth-century domestic manuals explain what to do if clothing caught fire. For an example, see Catherine Beecher, *The American Woman's Home* (New York: J. B. Ford, 1872), 352. Only after 1900 did relatively reliable statistics on deaths from burns and scalding from causes other than conflagrations become available. These reveal that fire-related injuries remained a significant cause of death well into the twentieth century. The death rate declined very slowly over the next forty years, despite the gradual disappearance of heating and cooking with fire and the advent of the public health and safety movements. See Sixteenth Census of the United States: 1940, *Vital Statistics Rates in the Unites States, 1900–1940* (Washington, DC: Government Printing Office, 1943), 220, 227, 237. For comparison, the death rate in 1900 for burns was 7.5 per 100,000. The rate for railroad accidents was 9.7 per 100,000. The census did not distinguish between domestic and industrial accidents, but it did treat burns from corrosive substances as a separate category.

14. *Drinker Diary*, Apr. 1797, 910.

15. Carl Bridenbaugh, *Cities in the Wilderness: The First Century of Urban Life in America, 1625–1742* (New York: Knopf, 1955), 207. Bridenbaugh writes that chimney fires were the "most common cause of fire." This is impossible to prove since no one kept systematic records of the causes of fires. However, the sources do suggest that many people believed this to be true.

16. Billy G. Smith, *The "Lower Sort": Philadelphia's Laboring People, 1750–1800* (Ithaca, NY: Cornell University Press, 1990), 156.

17. John E. Crowley, *The Invention of Comfort: Sensibilities & Design in Early Modern Britain & Early America* (Baltimore: Johns Hopkins University Press, 2001), 172; Priscilla J. Brewer, *From Fireplace to Cookstove: Technology and the Domestic Ideal in America* (Syracuse, NY: Syracuse University Press, 2000), 16; Benita Cullingford, *British Chimney Sweeps: Five Centuries of Chimney Sweeping* (Chicago: Ivan R. Dee, 2000), 6–7.

18. *Drinker Diary*, July 1767, 137.

19. Cullingford, *British Chimney Sweeps*, 48.

20. *Drinker Diary*, Jan. 1780, 365.

21. *Drinker Diary*, Nov. 1778, 335; Dec. 1778, 336.

22. Paul A. Gilje and Howard B. Rock, "'Sweep O! Sweep O!': African-American Chimney Sweeps and Citizenship in the New Nation," *William and Mary Quarterly* 51 (July 1994): 507–38.

23. Most famously, Frederick Douglass names "sweeping the chimney" as one of the ways he earned money in New Bedford after claiming his freedom. Frederick Douglass, *Narrative of the Life of Frederick Douglass* (Cambridge, MA: Harvard University Press, 2008), 113.

24. Richard S. Newman, *Freedom's Prophet: Bishop Richard Allen, the AME Church, and the Black Founding Fathers* (New York: New York University Press, 2008), 56–57.

25. Cullingford, *British Chimney Sweeps*, 28; Randall Moale, receipt (1837), "Bills and Receipts," Col. 493, 76x88.2, The Winterthur Library: Joseph Downs Collection of Manuscripts

and Printed Ephemera; De La Warr, Account Book, 1777–78, Doc. 67; Winterthur: Downs Collection.

26. *Drinker Diary*, Dec. 1773, 196. See also George Lewis Phillips, *American Chimney Sweeps: An Historical Account of a Once Important Trade* (Trenton, NJ: Past Times Press, 1957), 12–13.

27. Cullingford, *British Chimney Sweeps*, 6.

28. "An Act for Preventing Accidents That Happen by Fire" [1701], *The Statutes at Large of Pennsylvania*, vol. 3 (Harrisburg, PA: Clarence M. Buser, 1896), 163.

29. Donald J. Cannon, *Heritage of Flames: The Illustrated History of Early American Firefighting* (Garden City, NY: Doubleday, 1977), 74.

30. Phillips, *American Chimney Sweeps*, 22.

31. "Mr. Franklin," *Pennsylvania Gazette*, Feb. 11, 1734/35.

32. In 1762, nine pence would buy a little more than two pounds of beef. The average wage for a laborer was 46 pence (3.8 shillings). B. Smith, *"Lower Sort,"* Appendix F, Table F-2.

33. "Philadelphia, October 25," *Pennsylvania Gazette*, Oct. 25, 1775.

34. Novak, *People's Welfare*, 58.

35. "Act for Preventing Accidents," 252–53.

36. Ibid., 163; Bridenbaugh, *Cities in the Wilderness*, 58, 209, 366.

37. "Act for Preventing Accidents," 253.

38. Jill Lepore, *New York Burning: Liberty, Slavery, and Conspiracy in Eighteenth-Century Manhattan* (New York: Alfred A. Knopf, 2005), xii, xvi.

39. For newspaper accounts of slave arsonists, see also "Boston, January 6," *Pennsylvania Gazette*, Feb. 15, 1738; "Annapolis, February 7," *Pennsylvania Gazette*, Mar. 5, 1754; "Newport in Rhode Island," *Pennsylvania Gazette*, Mar. 11, 1762. On fires set to conceal a crime, see "New York, December 5," *Pennsylvania Gazette*, Dec. 12, 1771; "New York, December 13," *Pennsylvania Gazette*, Dec. 21, 1796. Elizabeth Drinker also made note of such incidents. See *Drinker Diary*, Dec. 28, 1896, 874; Feb. 1, 1897, 893; May 1, 1897, 913.

40. *Drinker Diary*, Apr. 1759, 17.

41. A. Roger Ekrich, *At Day's Close: Night in Times Past* (New York: W. W. Norton, 2005), 75–77. The watch was one of Benjamin Franklin's first targets for reform because citizens could buy their way out of the responsibility and because many voluntary watchmen did not take their responsibilities seriously enough for his taste. See Benjamin Franklin, *The Autobiography of Benjamin Franklin* (New York: P. F. Collier and Son, 1909), 103. On the organization of the watch, see Edward P. Allinson et al., *Philadelphia, 1681–1887: A History of Municipal Development* (Philadelphia: Allen, Lane and Scott, 1887), 35.

42. Cannon, *Heritage of Flames*, 75, 81.

43. *Drinker Diary*, May 1784, 422.

44. Robert S. Holzman, *The Romance of Firefighting* (New York: Harper and Brothers, 1956); Winer, "Development and Meaning of Firefighting," 48.

45. [no title], *Pennsylvania Gazette*, Apr. 18 to Apr. 25, 1734; *Pennsylvania Gazette*, June 6, 1734; for another instance of theft, see *Pennsylvania Gazette*, Apr. 30, 1730.

46. *Drinker Diary*, Dec. 1782, 407.

47. Cannon, *Heritage of Flames*, 18–19.

48. Ibid., 24–25.

49. "Philadelphia, April 10," *Pennsylvania Gazette*, Apr. 10, 1735.

50. "To the Publisher of the GAZETTE," *Pennsylvania Gazette*, Dec. 20, 1733.

51. Ibid.

52. Cannon, *Heritage of Flames*, 89–90.

53. Franklin, *Autobiography*, 104; Edmund S. Morgan, *Benjamin Franklin* (New Haven, CT: Yale University Press, 2002), 54, 57.

54. "Articles of the Union Fire Company" [1736], in *The Papers of Benjamin Franklin*, ed. Leonard Woods Labaree, William Bradford Wilcox, and Barbara Oberg (New Haven, CT: Yale University Press, 1959), 2:150–53 (hereafter *BFP*).

55. Ibid., 2:375.

56. Union Fire Company, *Minute Book*, 1745; Jan. 3, 1757; Jan. 5, 1778, reprinted in *Pennsylvania Magazine of History and Biography* 27 (Fall 1903): 476–81.

57. Philadelphia had actually purchased its first engine in 1719. It had proved less than satisfactory. Cannon, *Heritage of Flames*, 77.

58. Winer, "Development and Meaning of Firefighting," 65–92.

59. Union Fire Company, *Minute Book*, May 8, 1747; Nov. 2, 1747; July 2, 1749; Aug. 2, 1749; Mar. 7, 1750; Dec. 3, 1759.

60. Nicholas B. Wainwright, *A Philadelphia Story: The Philadelphia Contributionship for the Insurance of Houses from Loss by Fire* (Philadelphia: Contributionship, 1952), 21.

61. *Drinker Diary*, Oct. 1760, 77.

62. Wainwright, *Philadelphia Story*, 24. See also *The Philadelphia Contributionship for the Insurance of Houses by Fire: 250th Anniversary* (Philadelphia: The Company, 2002).

63. Carol Wojtowicz Smith, *The Philadelphia Contributionship for the Insurance of Houses from Loss of Fire, 1752–2002* (Philadelphia: Signature, 2002), 13–14. Historians of probability theory have noted that, for reasons that are "somewhat mysterious," eighteenth-century mathematicians did not see fires as occurring in regular patterns that could be analyzed to create predictions about probability for the purposes of insurance. They also note that insurance companies of all kinds ignored mathematicians' insights into probability, preferring to rely on qualitative assessments of risk. Gerd Gigerenzer et al., *The Empire of Chance: How Probability Changed Science and Everyday Life* (Cambridge: Cambridge University Press, 1989), 22, 25–26. Lorraine Daston has called this the anti-statistical bias of eighteenth-century insurance. See Lorraine J. Daston, "The Domestication of Risk: Mathematical Probability and Insurance, 1650–1830," in *The Probabilistic Revolution*, ed. Lorenz Krüger et al. (Cambridge, MA: MIT Press, 1987), 240.

64. A series of conferences have recently invited scholars to link this phenomenon to changing ideas of permanence in the "long eighteenth century." Most relevant is the work of economic historian Robin Pearson, "Urban Renewal or *Plus ca Change?* The Impact of Fire and Fire Insurance on Eighteenth-Century English Towns," unpublished paper presented at "Permanence and the Built Environment in the Eighteenth-Century Atlantic World," Oct. 4, 2008, at the Huntington Library.

65. David G. Schwartz, *Roll the Bones: The History of Gambling* (New York: Gotham Books, 2006), 93, 111, 127; Neal Elizabeth Millikan, "'Willing to Be in Fortune's Way': Lotteries in the Eighteenth-Century British North American Empire" (PhD diss., University of South Caro-

lina, 2008). Jackson Lears has provocatively tied this impulse to the history of Protestantism in America in *Something for Nothing: Luck in America* (New York: Penguin Books, 2003).

66. Wainwright, *Philadelphia Story*, 25.

67. For more on firemarks, see Anthony N. B. Garvan and Carol A. Wojtowicz, *Catalogue of the Green Tree Collection* (Philadelphia: Mutual Insurance, 1977).

68. C. Smith, *Philadelphia Contributionship*, 13.

69. Sharon V. Salinger, "Spaces, Inside and Outside in Eighteenth-Century Philadelphia," *Journal of Interdisciplinary History* 26 (Summer 1995): 3.

70. Ibid., 4; Elizabeth Gray Kogen Spera, "Building for Business: The Impact of Commerce on the City Plan and Architecture of the City of Philadelphia 1750–1800" (PhD diss., University of Pennsylvania, 1980), 190–91.

71. Wainwright, *Philadelphia Story*, 62; Spera, "Building for Business," 179.

72. Salinger, "Spaces, Inside and Outside," 3. Spera, "Building for Business," 180.

73. Salinger, "Spaces, Inside and Outside," 7.

74. Ibid., 197.

75. Ibid., 199.

76. Economists trace current interest in moral hazard to the work of Kenneth J. Arrow and Mark V. Pauly in the 1960s and 1970s. See Mark V. Pauly, "The Economics of Moral Hazard: Comment," *American Economic Review* 58 (June 1968): 531–37; Kenneth J. Arrow, "Insurance, Risk, and Resource Allocation," in *Essays in the Theory of Risk Bearing* (Chicago: Markham, 1971); see also Steven Shavell, "On Moral Hazard and Insurance," *Quarterly Journal of Economics* 93 (Nov. 1979): 541. For a contemporary definition of fraud, see "What Is Insurance Fraud?," www.insurance.ca.gov/0300-fraud/0100-fraud-division-overview/0100-what-is-insurance-fraud/ (accessed Mar. 8, 2011).

77. Tom Baker, "On the Genealogy of Moral Hazard," *Texas Law Review* 75 (1996): 237–92. The most notorious historical example of moral hazard in the older sense of the term was the nineteenth-century phenomenon of people killing their children for life insurance money—a hazard so heinous that many states passed laws in the nineteenth century forbidding parents from insuring the lives of their children. See Viviana A. Zelizer, "The Price and Value of Children: The Case of Children's Insurance," *American Journal of Sociology* 86 (Mar. 1981): 1036–56.

78. "Deed of Settlement of the Philadelphia Contributionship" [1752], *BFP*, 285. Franklin also advertised some of the restrictions in the *Pennsylvania Gazette*. See, for example, Jan. 8, 1754.

79. Massachusetts Mutual Fire Company, *Report of a Committee, Chosen to Digest a Plan, and Form Rules and Regulations for a Mutual Fire Insurance Company* (Boston: Thomas Fleet, 1797), 10. British fire insurance companies were more likely to spell out provisions against fraud. See "Proposals from the Sun Fire-Office in Threadneedle Street" (London, July 28, 1721); London Assurance of Houses and Goods from Fire, "Instructions for the Company's Agents" (London, 1730), 2.

80. Most of the evidence for eighteenth-century fraud related to fire insurance is indirect, consisting of statements about what could happen. See Robin Pearson, "Moral Hazard and the Assessment of Risk in Eighteenth and Early Nineteenth Century Britain," *Business History Review* 76 (Spring 2002): 1–35. A few eighteenth-century British court cases were written about by contemporaries. Sun Fire Office, *Roger Lynch and John Lynch, appellts. Robert Dalzel, Henry Cartwright, and John Everett, Esqs. Respts. The Respondents Case* (London, 1730); James Allan

Park, *A System of the Law of Marine Insurance with Three Chapters on Bottomry; on Insurance on Lives, and on Insurance against Fire* (London: Thomas and Andrews, 1799). Cases involving fire insurance became increasingly common in nineteenth-century American case law. See, for example, Fuller v. Boston Mutual Fire Insurance Co., 4 Met. 206; Borden v. Hingham Mutual Fire Insurance Co., 18 Pick. 523; Jacob M. Phillips v. The Merrimack Mutual Fire Insurance Company, 64 Mass. 355; 1852 Mass. LEXIS 167; 10 Cush. 355. Many involved shopkeepers allegedly overstating the amount of stock on hand before a fire.

81. *Drinker Diary*, Dec. 26, 1794, 634.

82. "For the Philadelphia Gazette," *Philadelphia Gazette*, Jan. 7, 1795; "Domestic Occurrences," *Monthly Register*, Jan. 1795, 61; Martin P. Snyder, *City of Independence: Views of Philadelphia before 1800* (New York: Praeger, 1975), 189–90.

83. For an extensive overview of these changes, see Mark Tebeau, *Eating Smoke: Fire in Urban America, 1800–1950* (Baltimore: Johns Hopkins University Press); and Amy Greenberg, *Cause for Alarm: The Volunteer Fire Department in the Nineteenth-Century City* (Princeton, NJ: Princeton University Press, 1998).

Chapter 2: The Uncertainties of Disease

1. Billy G. Smith, "Death and Life in a Colonial Immigrant City: A Demographic Analysis of Philadelphia," *Journal of Economic History* 38 (Dec. 1977): 879; Henry A. Gemery, "The White Population of the Colonial United States, 1607–1790," in *A Population History of North America*, ed. Michael R. Haines and Richard H. Steckel (Cambridge: Cambridge University Press, 2000), 165. Much less is known about mortality patterns among blacks. For an overview, see Lorena Walsh, "African American Population of the Colonial United States," in *A Population History*, 206–9. Maris A. Vinovskis, "Mortality Rates and Trends in Massachusetts before 1860," *Journal of Economic History* 32 (Mar. 1972): 199. Both Smith and Vinovskis also provide a detailed discussion of the methodological problems of colonial demography. For a survey of overall patterns from colonization through the eighteenth century, see Gerald Grob, *The Deadly Truth: A History of Disease in America* (Cambridge, MA: Harvard University Press, 2002).

2. See Rebecca J. Tannenbaum, *The Healer's Calling: Women and Medicine in Early New England* (Ithaca, NY: Cornell University Press, 2002); and James H. Cassedy, *Medicine in America: A Short History* (Baltimore: Johns Hopkins University Press, 1991), 10.

3. Richard Harrison Shryock, *Medicine and Society in America, 1660–1860* (New York: New York University Press, 1960), 49–54.

4. Herbert Leventhal, *In the Shadow of the Enlightenment: Occultism and Renaissance Science in Eighteenth-Century America* (New York: New York University Press, 1976); Richard Godbeer, *The Devil's Dominion: Magic and Religion in Early New England* (Cambridge: Cambridge University Press, 1992).

5. Roy Porter, "Introduction," *Patients and Practitioners: Lay Perceptions of Medicine in Pre-Industrial Society* (Cambridge: Cambridge University Press, 1985), 14.

6. Otho T. Beall Jr. and Richard H. Shryock, "Cotton Mather: First Significant Figure in American Medicine," *Proceedings of the American Antiquarian Society* 63 (Apr. 15, 1953): 68.

7. John B. Blake, *Public Health in the Town of Boston, 1630–1822* (Cambridge, MA: Harvard University Press, 1959), 10.

8. There is some evidence that inoculation was also an endemic practice in Europe. Correspondents of James Jurin, secretary of the Royal Society in the early eighteenth century, claimed that the practice was common in Wales. One also claimed that schoolboys self-inoculated using their penknives to scrape a raw spot on their hands into which they inserted pus from an infected person. James Jurin, *A Letter to Caleb Cotesworth, M.D.* (London: W. and J. Innys, 1723), 25–29.

9. Cotton Mather, *The Angel of Bethesda* (Barre, MA: American Antiquarian Society and Barre Publishers, 1972), 112.

10. Blake, *Public Health*, 33.

11. Ibid., 54; Ian Glyn and Jenifer Glyn, *The Life and Death of Smallpox* (New York: Cambridge University Press, 2004), 60.

12. Elizabeth Fenn, *Pox Americana: The Great Smallpox Epidemic of 1775–82* (New York: Hill and Wang, 2001), 16; Glyn and Glyn, *Life and Death*, 2–4.

13. Fenn, *Pox Americana*, 15.

14. *New York Weekly Journal*, Feb. 6, 1737.

15. Donald R. Hopkins, *Princes and Peasants: Smallpox in History* (Chicago: University of Chicago, 1983), 6.

16. Ibid., 235, 237.

17. Blake, *Public Health*, 20.

18. Hopkins, *Princes and Peasants*, 4, 50. Fenn, *Pox Americana*, includes a discussion of current scientific and historical understandings of why Native Americans were so susceptible (24–27).

19. Mather, *Angel of Bethesda*, 93.

20. Ibid., 94.

21. Perry Miller, *The New England Mind from Colony to Province* (Cambridge, MA: Harvard University Press, 1953), 349.

22. Peter Galison, *Image and Logic: A Material Culture of Microphysics* (Chicago: University of Chicago Press, 1997), 47. Galison credits anthropological linguists with originating the term.

23. "An Extract of Several Letters from Cotton Mather, D.D. to John Woodward, M.D. and Richard Waller, Esq.; S.R. Sect.," *Philosophical Transactions* 29 (1714–16): 63, 67.

24. Emanuel Timoni and John Woodward, "An Account, or History, of the Procuring the Small Pox by Incision, or Inoculation; As It Has for Some Time Been Practised at Constantinople," *Philosophical Transactions* 29 (1714–16): 72.

25. Hopkins, *Princes and Peasants*, 248. This story is recounted in a number of sources, but Hopkins explains that it comes from a letter Mather wrote Woodward.

26. Cited in Blake, *Public Health*, 54. For a full accounting of Mather's extensive correspondence with Woodward, see George Lyman Kittredge, "Cotton Mather's Scientific Communications to the Royal Society," *American Antiquarian Society Proceedings* 26 (Apr. 1916): 18–57.

27. Boylston was, as one biographer described him, "the son of a physician, taught by his father and without a medical degree." This kind of training was not uncommon well into the nineteenth century but would not have given Boylston the high status and authority of a university degree. Henry R. Viets, "Some Features of the History of Medicine in Massachusetts during the Colonial Period," *Isis* 23 (Sept. 1935): 397.

28. Zabdiel Boylston, *Historical Account of the Small-Pox Inoculated in New-England* (Lon-

don: S. Chandler, at Cross-Keys in the Poultry, 1729), 1–2. The first inoculations took place on June 26, 1721. The epidemic began in April.

29. Ibid., 18.

30. Ibid., 9–10.

31. Aug. 1, 1721, *Diary of Cotton Mather, 1709–1724*, Collections of the Massachusetts Historical Society, vol. 78 (Boston: Massachusetts Historical Society, 1912), 635.

32. Ibid., Aug. 15, 1721, 637–38.

33. Ibid., [various entries], Aug. 25–31, 639–43.

34. The overall population figure is Douglass's, calculated after the epidemic. James H. Cassedy, *Demography in Early America: Beginnings of the Statistical Mind, 1600–1800* (Cambridge, MA: Harvard University Press), 139.

35. Gordon W. Jones, "Introduction," in Mather, *Angel of Bethesda*, xxxiii.

36. "To the Author of the Boston NEWS-LETTER," *Boston News-Letter*, July 17–24, 1721.

37. John B. Blake, "The Inoculation Controversy in Boston: 1721–1722," *New England Quarterly* 25 (Dec. 1952): 493–94.

38. "A Project for Reducing the Eastern Indians by Inoculation," *New-England Courant*, Aug. 7, 1721. Yaws was a skin disease related to syphilis and endemic to Africa. It figures in debates about whether syphilis originated in the New World or the Old. See Alfred Crosby, *The Columbian Exchange: Biological and Cultural Consequences of 1492* (Westport, CT: Greenwood Publishing, 1972). The British army's presentation of blankets obtained at the Fort Pitt smallpox hospital in 1763 is the most famous example. The details remain a matter of controversy among historians. See Elizabeth A. Fenn, "Biological Warfare in Eighteenth-Century America: Beyond Jeffrey Amherst," *Journal of American History* 86 (Mar. 2000): 1552–80.

39. Quoted in Blake, "Inoculation Controversy," 493.

40. Nov. 14, 1721, *Diary of Cotton Mather*, 657–58.

41. [no title], *New-England Courant*, Nov. 13–20, 1721. Franklin claimed that "the above Account we receiv'd from the Doctor's [Mather's] own hand."

42. Cassedy, *Demography*, 135, 139. The number of infected people seems suspiciously precise but is probably a good approximation since the percentage of fatalities is consistent with other smallpox epidemics.

43. Patricia Cline Cohen, *A Calculating People: The Spread of Numeracy in Early America* (New York: Routledge, 1999), 100.

44. Blake, *Public Health*, 62.

45. "London, June 17, 1721," *Boston News-Letter*, Oct. 16–23, 1721.

46. Genevieve Miller, *The Adoption of Inoculation for Smallpox in England and France* (Philadelphia: University of Pennsylvania Press, 1957), 70–89.

47. Genevieve Miller, "Smallpox Inoculation in England and America: A Reappraisal," *William and Mary Quarterly* 13 (Oct. 1956): 479.

48. Ibid., 486; Ola Elizabeth Winslow, *The Destroying Angel: The Conquest of Smallpox in Colonial Boston* (Boston: Houghton Mifflin, 1974), 66–67.

49. James Jurin, *An Account of the Success of Inoculating the Small Pox in Great Britain* (London: J. Peele, 1724), 18–28, 29.

50. Ibid., 32.

51. G. Miller, *Adoption of Inoculation*, 83.

52. Cohen, *Calculating People*, 87.

53. William Douglass, *A Dissertation Concerning Inoculation of the Small-Pox* (Boston: D. Henchman in Cornhill, 1730), iii.

54. *Drinker Diary*, Oct. 26, 1762, 97; Aug. 22, 1759, 31; Sept. 11, 1759, 31.

55. Sarah Blank Dine, "Inoculations, Patients, and Physicians: The Transformation of Medical Practice in Philadelphia, 1730–1810," *Transactions and Studies of the College of Physicians of Philadelphia* 20 (1998): 71.

56. Billy G. Smith, *The "Lower Sort": Philadelphia's Laboring People, 1750–1800* (Ithaca, NY: Cornell University Press, 1990), 48.

57. Dine, "Inoculations, Patients, and Physicians," 70.

58. Sarah Blank Dine, "Diaries and Doctors: Elizabeth Drinker and Philadelphia Medical Practice, 1760–1810," *Pennsylvania History* 68 (Autumn 2001): 416.

59. The account of the inoculation is *Drinker Diary*, Apr. 5, 1760, 53. For other visits, see Sept. 18, 1759, 32; Nov. 13, 1759, 39; Sept. 11, 1760, 31; Sept. 13, 1760, 31; Sept. 18, 1760, 32.

60. Drinker occasionally attended scientific demonstrations during this period. On February 8, 1760, she recorded that she "spent this afternoon, with Molly Foulk at the Widow Bringhursts, where we were entertained with diverse objects in a microscope and with several experiments in Electricity." *Drinker Diary*, Feb. 8, 1760, 47.

61. Glynis M. Breakwell, *The Psychology of Risk* (Cambridge: Cambridge University Press, 2007), 99–100.

62. Benjamin Rush, *The Autobiography of Benjamin Rush: His "Travels Through Life" together with His Commonplace Book for 1789–1813* (Princeton, NJ: Princeton University Press, 1948), 80.

63. "Philadelphia," June 21, 1860, *Pennsylvania Gazette*.

64. *Drinker* Diary, Jan. 6, 1763, 99; Feb. 11, 12, 18, 21, 1763, 107.

65. *Drinker Diary*, Dec. 16, 1765, 125.

66. *Drinker Diary*, Aug. 28, 1765, 123.

67. Dine, "Inoculation, Patients, and Physicians," 72.

68. Blake, *Public Health*, 111; Fenn, *Pox Americana*, 42, 92–95, 98–103.

69. Elizabeth Fenn views inoculation as remaining relatively rare. See Fenn, *Pox Americana*, 41. In contrast, Sarah Blank Dine portrays the practice as widely adopted in Philadelphia by the 1790s. See Dine, "Inoculation, Patients, and Physicians," 81.

70. Sara Stidstone Gronim, "Imagining Inoculation: Smallpox, the Body, and Social Relations of Healing in the Eighteenth Century," *Bulletin of the History of Medicine* 80 (2006): 247–68.

71. Dine, "Inoculation, Patients, and Physicians," 79–80.

72. John Blake, *Benjamin Waterhouse and the Introduction of Vaccination* (Philadelphia: University of Pennsylvania Press, 1957), 11, 37.

73. Dorothy Porter and Roy Porter, "The Politics of Prevention: Anti-Vaccinationism and Public Health in Nineteenth-Century England," *Medical History* 32 (1998): 231; Nadja Durbach, *Bodily Matters: The Anti-Vaccination Movement in England, 1853–1907* (Durham, NC: Duke University Press, 2005).

74. James Keith Colgrove, "Between Persuasion and Compulsion: Smallpox Control in Brooklyn and New York," *Bulletin of the History of Medicine* 78 (Summer 2004): 354; Michael Willrich, "'The Least Vaccinated of Any Civilized Country': Personal Liberty and Public Health in the Progressive Era," *Journal of Policy History* 20 (Jan. 2008): 76–93; Martin Kaufman,

"The American Anti-Vaccinationists and Their Arguments," *Bulletin of the History of Medicine* 4 (Sept.–Oct. 1967): 465–66.

75. Kaufman, "American Anti-Vaccinationists," 464, 467. Since the vaccination process involved creating an open wound and, before the era of germ theory, physicians often did not sterilize instruments, it is quite possible that the process passed along blood-borne diseases or created an opportunity for infections such as *Staphylococcus*.

76. Ibid., 464.

77. Baruch Fischhoff, "Risk Perception and Communication Unplugged: Twenty Years of Process," *Risk Analysis* 15 (Summer 1995): 139.

78. For a sense of the debates about vaccination, see "Given the Risks, Vaccinations Should Not Be Mandated" and "Childhood Vaccinations Are Important for Public Health," in *Should Vaccinations Be Mandatory?*, ed. Noël Merino (Detroit: Greenhaven, 2010).

Chapter 3: Doing Something about the Weather

1. "Philadelphia—November 23," *Pennsylvania Gazette*, Nov. 23, 1732. The "cold" was probably an influenza epidemic. Franklin noted that it had started earlier in Boston. A few weeks later he noted that "many elderly people die among us of Colds, and the Pleurisy has taken off several Young People lately." *Pennsylvania Gazette*, Dec. 7, 1732.

2. "Philadelphia, July 11," *Pennsylvania Gazette*, July 11, 1734. A week later, Franklin again reported, "By all accounts from the Country, it appears that this Harvest by Reason of the Extream Heat, has been one of the hardest for the Workmen to go thro', that has ever been known here." "Philadelphia, July 18," *Pennsylvania Gazette*, July 18, 1734.

3. Peter Eisenstadt, "The Weather and Weather Forecasting in Colonial America" (PhD diss., New York University, 1990); Kristine Harper, *Weather by the Numbers: The Genesis of Modern Meteorology* (Cambridge, MA: MIT Press, 2008); Paul N. Edwards, *A Vast Machine: Computer Models, Climate Data, and the Politics of Global Warming* (Cambridge, MA: MIT Press, 2010).

4. *Drinker Diary*, May 29, 1774.

5. For an example of historians' efforts to reconstruct the history of colonial weather based on these accounts, see David M. Ludlum, *Early American Winters, 1604–1820* (Boston, MA: American Meteorological Society, 1966).

6. Some would have also recognized the biblical allusion to the verse in Matthew in which Jesus says, "When it is evening, ye say it will be fair weather, for the sky is red; and in the morning it will be foul weather today, for the sky is read and lowering." Matthew 16:2, 3. Cited in Edward B. Garriot, *Weather Folk-lore and Local Weather Signs* (Washington, DC: Government Printing Office, 1903), 12.

7. *Drinker Diary*, Aug. 3, 1805.

8. [fig. Calculations for the Month of June 1792], in Silvio Bedini, *The Life of Benjamin Banneker: The First African-American Man of Science*, 2nd ed. (Baltimore: Maryland Historical Society, 1999), 187.

9. Eisenstadt, "Weather and Weather Forecasting," 24–29.

10. Jane M. Hatch and George William Douglas, *The American Book of Days* (New York: Wilson, 1978), 136–38.

11. Eisenstadt, "Weather and Weather Forecasting," 73–74, 70.

12. Benjamin Franklin, *The Complete Poor Richard Almanacks Published by Benjamin Franklin, Reproduced in Facsimile with an Introduction by Whitfield J. Bell, Jr.*, vol. 1 (Barre, MA: Imprint Society, 1970), 99.

13. Eisenstadt, "Weather and Weather Forecasting," 127–28.

14. Jan Golinski, *British Weather and the Climate of Enlightenment* (Chicago: University of Chicago Press, 2007), 79.

15. Thomas Mann Randolph Jr. to Thomas Jefferson (May 3, 1790), cited in James Roger Fleming, *Meteorology in America, 1800–1870* (Baltimore: Johns Hopkins University Press, 1990), 1.

16. Golinski, *British Weather*, 109, 194.

17. "Brief History of the Barometer," www.barometer.ws/history.html, accessed May 24, 2011.

18. Editorial, Mark Twain [Samuel Langhorne Clemens] (1835–1910), *Hartford Courant* (Aug. 27, 1897), www.bartleby.com/66/9/61909.html, accessed Oct. 9, 2007.

19. Charles Woodmason, "Poetical Epistle" (1754), in *The Papers of Benjamin Franklin*, ed. Leonard Woods Labaree, William Bradford Willcox, and Barbara Oberg (New Haven, CT: Yale University Press, 1959), 5:61 (hereafter *BFP*). Woodmason was not the only fan to praise Franklin's invention in verse. See also "Nathaniel Evans: Verses Addressed to Benjamin Franklin" (1765), *BFP*, 10:425.

20. For a more detailed scientific explanation of how lightning rods work, see Basil Schonland, *The Flight of the Thunderbolts* (Oxford: Clarendon, 1964), 18–19; Oliver J. Lodge, *Lightning Conductors and Lightning Guards* (London: Whittaker, 1892), 366–67.

21. Schonland, *Flight of the Thunderbolts*, 23.

22. *Gentleman's Magazine* 39 (1769): 457, cited in E. Philip Krider, "Lightning Rods in the 18th Century," unpublished conference paper, Friends of the Franklin Papers, Oct. 10, 1995.

23. Benjamin Franklin to Horace-Benedict de Sausure, Oct. 8, 1772, *BFP*, 19:325.

24. No architectural examples of eighteenth-century rods survive intact, to my knowledge, although some researchers have found traces, such as internal mountings in Independence Hall, Philadelphia. E. Philip Krider, personal communication, Nov. 2007. Descriptions from Franklin's correspondence afford the best explanation of how and where these early systems were installed. "Charlestown," Nov. 1, 1760, 10:53–54; "Of Lightning, and the Method (Now Used in America) of Securing Buildings and Persons from Its Mischievous Effects," (1769), *BFP*, 14:260–64. Franklin also provided instructions to some correspondents; see, for instance, "To David Hume," Jan. 21, 1762, *BFP*, 10:263.

25. Randolph Shipley Klein, *Portrait of an Early American Family: The Shippens of Pennsylvania across Five Generations* (Philadelphia: University of Pennsylvania Press, 1975), 165, 168; Arthur Lee, "An Account of the Effects of Lightning on Two Houses in the City of Philadelphia," *Memoirs of the American Academy of Arts and Sciences* 1 (1783): 248–49. The Shippen family was also among the Philadelphia Contributionship's earliest subscribers for fire insurance. www.philadelphiabuildings.org/contributionship/search.cfm.

26. J. A. Leo Lemay, *Ebenezer Kinnersley, Franklin's Friend* (Philadelphia: University of Pennsylvania Press, 1964), 73–74.

27. Lee, "Account of the Effects of Lightning," 248–49.

28. See Eleanor M. Tilton, "Lightning-Rods and the Earthquake of 1755," *New England*

Quarterly 13 (Mar. 1940): 86. See also I. B. Cohen, "Prejudice against the Introduction of the Lightning Rod," *Journal of the Franklin Institute* 253 (1952): 393–440.

29. "A Letter to the Boston Gazette" (Jan. 1756), cited in Tilton, "Lightning-Rods," 91.

30. "From John Winthrop," Jan. 6, 1768, *BFP*, 15:14; "To John Winthrop," July 2, 1768, *BFP*, 15:168.

31. Benjamin, Franklin, *Poor Richard Improved: Being an Almanack and Ephemeris* (Philadelphia: B. Franklin and D. Hall, 1853), n.p.

32. The author of "Lightning Rods" in the *Genesee Farmer* (Dec. 22, 1838), 401, hoped that readers would not be deterred from erecting rods because they could not obtain glass insulators. Homemade wooden ones would do just as well. In a transitional article, *Scientific American* touted the virtues of iron wire manufactured by Cooper and Hewitt "which answers for every purpose for lightning conductors." "Lightning Rods for Houses," *Scientific American* 7 (July 10, 1852): 344.

33. J. D. B. De Bow, *Statistical View of the United States . . . Being a Compendium of the Seventh Census* (1850), 5:119.

34. Charles Sellers, *The Market Revolution: Jacksonian America, 1815–1846* (New York: Oxford University Press, 1991).

35. David Jaffee, "The Village Enlightenment in New England, 1760–1820," *William and Mary Quarterly* 47 (July 1990): 327–28. Although Jaffee focuses on New England, the phenomenon was also widespread in the Mid-Atlantic region, especially Pennsylvania, New York, and Northern Delaware, and traveling west. In his study of Milwaukee newspapers between 1837 and 1846, Donald Zochert identified many articles about science, mostly republished from East Coast newspapers and journals. See "Science and the Common Man," in *Science in America since 1820*, ed. Nathan Reingold (New York: Science History, 1976), 7–32.

36. Melville's original readers would have recognized the literal basis of the story. At the time of its publication, a growing number of lightning rod salesmen wandered the villages and back roads of America pitching their wares to both the wary and the unsuspecting. Melville himself supposedly encountered such a character during his Pittsfield stay. See Alan Moore Emery, "Melville on Science: 'The Lightning-Rod Man,'" *New England Quarterly* 56 (Dec. 1983): 555–68; and Lea Bertani Vozar Newman, *A Reader's Guide to the Short Stories of Herman Melville* (Boston: G. K. Hall, 1986), 269–70. Traces of real lightning rod salesmen are rare in the historical record. For a discussion of a notable exception, see Elizabeth Cavichi, "Earth Grounds and Heavenly Spires: Lightning Rod Men, Patent Inventors, and Telegraphers," in *Playing with Fire: Histories of the Lightning Rod*, ed. Peter Heering, Oliver Hochadel, and David J. Rhees (Philadelphia: American Philosophical Society, 2009), 182–83.

37. Herman Melville, "The Lightning-Rod Man," in *The Piazza Tales and Other Prose Pieces, 1839–1860* (Evanston: Northwestern University Press, 1987), 119.

38. Ibid.

39. Ibid., 124.

40. The narrator asks, "Of what use was your rod then?" True to the rhetoric of the time, the salesman responds, "Of life-and-death use." He blames the failure on his workman's failure to properly install the device (he failed to insulate it from other metal fittings on the roof). "Not my fault but his," he explains. Ibid., 11. Discussions of failures begin with Franklin and are common across the nineteenth century. Improper installation is the most frequent explana-

tion. See, for example, Benjamin Franklin to David Hume, Jan. 21, 1762, *BFP*, 10:19; *Otis' Patent Lightning Conductors* (New York: Lyon's Manufacturing Company), 6–7. Arthur R. Bostwick, citing the work of physicist Oliver Lodge at the end of the nineteenth century, refutes these arguments, stating instead that earlier failures were a result of incomplete understanding of the physics of lightning. "The Modern View of Lightning Rods," *Literary Digest*, Oct. 27, 1894, 764.

41. Sally Kohlstedt, "Parlors, Primers, and Public Schooling: Education for Science in Nineteenth-Century America," *Isis* 81 (Sept. 1990): 424–45.

42. The *Pittsfield Sun* for August and September 1853 (the period during which Melville lived in the area) contains no advertisements for lightning rods. It did publish a number of articles about buildings and people being hit by lightning. See *Pittsfield Sun*, Aug. 17, 1853, Sept. 1, 1853.

43. See, for example, "Lightning Rods," *New York Religious Chronicle* 3 (July 9, 1825): 111, which discusses a paper on lightning rods published in Silliman's *Journal of Science*. An article in the *Christian Register* 10 (Feb. 12, 1831): 28, explains the lightning rod construction theories of Londoner John Murray as described in a paper on "Atmospheric Electricity."

44. Melville, "Lightning-Rod Man," 120–24, 122.

45. Richard D. Birdsall, *Berkshire County: A Cultural History* (New Haven, CT: Yale University Press, 1959), 30, 154, 171.

46. "Lightning Rods," *Genesee Farmer* 2 (Aug. 25, 1832): 269.

47. "Lightning Rods," *Genesee Farmer* 2 (Sept. 8, 1832): 284–85; "Lightning Rods," *Genesee Farmer* 2 (Sept. 22, 1832): 299.

48. "Lightning Rods for Barns," *Genesee Farmer* 3 (Sept. 16, 1837): 290.

49. "Danger to Barns from Lightning," *Genesee Farmer* 3 (Sept. 23, 1837): 299.

50. I searched the APS Index to Early American Periodicals for "Lightning Rods" and found nearly forty articles for the period between 1820 and 1850. See, for example, "Proper Construction of Lightning Rods," *Cultivator* 4, new series (May 1847): 161; "A Word about Lightning Rods," *Scientific American* 3 (Aug. 26, 1848): 387. The articles in religious papers are particularly interesting and are explained in part by Herbert Hovenkamp's thesis that science and religion were not necessarily viewed as antithetical by these antebellum enlightened Protestants. See Herbert Hovenkamp, *Science and Religion in America, 1800–1860* (Philadelphia: University of Pennsylvania Press, 1978).

51. *Scientific American*, for instance, moved away from giving useful advice toward evaluating the virtues of various patent systems and commercially available products. See "Lightning Rods for Houses," *Scientific American* 7 (July 10, 1852): 344; and "More about Lightning Rods," *Scientific American* 8 (July 9, 1853): 341.

52. Although her primary focus is women, Catherine Kelly provides a useful summary of these tensions. See Catherine E. Kelly, *In the New England Fashion: Reshaping Women's Lives in the Nineteenth Century* (Ithaca, NY: Cornell University Press, 1999), esp. 9–10, and chap. 8.

53. "Lightning Rods," *Scientific American* 3 (Apr. 15, 1848): 237.

54. "More About Lightning Rods," *Scientific American* 8 (July 9, 1853): 341.

55. *Otis' Patent Lightning Conductors*, 23; Lucius Lyon, *A Treatise on Lightning Conductors* (New York: G. P. Putnam, 1853). Interestingly, the literary critic Emery (op. cit. footnote 6) mistook Lyon's treatise for a legitimate scientific work.

56. *Otis' Patent Lightning Conductors*, 24.

57. H. F. Morrow, for example, set up a shop in Chester, Pennsylvania, to sell and install

rods. After a brief while, he gave his rod the name "Morrow Corrugated Rod," implying that it was a patented rod. See [no title], *Delaware County American*, May 24, 1865; "Copper Lightning Rod," *Delaware County American*, May 19, 1865; "Busy," *Delaware County American*, July 7, 1865. One of Lyon's direct competitors, the Quimby Company, installed most of their rods within a fifty-mile radius of Manhattan using their own mechanics but would ship systems and directions to more distant locations. They seem to have found a market in Caribbean and Central American sugar plantations. See *Circular of A.M. Quimby & Son, Dealers in Quimby's Improved Lightning Rods for Houses and Vessels* (New York: Baker, Godwin, 1854).

58. "My workman was heedless. In fitting the rod at top to the steeple, he allowed a part of the metal to graze the tin sheeting. Hence the accident. Not my fault, but his." Melville, "Lightning-Rod Man," 120.

59. See Timothy B. Spears, *One Hundred Years on the Road: The Traveling Salesman in American Culture* (New Haven, CT: Yale University Press, 1995).

60. John Phin, *Plain Directions for the Construction and Erection of Lightning Rods* (New York: Industrial, 1879), v.

61. Ibid. (italics in the original).

62. Otis's pamphlet also contains what could be read as a frank admission of the problem of how these devices were sold. "Intelligent local agents are being appointed in every town and county in the United States, to avoid the impositions of irresponsible lightning-rod peddlers, whose flagrant misrepresentations have defrauded thousands in different parts of our land, besides increasing the exposure of their families and property to the fury of the lightning blast." *Otis' Patent Lightning Conductors*, 32.

63. In addition to the Otis pamphlet, see C. J. Hubbell, *The Hubbell Patent System* (Washington, DC: C. W. Brown, Printer, 1885); Major H. G. Denniston, *Lightning Rods: Their Adoption and Value When Properly Constructed and Scientifically Applied* (New York: Williams, 1889); *Circular of A.M. Quimby & Son.*

64. *North American Lightning Rod Company Catalog* (Philadelphia: Rand-McNally, n.d.), 6–7.

65. Invoices in the Warshaw Collection files suggest that customers paid about twenty cents a foot between the 1860s and 1890s, buying several hundred dollars worth of rod, connections, and grounds for a house and outbuildings. See W. H. Demorest, "Invoice" (Oct. 1, 1866); New York Star Lightning Co., "Invoice" (Sept. 13, 1884); Atlantic Lightning Rod Company (1893) Lightning Rods—Box 1, Warshaw Collection, NMAH. The lightning-rod man in Melville's story wanted a dollar a foot—which seems to me dramatic license, although prices may have dropped over the century as production became cheaper and more efficient. For house prices, see Cooperative Building Plan Association, *Complete Collection of Shoppel's Modern Houses* (New York: Co-operative Building Plan Association, ca. 1885).

66. As early as 1860, the United States Census of Manufactures identified 164 manufacturers. *Eighth Census of the United States* (1860), 666–67.

Chapter 4: Animal Risk for a Modern Age

1. On domestication, see Edward Hyams, *Animals in the Service of Man: 10,000 Years of Domestication* (London: Dent, 1972); Frederick Everard Zeuner, *A History of Domesticated*

Animals (London: Hutchinson, 1963). Recent scholarship on the history of domestication suggests that animals and people domesticated each other. See, for example, Stephen Budiansky, *The Covenant of the Wild: Why Animals Chose Domestication* (New York: William Morrow, 1992).

2. On animals as technology, see Philip Scranton and Susan R. Schrepfer, *Industrializing Organisms: Introducing Evolutionary History* (New York: Routledge, 2004).

3. Hendrik Hartog, "Pigs and Positivism," *Wisconsin Law Review* 4 (1985): 900–903; Peter Karsten, "Cows in the Corn, Pigs in the Garden, and 'The Problem of Social Costs': 'High' and 'Low' Legal Cultures of the British Diaspora Lands in the 17th, 18th, and 19th Centuries," *Law and Society Review* 32 (1998): 67–68: Martin L. Primack, "Farm Fencing in the Nineteenth Century," *Journal of Economic History* 29 (June 1969): 287–91.

4. Ann Norton Greene, *Horses at Work: Harnessing Power in Industrial America* (Cambridge, MA: Harvard University Press, 2008), 72.

5. A few cities kept track of traffic fatalities from horse-drawn vehicles. For New York, see Clay McShane, *Down the Asphalt Path: The Automobile and the American City* (New York: Columbia University Press, 1994), 49.

6. See, for example, "By the Mayor, Aldermen, and Citizens of Philadelphia: An Ordinance for the Regulation of the Drivers of Carriages and Horses in and through the Streets of the City of Philadelphia," *Pennsylvania Gazette*, June 23, 1790; "October Sessions," *South Carolina Gazette*, Oct. 27, 1772.

7. Enlightenment-influenced reformers did make a sustained effort to change vernacular practices with regard to the issue of animal cruelty, but they did not frame work in terms of risk. See Bernard Oreste Unti, "The Quality of Mercy: Organized Animal Protection in the United States, 1866–1930" (PhD diss., American University, 2002); Keith Thomas, *Man and the Natural World: A History of the Modern Sensibility* (New York: Pantheon Books, 1983), 149.

8. Thomas Bewick's late eighteenth-century engraving "The Frightened Mother" neatly captures this in his portrait of a mother desperately yelling at a toddler to stop pulling on a pasture horse's tail. Robb Sagendorph, *America and Her Almanacs: Wit, Wisdom, and Weather* (Dublin, NH: Yankee, 1970), 181.

9. See David W. Anthony, "The Domestication of the Horse," in *Equids in the Ancient World*, vol. 2, ed. Richard H. Meadow and Hans-Peter Uerpmann (Wiesbaden: Reichert, 1991); David W. Anthony and Corcas R. Brown, "The Origins of Horseback Riding," *Antiquity* 65 (1991): 22–38; Marsha L. Levine, "Dereivka and the Problem of Horse Domestication," *Antiquity* 64 (1990): 727–40.

10. For a description of how to train horses to face these and other challenges, see Henry, Earl of Pembroke, *Military Equitation; or, A Method of Breaking Horses and Teaching Soldiers to Ride*, 3rd ed. (London: E. Easton, 1778), 27 and passim.

11. For a detailed analysis of horses (and mules) versus oxen, see Greene, *Horses at Work*, 27–32.

12. Jack Larkin, *The Reshaping of Everyday Life, 1790–1840* (New York: Harper and Row, 1988), 215.

13. Greene, *Horses at Work*, 2.

14. "Petersburg, May 20," *Pennsylvania Gazette*, June 2, 1790.

15. Rebecca Solnit, *River of Shadows: Eadweard Muybridge and the Technological Wild West* (New York: Viking, 2003), 38–39.

16. [Interview with Mazique Sanco], "WPA Slave Narrative Project, Texas Narratives, Volume 16, Part 4," http://memory.loc.gov/cgi-bin/query/D?mesnbib:2:./temp/~ammem_TzSH:: (accessed Apr. 20, 2012).

17. "Horace Greeley," *North Star*, July 20, 1849.

18. "Philadelphia, April 8," *Philadelphia Gazette*, Apr. 8, 1731.

19. "Frightful Accident on the Delaware," *Media Advertiser*, Jan. 30, 1856.

20. Clay McShane and Joel Tarr, *The Horse in the City: Living Machines in the Nineteenth Century* (Baltimore: Johns Hopkins University Press, 2007), 39.

21. Michael N. Searles, "Taking Out the Buck and Putting In a Trick: The Black Working Cowboys' Art of Breaking and Keeping a Good Cow Horse," *Journal of the West* 44 (2005): 53–60; Sara R. Massey, "Black Cowboys: Wrangling the Numbers," *Journal of South Texas* 17 (2004): 15–30.

22. "Horse Breaking," *Spirit of the Times* 14 (Nov. 23, 1844): 465.

23. These terms are still widely used among horse people, although some trainers, notably the well-known "horse-whisperer" Monty Roberts, prefer the term "starting" a young horse. Monty Roberts, *The Man Who Listens to Horses* (New York: Random House, 1997).

24. Leroy Judson Daniels, *Tales of an Old Horsetrader: The First Hundred Years* (Iowa City: University of Iowa Press, 1987), 6.

25. Ibid., 4. The other writer was Fairman Rogers, "Horsemanship, and the Methods of Training the Horse to Obey His Rider," *United States Service Magazine* 1 (Mar. 1864): 265. Rogers also decried these methods. He described the results as producing a horse that had a "sulky and wicked disposition, to improve which much careful subsequent handling was required."

26. "Time of Putting Colts to Work," *Southern Planter* 4 (Apr. 1848): 114.

27. Horses raced under saddle were the partial exception because they were sometimes raced very young before they had reached their full body mass and had begun to slow down.

28. Daniels, *Tales of an Old Horsetrader*, 5–6.

29. Clive Richardson, *The Horse Breakers* (London: J. A. Allen, 1998), 142, 246.

30. John S. Rarey, "The American Art of Taming Horses," *Southern Planter* 10 (Oct. 1858): 3. For more on Rarey and other horse tamers, see Richardson, *Horse Breakers*, 159–203.

31. "Horse Training," *Ohio Farmer* 20 (May 18, 1872): 308.

32. Merritt W. Harper, *The Training and Breaking of Horses* (New York: MacMillan, 1918), 13–15.

33. Lawrence M. Friedman, *A History of American Law*, 2nd edition (New York: Simon and Schuster, 1985), 264–65.

34. Clay McShane, "Gelded Age Boston," *New England Quarterly* 74 (Spring 2001): 284.

35. "Adventures in Search of a Horse," *New York Mirror* 14 (Mar. 18, 1837): 302.

36. On taking victims to the hospital, see, for example, "Fatal Accident in Central Park," *New York Times*, July 8, 1868, 4.

37. Conn v. Hunsberger, No. 231, Supreme Court of Pennsylvania, 224 Pa. 154; A. 324; 1909 Pa. LEXIS 744. Over the course of the nineteenth century, the courts seem to have taken an increasingly firm stand on this responsibility. Earlier in the century, most cases involved stable owners suing renters who abused horses or destroyed equipment. See, for example, Hughes

v. Boyer, Supreme Court of Pennsylvania, Middle District, Harrisburg, 9 Watts 556; 1840 Pa. LEXIS 71.

38. "A Day to be Out of Doors," *New York Times*, June 2, 1890.

39. T. H. Breen, "Horses and Gentlemen: The Cultural Significance of Gambling among the Gentry of Virginia," *William and Mary Quarterly* 34 (Apr. 1977): 249.

40. For more on the transformation of harness racing into a professional sport, see Melvin L. Adelman, "The First Modern Sport in America: Harness Racing in New York City, 1825–1870," *Journal of Sport History* 8 (Spring 1981): 5–32.

41. "Bonner's Stables," *Massachusetts Ploughman* 4 (Oct. 11, 1884): 44; "Robert Bonner and His Trotting Horses," *American Farmer* 73 (Aug. 1, 1892): 15.

42. The stable was on Fifty-Fifth Street. James D. McCabe Jr., *Lights and Shadows of New York Life* (New York: Farrar, Straus and Giroux, 1970), 758.

43. Harlem Lane was renamed St. Nicholas Avenue in the late nineteenth century. For a description of the lane and horse culture of New York in the 1850s, see Lloyd Morris, *Incredible New York: High Life and Low Life from 1850 to 1950* (Syracuse, NY: Syracuse University Press, 1996), 95.

44. "Our Drives," *Brooklyn Eagle*, July 17, 1875.

45. Flora Temple, the most famous trotter of the 1850s, set a record in 1855 of a mile in 2 minutes, 23.5 seconds (26.9 mph). Dwight Akers, *Drivers Up: The Story of American Harness Racing* (New York: G. P. Putnam's Sons, 1947), 88.

46. E. B. Abercrombie, "Trotting Road Teams and Their Drivers," *Outing* 29 (Oct. 1896): 8.

47. "Improving the Breed of Horses," *New York Evangelist*, Mar. 24, 1887, 7.

48. "Mr. Bonner Buys Maude S.," *New York Times*, Aug. 20, 1884.

49. Akers, *Drivers Up*, 131.

50. P. T. Barnum to Robert Bonner, July 26, 1871; P. T. Barnum to Robert Bonner, June 15, 1874, Papers of Robert Bonner (hereafter Bonner Papers), Series II, Box 7, New York Public Library.

51. James B. Townsend, "The American Trotting Horse," *Frank Leslie's Popular Monthly* 20 (Nov. 1885): 540.

52. [no title], *Albion* 47 (July 24, 1869): 30.

53. "Mr. Bonner's Breeding and Training Farm," *New York Evangelist*, June 17, 1875, 7.

54. "Mr. Bonner Buys Maud S.," *Maine Farmer*, Aug. 28, 1884, 1.

55. William Keetch to Robert Bonner, Apr. 4, 1873, Series II, Box 7, Bonner Papers.

56. The *New York Times* reported that the "city ordinance was amended so as to prohibit the driving of any cart, carriage, horse car, or other vehicle at a greater speed than at the rate of five miles per hour." "To Stop Reckless Driving," *New York Times*, May 30, 1888.

57. "Grant Arrested for Fast Driving," *Daily National Intelligencer*, Apr. 9, 1866.

58. [no title], *New York Times*, Nov. 30, 1880.

59. [Receipt], Apr. 27, 1869, Series II, Box 7, Bonner Papers.

60. "A Warning to Fast Drivers of Fast Horses," *New York Times*, Dec. 22, 1865.

61. "Surely a Winning Fight," *New York Times*, Mar. 23, 1892.

62. "The Speedway," *New York Times*, Apr. 2, 1892.

63. "Heard by the Park Board," *New York Times*, May 24, 1892.

64. "The Harlem River Speedway, New York City," *Scientific American* 126 (Feb. 6, 1897): 89.

Chapter 5: Railroads, or Why Risk in a System Is Different

1. Precise numbers of railroad-related fatalities are impossible to obtain for most of the nineteenth century. Mark Aldrich has compiled data for the period 1846–1900 from available sources. His analysis shows that although the absolute number of fatalities increased during this period, the fatality rate as measured by miles traveled declined. Mark Aldrich, *Death Rode the Rails: American Railroad Accidents and Safety, 1828–1965* (Baltimore: Johns Hopkins University Press, 2006), Appendix I, Table A1.2. In 1899 (one of the first years in which statistics are relatively reliable), a total of 7,123 fatalities and 44,620 injuries were reported to the Interstate Commerce Commission. Although passenger train wrecks attracted the most attention, passengers did not account for the majority of accidents or fatalities. Employees constituted by far the largest number of casualties overall, about 72%, but only 31% of the fatalities. Passengers made up 7% of the total casualties but only 3.3% of the fatalities. "Others"—a vague category mostly describing people hit while standing next to or crossing the tracks—provided only 21% of the total casualties but 65% of the fatalities. The majority of "others" were considered by the railroads to be "trespassers," meaning they were not killed at a crossing or at a station, but a not insubstantial 443 people were killed and 3,306 injured at stations. In contrast, only 239 passengers were killed. *Fourteenth Annual Report of the Interstate Commerce Commission* (Washington, DC: U.S. Government Printing Office, 1901), 73–74. In recent years, almost all railroad fatalities happen at crossroads or to trespassers. See note 9 below.

2. "Another Fatal Accident," *Torch Light and Public Advertiser*, Aug. 17, 1837.

3. "Dreadful Rail-Road Disaster and Loss of Lives," *Richmond Enquirer*, Aug. 18, 1837; "The Dreadful Accident on the Roanoke and Portsmouth Railroad," *New York Spectator*, Sept. 14, 1837; "Another Fatal Accident."

4. "The Late Rail Road Accident," *New York Spectator*, Sept. 18, 1837.

5. For an explanation of "socio-technological systems" as a concept, see Thomas P. Hughes, "The Evolution of Large Technological Systems," in *The Social Construction of Technological Systems: New Directions in the Sociology and History of Technology*, ed. Wiebe E. Bijker, Thomas P. Hughes, and Trevor Pinch (Cambridge, MA: MIT Press, 1878), 51–82.

6. Charles Perrow, *Normal Accidents: Living with High-Risk Technologies* (Princeton, NJ: Princeton University Press, 1984), is one of the most influential statements on the distinctive characteristics of risk in complex socio-technological systems. In practice, nineteenth-century railroads often functioned as what Perrow and others have called "loosely coupled" systems (89–93).

7. The peak year for railroad fatalities was 1913. In that year, the Interstate Commerce Commission reported that 10,550 people were killed and 86,668 injured in accidents involving the operation of steam railways. *Twenty-Fourth Annual Report of the Interstate Commerce Commission* (Washington, DC: U.S. Government Printing Office, 1914), 53. A total of 750 people died in railroad accidents between January and October 2004; 94.27% were killed at highway-rail crossings or while trespassing on the tracks. Federal Railroad Administration, Office of Safety Analysis, "Accident/Incident Overview," http://safetydata.fra.dot.gov/officeofsafety/public site/summary.aspx (accessed Apr. 22, 2012).

8. Aldrich, *Death Rode the Rails*, 3.

9. The most influential students of railroad safety have framed their analysis in terms of

political economy, arguing that powerful railroad companies fought for the right to decide how to implement safety measures in ways that would be most economically beneficial to them. For a synthetic overview, see Aldrich, *Death Rode the Rails*, especially the "Preface" and "Notes on Sources." Legal scholarship on this "economic-oriented" approach is often identified with James Willard Hurst, *Law and Social Order in the United States* (Ithaca, NY: Cornell University Press, 1977); and Morton J. Horwitz, *The Transformation of American Law, 1780–1860* (Cambridge, MA: Harvard University Press, 1977). For a recent critique, see Peter Karsten, *Heart versus Head: Judge-Made Law in Nineteenth Century America* (Chapel Hill: University of North Carolina Press, 1997), 3.

10. For more on the adoption of safety devices see Steven W. Usselman, *Regulating Railroad Innovation: Business, Technology, and Politics in America, 1840–1920* (Cambridge: Cambridge University Press, 2002); and Charles Clark, "The Railroad Safety Movement in the United States: Origin and Development, 1893–1913" (PhD diss., University of Illinois, 1966). Much of the older literature on railroad safety focuses on train wrecks. See, for example, Robert B. Shaw, *Down Brakes: A History of Railroad Accidents, Safety Precautions and Operating Practices in the United States of America* (London: P. R. Macmillan, 1961).

11. W. L. Garrison, "Railroad Traveling," *North Star*, Nov. 30, 1849. For other narratives illustrating this enthusiasm, see H. Roger Grant, ed., *We Took the Train* (Dekalb: Northern Illinois Press, 1990).

12. Garrison, "Railroad Traveling." In the 1850s, railroad advocates attempted to quantify the difference in safety between different modes of transportation. They claimed that between 1853 and 1859, 1,109 had died in railroad accidents versus 2,304 on steamboats. John F. Stover, *Iron Road to the West: American Railroads in the 1850s* (New York: Columbia University Press, 1978), 210. These numbers probably represent perception rather than any kind of accurate census.

13. John F. Stover, *American Railroads* (Chicago: University of Chicago Press, 1961), 26–27.

14. For detailed discussion of early railroad technology, see the work of John H. White, beginning with *American Locomotives: An Engineering History, 1830–1880* (Baltimore: Johns Hopkins University Press, 1968); *The American Railroad Freight Car: From the Wood-Car Era to the Coming of Steel* (Baltimore: Johns Hopkins University Press, 1993); and *The American Railroad Passenger Car* (Baltimore: Johns Hopkins University Press, 1978).

15. Shaw, *Down Brakes*, 34–39.

16. Walter Licht, *Working for the Railroad: The Organization of Work in the Nineteenth Century* (Princeton, NJ: Princeton University Press, 1983), 37.

17. Some states began requiring testing for physical impairments as early as the 1880s. See Mark Aldrich, "Train Wrecks to Typhoid Fever: The Development of Medicine Organizations, 1850 to World War I," *Bulletin of the History of Medicine* 75 (2001): 274–75.

18. Licht, *Working for the Railroad*, 37, 38, 43.

19. "Stealing a Locomotive," *Saturday Evening Post*, Oct. 4, 1856, 3.

20. Massachusetts chief justice Lemuel Shaw remarked that early legislation treated the railroad as "an iron turnpike, upon which individuals and transportation companies were to enter and run with their own cars and carriages, paying a toll to the corporation for the use of the road." Quoted in Leonard W. Levy, *The Law of the Commonwealth and Chief Justice Shaw* (Oxford: Oxford University Press, 1957), 141.

21. "Railroad Accidents," *American Railroad Journal* 18 (May 8, 1845).

22. Henry D. Thoreau, *Walden: A Fully Annotated Edition*, ed. Jeffrey S. Cramer (New Haven, CT: Yale University Press, 2004), 112.

23. For Thoreau's use of the metaphor "get off the tracks," see, for instance, ibid., 115; for derailment, 95.

24. John R. Stilgoe, *Metropolitan Corridor: Railroads and the American Scene* (New Haven, CT: Yale University Press, 1983), 168–69.

25. Legal historian William J. Novak argues that *Thorpe v. Rutland and Burlington Railroad* (1855), which upheld the right of the state of Vermont to require railroads to fence their tracks, is one of the most important police power cases of the antebellum era. See William J. Novak, *The People's Welfare: Law and Regulation in Nineteenth-Century America* (Chapel Hill: University of North Carolina Press, 1996), 109.

26. Charles Dickens, *American Notes for General Circulation* (Harmondworth, Middlesex: Penguin Books, 1972), 113.

27. Isaac F. Redfield (Chief Justice of Vermont), *A Practical Treatise upon the Law of Railroads*, 2nd ed. (Boston: Little, Brown, 1858), 364–65.

28. Dionysius Lardner, *Railway Economy: A Treatise on the New Art of Transport, Its Management, Prospects, and Relations* (New York: Harper and Brothers, 1850), 284.

29. For an extended discussion of legal cases resulting from passengers being forced to disembark from moving trains, see Barbara Young Welke, *Recasting American Liberty: Gender, Race, Law, and the Railroad Revolution, 1865–1920* (Cambridge: Cambridge University Press, 2001).

30. See, for example, Gustavas Nicolls to George Tucker, Oct. 27, 1853, Sept. 3, 1851; Gustavus Nicolls to James Milledland, Aug. 23, 1853, Aug. 23, 1853; Gustavus Nicolls to Peter Adams, Nov. 1, 1851, Box 104–1, Reading Railroad Papers, Hagley Museum and Library.

31. Mark Aldrich has suggested that in a statistical sense "worker and passenger safety was probably improving" during the collision crisis. The crisis may actually have been one of risk perception. Aldrich, *Death Rode the Rails*, 38.

32. For fuller descriptions, see Robert B. Shaw, *A History of Railroad Accidents, Safety Precautions, and Operating Practices* (Pottsdam, NY: Northern Press, 1978), 45, 127, 230, and passim.

33. "Railway Accidents," *American Railroad Journal* 26 (Oct. 1, 1853): 539.

34. "Another Case of Wholesale Murder," *National Era* 7 (May 12, 1853): 74; "Railroad Murders," *National Era* 7 (Aug. 18, 1853): 131.

35. Aldrich, *Death Rode the Rails*, Appendix I, Table A1.2, Appendix II, Table A2.1.

36. Glynnis M. Breakwell, *The Psychology of Risk* (Cambridge: Cambridge University Press, 2007), 29; Perrow, *Normal Accidents*, 326.

37. Railroad charters issued by states in the early years of the railroad often contained verbiage about responsibility to protect public safety, but these contractual obligations were not often enforced and often unenforceable because of vague wording. Sarah H. Gordon, *Passage to Union: How Railroads Transformed American Life, 1829–1929* (Chicago: Ivan R. Dee, 1996), 57.

38. Novak, *People's Welfare*, 108–9.

39. Aldrich, *Death Rode the Rails*, 305. For more on the commission, see Thomas K. McCraw, *Prophets of Regulation: Charles Francis Adams, Louis D. Brandeis, James M. Landis, Alfred E. Kahn* (Cambridge, MA: Belknap Press of Harvard University Press, 1984); Edward Chase Kirkland, *Men, Cities, and Transportation: A Study in New England History, 1820–1900*, vol. 2 (Cambridge,

MA: Harvard University Press, 1948); William A. Crafts, "The Second Decade of the Massachusetts Railroad Commission," *Railroad Gazette* 28 (Aug. 4, 1898): 551.

40. Massachusetts Railroad Commission, *Third Annual Report of the Board of Railroad Commissioners* (Boston: Wright and Potter, 1872), cxxvii–cxxix. Adams complained that the workers dismantled the guards.

41. Aldrich, *Death Rode the Rails*, 194.

42. White, *American Railroad Freight Car*, 498.

43. Steven Usselman's work on safety devices has been particularly influential. See Usselman, "Air Brakes for Freight Trains: Technological Innovation in the American Railroad Industry, 1869–1900," *Business History Review* 58 (Spring 1984): 30–50; and Usselman, *Regulating Railroad Innovation*. See also Clark, "Railroad Safety Movement"; White, *American Railroad Freight Car*; and Aldrich, *Death Rode the Rails*.

44. In the 1870s, George Westinghouse charged four hundred dollars to install a compressor on each engine plus fifty dollars per car for additional air brake equipment. To equip every one of the estimated 480,190 freight cars and 17,084 locomotives in use on American railroads in 1879 would have cost nearly twenty-five million dollars. Albert Fishlow, "Technological Change in the Railroad Sector, 1840–1910," unpublished paper, 1963, Hagley Museum and Library, 19.

45. Aldrich, *Death Rode the Rails*, 184.

46. Master Car Builders Association, *Thirteenth Annual Report* (New York: The Company, 1879), 104.

47. Edward C. Kirkland, *Charles Francis Adams, Jr., 1835–1915: The Patrician at Bay* (Cambridge, MA: Harvard University Press, 1965), 14, 37, 39.

48. J. J. Thomas, *Fifty Years on the Rail* (New York: Knickerbocker Press, 1912), 60.

49. Support was stronger for the automatic coupler than air brakes. See Clark, "Railroad Safety Movement," 185.

50. James H. Ducker, *Men of the Steel Rails: Workers on the Atchison, Topeka & Santa Fe Railroad, 1869–1900* (Lincoln: University of Nebraska Press, 1983), 121.

51. H. S. Haines, "Railway Accidents: Their Causes and Prevention," *Railroad Gazette* 25 (June 30, 1893): 484. Haines pointed out that the ICC only counted injuries or fatalities. The industry figures that he used only included accidents that damaged equipment.

52. Mark Aldrich, *Safety First: Technology, Labor, and Business in the Building of American Work Safety, 1870–1939* (Baltimore: Johns Hopkins University Press, 1997), 169.

53. For a similar example in the commercial airline industry, see Perrow, *Normal Accidents*, 162.

54. Aldrich, *Safety First*, 178. This phenomenon was also observed with the introduction of antilock brakes on automobiles in the 1990s. See Clifford M. Winston, Vikram Maheshri, and Fred Mannering, "An Exploration of the Offset Hypothesis Using Disaggregate Data: The Case of Airbags and Antilock Brakes," *Journal of Risk and Uncertainty* 32 (Spring 2006): 83–99.

55. Haines, "Railway Accidents," 485.

56. Paul Michel Taillon, "'To Make Men out of Crude Material': Work Culture, Manhood, and Unionism in the Railroad Running Trades," in *Boys and Their Toys? Masculinity, Technology, and Class in America*, ed. Roger Horowitz (New York: Routledge, 2001), 35.

57. This informal system has been noted by various historians of railroad labor. See, for

example, Taillon, "'To Make Men Out of Crude Material,'" 39; Licht, *Working for the Railroad*, chap. 2; John Williams-Searle, "Broken Brotherhood: Disability, Manliness, and Safety on the Rails, 1868–1908" (PhD diss., University of Iowa, 2004).

58. Thomas, *Fifty Years on the Rail*, 164–65.

59. Ibid., 21.

60. Ibid., 22.

61. Ibid., 39, 41.

62. Ibid., 53.

63. Ibid., 77.

64. On twentieth-century drinking, see John W. Orr, *Set Up Running: The Life of a Pennsylvania Railroad Engineman* (University Park: Pennsylvania State Press, 2001), 15, 21. For a nineteenth-century railroad man's temperance narrative, see Neason Jones, *Tom Keenan: Locomotive Engineer* (London: Fleming H. Revell, 1903).

65. [no title], *Railroad Gazette* 47 (July 9, 1909): 41.

66. "Some Points of Discipline," *Railroad Gazette* 22 (Feb. 14, 1890): 113; "Joint Responsibility of Conductor and Engineer," *Railroad Gazette* 22 (Aug. 22, 1890): 587–88.

67. R. L. Caincross, "Railroad Brotherhoods and Discipline," *Railroad Gazette* 46 (Feb. 26, 1909): 407. In response, K. A. Gohering wrote to the editor claiming that employees of the Ann Arbor Railroad used union meetings to talk about "accidents, rules, and other matters pertaining to the best interests of the company." See "Business-like Cooperation by Employees," *Railroad Gazette* 46 (Mar. 26, 1909): 690.

68. "Some Ethical Points of Discipline," *Railroad Gazette* 24 (July 29, 1892): 565.

69. Shelton Stromquist, *A Generation of Boomers: The Pattern of Railroad Labor Conflict in Nineteenth-Century America* (Urbana: University of Illinois Press, 1987), 189–90.

70. Licht, *Working for the Railroad*, 113–17.

71. "Good Discipline Easily Attainable If You Want It," *Railroad Gazette* 25 (June 1, 1894): 381.

72. "Can Railroad Discipline Be Improved?," *Railroad Gazette* 24 (July 15, 1892): 531.

73. "Dangerous Retrenchment," *Railroad Gazette* 22 (Feb. 7, 1890): 94.

74. "Some Ethical Points of Discipline," *Railroad Gazette* 24 (July 29, 1892): 565.

75. Ibid.

76. Haines, "Railway Accidents," 486.

77. Ducker, *Men of the Steel Rails*, 35–37.

78. [illegible name] to Alphonse Feldpauche, Dec. 24, 1903, Box 419, The Association of the Transportation Officers of the Pennsyslvania Railroad (hereafter ATO), Pennsylvania Railroad Papers, Hagley Museum and Library.

79. Aldrich, "Train Wrecks to Typhoid Fever," 273.

80. S. E. Long to A. Feldpauche, Sept. 3, 1908, ATO, Box 419, PRR.

81. ATO, "Report of the Committee on Conducting Transportation on 'The Subject of Keeping a Full Record . . . ,'" Apr. 15, 1897, 4–5, Box 413, PRR; "The Brown System of Discipline," *Railway World*, Sept. 17, 1904, 1078; see also Stromquist, *Generation of Boomers*, 241. There is some confusion about when Brown created the system. The ATO report is probably most accurate because the anonymous author interviewed Brown himself.

82. Shelton Stromquist describes Brown as a "major step toward the internalization of discipline on an individual basis." See Stromquist, *Generation of Boomers*, 241.

83. Ibid., 103.

84. ATO, "Report of the Committee on Conducting Transportation," 5.

85. James O. Fagan, *Confessions of a Railroad Signalman* (Boston: Houghton Mifflin, 1908), 158.

86. Ibid., 160.

87. Ibid., 175.

88. Ibid., 22.

89. For more on the twentieth-century history of railroad safety, see Aldrich, *Safety First*, chap. 5; and Aldrich, *Death Rode the Rails*.

Chapter 6: The Professionalization of Safety

1. Morris J. Vogel and Charles E. Rosenberg, *The Therapeutic Revolution: Essays in the Social History of American Medicine* (Philadelphia: University of Pennsylvania Press, 1979), 234; Kermit Hall, *The Magic Mirror: Law in American History* (New York: Oxford University Press, 1989), 198, 231. See also Burton J. Bledstein, *The Culture of Professionalism: The Middle Class and the Development of Higher Education in America* (New York: Norton, 1976); and Andrew Delano Abbott, *The System of Professions: An Essay on the Division of Expert Labor* (Chicago: University of Chicago Press, 1988). Sociologists define professionalization as having three characteristics: "an ideology of expertise to be placed at the service of a wider public," "an interest in securing income and prestige for its practitioners," and "a social space with which the professional group could be identified." Patricia Lengermann and Gillian Niebrugge, "Thrice Told Tales: Narratives of Sociology's Relation to Social Work," in *Sociology in America: A History*, ed. Craig Calhoun (Chicago: University of Chicago Press, 2007), 80.

2. For more on mining and railroads, see William Graebner, *Coal-Mining Safety in the Progressive Period: The Political Economy of Reform* (Lexington: University Press of Kentucky, 1976); Mark Aldrich, *Safety First: Technology, Labor, and Business in the Building of American Work Safety, 1870–1939* (Baltimore: Johns Hopkins University Press, 1997); *Death Rode the Rails: American Railroad Accidents and Safety, 1828–1965* (Baltimore: Johns Hopkins University Press, 2006).

3. Terence Powderly claimed that the Knights were central to the creation of bureaus of labor statistics. Establishment of these agencies was one of the planks in the Knights 1878 constitution, but, as this chapter will show, a number of them preceded the emergence of the Knights as a public organization. See Terence Powderly, *Thirty Years of Labor* (Columbus, OH: Excelsior Publishing House, 1889), chap. 7.

4. Census of Manufactures, 1870, http://mapserver.lib.virginia.edu/php/newlong2.php (accessed Jan. 19, 2011).

5. On accidents in early American industry, see Anthony F. C. Wallace, *Rockdale: The Growth of an American Village in the Early Industrial Revolution* (New York: Knopf, 1978), 149–50; Daniel J. Walkowitz, *Worker City, Company Town: Iron and Cotton-Worker Protest in Troy and Cohoes, New York, 1855–84* (Urbana: University of Illinois Press, 1978), 108–9; Peter Way, *Common Labor: Workers and the Digging of North American Canals, 1780–1860* (Baltimore: Johns Hopkins

University Press, 1997). On the new risks of chemicals, see Christopher C. Seller, *Hazards of the Job: From Industrial Disease to Environmental Health Sciences* (Chapel Hill: University of North Carolina Press, 1997). For a discussion of the erosion of employer paternalism, see Jamie Bronstein, *Caught in the Machinery: Workplace Accidents and Injured Workers in Nineteenth-Century Britain* (Stanford, CA: Stanford University Press, 2008), 40–49, 171.

6. *Historical Statistics of the United States: From Earliest Times to the Present, Millennial Edition*, vol. 2, Part B (Cambridge: Cambridge University Press, 2000), 2–18. A total of 55.8% of free wage earners were found in agriculture. Less than 15% of the labor force was employed in manufacturing. Smaller percentages worked in domestic service, clerical sales and service, and the professions. The census did not count women's unpaid domestic labor as productive employment.

7. J. Lynn Barnard, *Factory Legislation in Pennsylvania: Its History and Administration* (Philadelphia: University of Pennsylvania, 1907), 6–7.

8. William Brock, *Investigation and Responsibility: Public Responsibility in the United States, 1865–1900* (Cambridge: Cambridge University Press, 1984), 149. For more on the role of statistical arguments in advocating for legislation, see Arwen Mohun, "On the Frontier of the Empire of Chance: Statistics, Accidents, and Risk in Industrializing America," *Science in Context* 18 (2005): 337–57.

9. The two men knew each other. Wright later asked Adams to consult on the Pullman strike. James Leiby, *Carroll Wright and Labor Reform: The Origins of Labor Statistics* (Cambridge, MA: Harvard University Press, 1960), 168–69.

10. Brock, *Investigation and Responsibility*, 150, 152.

11. See, for example, Carroll D. Wright, "The Work of the United States Bureau of Labor," *Proceedings at the Third Annual Session of the National Commission of Bureaus of Statistics of Labor in the United States* (Boston, 1885): 125–26.

12. Martin Ambrose Foran, "Speech of Hon. M.A. Foran of Ohio, in the House of Representatives, April 19, 1884," *Bureau of Labor Statistics* (Washington, 1884), 3, 5.

13. William Forbath, *Law and the Shaping of the American Labor Movement* (Cambridge, MA: Harvard University Press, 1991), 17.

14. William I. Trattner, *Crusade for the Children: A History of the National Child Labor Committee and Child Labor Reform in America* (Chicago: Quandrangle Books, 1970), 29–33.

15. For a detailed list of departments and programs, see William Franklin Willoughby, *Inspection of Factories and Workshops* (Department of Social Economy for the United States Commission to the Paris Exposition, 1901).

16. Department of the Interior, Census Office, *Report on the Manufactures of the United States at the Tenth Census* (Washington, DC: U.S. Government Printing Office, 1883), 15. For a detailed account of Ohio's industrialization, see Philip D. Jordan, *Ohio Comes of Age, 1873–1900* (Columbus, OH: Ohio State Archeological and Historical Society, 1943), chap. 7.

17. International Association of Factory Inspectors, *Journal of the Proceedings of the International Association of Factory Inspectors of America*, Fifteenth Annual Convention, Held at Niagara Falls, New York, Sept. 24–27, 1901, n.p. (hereafter *Journal of the Proceedings*).

18. Ibid.

19. Willoughby, *Inspection of Factories and Workshops*, 4.

20. Jonathan Garlock, *Guide to the Local Assemblies of the Knights of Labor* (Westport, CT: Greenwood Press, 1982).

21. Willoughby, *Inspection of Factories and Workshops*, 5 and passim.

22. Ohio. Department of Inspection of Workshops, Factories, and Public Buildings, *First Annual Report of the Department of Inspection of Workshops, Factories, and Buildings* (Columbus, 1885), 4.

23. Ibid., 6.

24. Ibid., 24.

25. Ohio. Department of Inspection of Workshops, Factories, and Public Buildings, *Fifth Annual Report of the Department of Inspection of Workshops, Factories, and Buildings* (Columbus, 1889), 35.

26. Ohio. Department of Inspection of Workshops, Factories, and Public Buildings, *Second Annual Report of the Department of Inspection of Workshops, Factories, and Buildings* (Columbus, 1886) 73.

27. Helena Bergman, "That Noble Band? Male and Female Factory Inspectors in the Turn-of-the-Century United States" (MA thesis, SUNY Binghamton, 1995).

28. *Journal of the Proceedings*, Fifteenth Annual Convention, n.p. This volume reprinted selected proceedings from previous meetings.

29. Ibid., 17.

30. Ohio. Department of Inspection of Workshops, Factories, and Public Buildings, *Second Annual Report*, 38, 41. For more on fire escapes, see Sarah Wermeil, *The Fireproof Building: Technology and Public Safety in the Nineteenth-Century American City* (Baltimore: Johns Hopkins University Press, 2000), 190–212.

31. International Association of Factory Inspectors, *Journal of the Proceedings of the International Association of Factory Inspectors of America*, 1st Annual Meeting, 1886, 3. The association was international in that Canadians were welcome to join.

32. New York, Office of Factory Inspectors, *Annual Report of the Factory Inspectors of New York for the Year Ending in 1887* (2nd Annual), 32–38.

33. Ohio. Department of Inspection of Workshops, Factories, and Public Buildings, *Fifth Annual Report*, 34, 190, 192, 193–94, 197.

34. Lengermann and Niebrugge, "Thrice Told Tales," 100, 105.

35. Florence Kelley, *Notes of Sixty Years: The Autobiography of Florence Kelley*, ed. and intro. Kathryn Kish Sklar (Chicago: Charles H. Kerr, 1986), 86.

36. Viviana A. Rotman Zelizer, *Pricing the Priceless Child: The Changing Social Value of Children* (New York: Basic Books, 1985); Steven Mintz, *Huck's Raft: A History of Childhood* (Cambridge, MA: Harvard University Press, 2004), 181; Joel A. Tarr and Mark Tebeau, "Managing Danger in the Home Environment, 1900–1940," *Journal of Social History* 29 (Winter 1996): 797.

37. Donald L. Miller, *City of the Century: The Epic of Chicago and the Making of America* (New York: Simon and Schuster, 1996).

38. Kathryn Kish Sklar, *Florence Kelley and the Nation's Work* (New Haven, CT: Yale University Press, 1995), 142.

39. www.hullhouse.org/aboutus/history.html (accessed Jan. 12, 2010); Kelley, *Notes of Sixty Years*, 12.

40. Mary Jo Deegan, *Jane Addams and the Men of the Chicago School* (New Brunswick, NJ: Transaction Books, 1988).

41. Sandra D. Harmon, "Florence Kelley in Illinois," *Illinois State Historical Journal* 74 (Autumn 1981): 64; Kelley, *Notes of Sixty Years*, 80.

42. It prohibited "the manufacture of certain items of clothing in apartments, tenement houses, and living rooms, except by families living therein," and only if the workshops were kept clean, free of vermin and "infection or contagious matter." Children under fourteen years of age were prohibited from working in "any manufacturing establishment, factory, or workshop in the state." The hours clause read: "No female shall be employed in any factory or workshop more than eight hours in any one day or forty-eight in any one week." "Factories and Workshops," *Laws of the State of Illinois, Passed by the Thirty-Eighth General Assembly* (Springfield, IL: H. W. Rokker, 1893), 99, 100.

43. Illinois, *Third Annual Report of the Factory Inspectors of Illinois for the Year Ending December 15, 1895* (Springfield, IL: Ed F. Hartman, 1896), 5.

44. Illinois, *Second Annual Report of the Factory Inspectors*, 13.

45. Illinois, *Third Annual Report of the Factory Inspectors*, 15–16. On changing beliefs about child labor, see Zelizer, *Pricing the Priceless Child*, 56–60. Illinois, *Second Annual Report of the Factory Inspectors*, 25.

46. Illinois, *Second Annual Report of the Factory Inspectors*, 24.

47. Illinois, *Third Annual Report of the Factory Inspectors*, 11; Mintz, *Huck's Raft*, 181.

48. Illinois, *Second Annual Report of the Factory Inspectors*, 11. Other employers included the Lancaster Caramel Company, Spaulding & Merrick Tobacco Factory, and the Kimball and Company piano manufacturers.

49. Illinois, *Third Annual Report of the Factory Inspectors*, 16.

50. Ibid., 18.

51. Ibid, 16.

52. Ibid., 23.

53. Ibid., 26.

54. Ibid., 4–5.

55. William Hard, "Making Steel and Killing Men," *Everybody's Magazine* 17 (Nov. 1907): 585–86, 580.

56. John A. Fitch, *The Steel Workers* (New York: Charities Publication Committee, 1911), 57–71.

57. Hard, "Making Steel," 581.

58. Ibid., 588.

59. *Report on Conditions of Employment in the Iron and Steel Industry in the United States*, vol. 4, *Accidents and Accident Prevention*, Senate document 110, 62nd Congress, 1st Session (Washington, DC: U.S. Government Printing Office, 1913), 179.

60. For an extended analysis of why industrial employment became more dangerous in the late nineteenth century, see Aldrich, *Safety First*, 85–86 and passim.

61. Daniel Nelson, *Managers and Workers: Origins of the New Factory System in the United States, 1880–1920* (Madison: University of Wisconsin Press, 1975), 35.

62. David Brody, *Steelworkers in America: The Non-Union Era* (Cambridge, MA: Harvard University Press, 1960), 5–7, 22.

63. Crystal Eastman, *Work Accidents and the Law* (New York: Charities Publication Committee, 1910), 7.

64. Ibid., 108–9.

65. Ibid., 110.

66. Brody, *Steelworkers in America*, 155, 161, 147.

67. The literature on the history of workmen's compensation is vast. For a recent overview, see John Fabian Witt, *The Accidental Republic: Crippled Workingmen, Destitute Widows, and the Remaking of American Law* (Cambridge, MA: Harvard University Press, 2004), 125 and passim.

68. Ida Tarbell, *The Life of Elbert H. Gary* (New York: Appleton, 1925), 58, 88, 94.

69. This story has been told in various forms. See, for instance, Charles L. Close, "Welfare Work in the Steel Industry," *Address by Charles L. Close* (May 28, 1920); and David Beyer's account in Eastman, *Work Accidents and the Law*, 244–45.

70. Aldrich, *Safety First*, 91.

71. "Steel Mill City Locks Out Police," *Chicago Daily Tribune*, May 14, 1906.

72. "Novel Plot to Defraud," *Chicago Daily Tribune*, May 29, 1896.

73. Close, "Welfare Work," 5.

74. The most detailed account of these features is in safety engineer David Breyer's description of the safety program at American Steel and Wire, another U.S. Steel subsidiary written for the Pittsburgh survey. See Eastman, *Work Accidents and the Law*, 244–61.

75. Ibid., 245.

76. Ibid., 248–49.

77. Ibid.

78. Ibid., 262.

79. Aldrich, *Safety First*, 92.

80. Ibid., 94–104.

81. Like everything else in this story, practice was more complicated. Not all employers were required to or chose to buy compensation insurance. Workers continued to sue because workmen's compensation payouts were small and initially limited to certain types of injuries—mostly dismemberment. The premium structure convinced some employers not to rehire workers who had recovered from previous injuries because it would raise their overall payments.

82. Nelson, *Managers and Workers*, 151.

83. For instance, David Beyer, who wrote the appendix on the safety program at American Steel and Wire for Eastman's *Work Accidents and the Law*, became the chief safety engineer for Liberty Mutual Insurance Company in 1912; Aldrich, *Safety First*, 118. Lew Palmer, who we will get to know in the next chapter, began his career as an engineer at Homer and Laughlin Steel Company. In 1913, he was appointed Chief Inspector for the Pennsylvania State Department of Labor and Industry. He worked inspecting shipyards during World War I for the state of Pennsylvania in the late 1910s and was later employed by the Equitable Life Assurance Society. "Lew Palmer Dies," *Air Transport*, Apr. 1945, n.p.

84. Lew R. Palmer, "History of the Safety Movement," *Annals of the American Academy of Political and Social Science* 123 (Jan. 1926): 10.

85. Ibid., 10.

86. Robert L. Meyer, "A Look at the First Congress," *Safety and Health* 135 (Feb. 1987): 22.

87. The 1912 "Social Creed" set out by the Federal Council of the Churches of Christ in America did include "protection of the worker from dangerous machinery, occupational diseases, and mortality" as one of its sixteen principles. Charles Howard Hopkins, *The Rise of the Social Gospel in American Protestantism, 1865–1915* (New Haven, CT: Yale University Press, 1950), 317; Federal Council of the Churches of Christ in America, *The Church and Modern Industry* (Federal Council of the Churches of Christ in America, 1913).

88. Federal Council, *Church and Modern Industry*, 11.

89. Aldrich, *Safety First*, 110–11.

90. Ibid., 132.

91. Ibid., 118–20.

92. "Second Safety Congress of the National Council for Industrial Safety," Sept. 23 and 24, 1913, 136. NSC lore holds that it was Whittle's discussion of automobile accident statistics that caught Campbell's attention, but I could find no corroborating evidence of this. The Chicago connection is probably much more important. Robert Meyer, "One Giant Step for Safety," *Safety and Health*, Mar. 1987, 47.

93. "Much Needed Regulations," *Outlook*, Aug. 29, 1914, 1038.

Chapter 7: The Safety-First Movement

1. "Queries and Answers," *New York Times*, Sept. 19, 1915.

2. Dianne Bennett and William Graebner, "Safety First: Slogan and Symbol of the Industrial Safety Movement," *Journal of Illinois Historical Society* 68 (June 1975): 243–56; Roy Rutherford Bailey, *Sure Pop and the Safety Scouts* (Chicago: World Book Company, 1915); Paul B. Dickey, *Safety First: A Musical Comedy in Two Acts, Presented by the Princeton University Triangle Club* (Cincinnati, OH: J. Church, 1916); *Safety Last!*, directed by Fred C. Newmeyer and Sam Taylor (Los Angeles: Hal Roach Studios, 1923).

3. "Queries and Answers." The Safety First Federation was a short-lived organization headquartered in New York City. Surviving records are in the Charles Bernheimer Papers, Historical Society of New York.

4. Barbara Welke, *Recasting American Liberty: Gender, Race, Law, and the Railroad Revolution, 1865–1920* (Cambridge: Cambridge University Press, 2001), 36.

5. Edward C. Spring, assistant to the president of the Lehigh Valley Transit Co., was a key figure. He served as the first chairman of the NSC's public safety section and also played an important role in the American Museum of Safety's public safety efforts. *Proceedings of the National Safety Council* 5 (1916): 797.

6. Randolph Bergstrom, *Courting Danger: Injury and Law in New York City, 1870–1910* (Ithaca, NY: Cornell University Press, 1992), 19. Although it is a case study of a single city, Bergstrom's book remains definitive on the early twentieth-century shift in personal injury cases outside the workplace.

7. Barbara Welke quotes another source that puts the figure at 10% of companies' operating expenses, "second only to wages." Welke, *Recasting American Liberty*, 29.

8. "Review of the Denver Meeting of the Claims Agents Association," *Electric Railway Journal* 34 (Nov. 20, 1909): 1063.

9. Bergstrom, *Courting Danger*, 163.

10. E. F. Schneider, "Prevention of Accidents," *Electric Railway Journal* 35 (Apr. 2, 1910): 617.

11. "A Reduction in Costs of Injuries and Damages," *Electric Railway Journal* 37 (Feb. 4, 1911): 192.

12. "The Problem of Reducing Accidents-II," *Electric Railway Journal* 33 (Jan. 16, 1909): 98–99.

13. See, for example, "Accidents in New York in March," *Electric Railway Journal* 35 (Apr. 30, 1910): 800.

14. "Instruction on Accidents," *Electric Railway Journal* 31 (May 2, 1908): 752.

15. Peter D. Norton, *Fighting Traffic: The Dawn of the Motor Age in the American City* (Cambridge, MA: MIT Press, 2008), 4, 5, 71–79.

16. Dorothea Haven Scoville and Doris Long, *Safety for the Child: A Practical Guide for Home and School* (New York: Republic Book Company, 1921), 45.

17. Philip Davis, *Street-land: Its Little People and Big Problems* (Boston: Small, Maynard, 1915), 34.

18. In an exercise of legal logic similar to the fellow servant rule, streetcar companies were protected by a legal doctrine called "imputable negligence" that required a supervising adult to exercise diligence. Bergstrom, *Courting Danger*, 61.

19. "Campaign against Accidents in Chicago," *Electric Railway Journal* 33 (June 5, 1909): 1046.

20. "Look Out for the Little Ones," *Street Railway Journal* [name changed to *Electric Railway Journal* the following year] 34 (July 3, 1909): 54.

21. "Safety Habit Is Spreading," *Los Angeles Times*, June 4, 1911, described the efforts of Ella Flagg Young, superintendent of the Chicago public schools. On Brooklyn, see Welke, *Recasting American Liberty*, 37.

22. "Cincinnati Accident Bulletin for Children," *Electric Railway Journal* 35 (June 25, 1910): 1105.

23. See Julie Johnson-McGrath, "Speaking for the Dead: Forensic Pathologists and Criminal Justice in the United States," *Science, Technology, and Human Values* 20 (Autumn 1995): 441.

24. "Peter Hoffman, Ex-Coroner and Sheriff, Dies," *Chicago Daily Tribune*, July 31, 1948.

25. *Proceedings of the Second Safety Congress* (1913), 131; Peter Hoffman, *What You Must Know for Safety: A Manual of Information and Statistical Records Relating to the Conservation of Human Life* (Chicago: Public Safety Commission of Chicago and Cook County, 1915), 8. Some of the steel men at these gatherings must have been annoyed at this claim. George Whittle's version was that Richardson's company had itself been catechized in "safety first" by safety experts from Illinois Steel. See *Proceedings of the First Annual Safety Congress* (1912), 76.

26. Hoffman, *What You Must Know*, 41.

27. "Autos Run Down School Children; Two Near Death," *Chicago Tribune*, Sept. 13, 1913; "Public Safety Work Starts," *Chicago Tribune*, Sept. 19, 1913.

28. Hoffman, *What You Must Know*, 12, 26, 154.

29. Ibid., 50.

30. Ibid., 50, 65–66.

31. Norton, *Fighting Traffic*, 177.

32. *Proceedings of the National Safety Council* 7 (1918): 350.

33. Julian H. Harvey, *The Rochester Public Safety Campaign: A Report of Organized Public

Safety Activities in the City of Rochester, N.Y. (Chicago: National Safety Council, 1918), 19, 24, 30–33. On Rochester's history as a center of evangelicalism, see Paul E. Johnson, *A Shopkeeper's Millennium: Society and Revivals in Rochester, New York, 1815–1837* (New York: Hill and Wang, 1979).

34. See, for example, "Public Safety Campaigns," *Safety*, Dec. 1913, 14; "Buffalo Campaign," *Safety*, May 1914, 105; "Brooklyn's Safety Campaign," *Safety*, Mar. 1914, 53. By 1916, the federal government had even gotten involved, sending out a train with twelve cars of safety exhibits to tour the country, based on a series of exhibits in Washington, DC. See "The Safety Train," *Outlook*, May 31, 1916, 240.

35. Norton, *Fighting Traffic*, 33–34.

36. For Campbell's background, see Robert Meyer, "A Look at the First Congress," *Safety and Health*, Feb. 1987, 22; and Meyer, "One Giant Step for Safety," *Safety and Health*, Mar. 1987, 46.

37. Harvey, *Rochester Public Safety Campaign*, 5–8, 26.

38. In this period, these organizations not only provided an important networking function for people in the business community but also styled themselves as promoting reform of the rationalizing, conservation, and efficiency-oriented kind. They also strove to provide a self-governing alternative to state-sponsored regulation.

39. Harvey, *Rochester Public Safety Campaign*, 5, 7.

40. Ibid., 14.

41. "Some Interesting Facts regarding the Organization and Growth of the National Safety Council," Mar. 31, 1939, typescript, National Safety Council Library, Itasca, IL, 4.

42. On Eno, see Clay McShane, *Down the Asphalt Path: The Automobile and the American City* (New York: Columbia University Press, 1994), 185. On Campbell's connection to Eno, see *NSC Proceedings* 4 (1915): 13. On traffic "safeguarding" in Rochester, see Harvey, *Rochester Public Safety Campaign*, 28.

43. For more on Eno, see Norton, *Fighting Traffic*, 51.

44. Harvey, *Rochester Public Safety Campaign*, 33.

45. "Sixty-Two Autoists Appear before Traffic Court Judge," *Rochester Times-Union*, June 4, 1918.

46. Welke, *Recasting American Liberty*, 30.

47. How they came up with the 70% figure and whether or not they actually believed themselves remain a mystery.

48. Harvey, *Rochester Public Safety Campaign*, 22.

49. Ibid., 26.

50. Ibid., 32.

51. Ibid., 31.

52. Ibid., 19.

53. "Big Decrease in Accidents," *Rochester Times-Union*, Aug. 6, 1918.

54. Harvey, *Rochester Public Safety Campaign*, 43.

55. Ibid., 44.

56. Mary Martha Butt Sherwood, *Duty Is Safety; or, Troublesome Tom* (Philadelphia: G. S. Appleton, 1850).

57. For an example of a checklist to be sent home with children, see Scoville and Long, *Safety for the Child*, 337.

58. On the gendering of accidents, see *Proceedings of the National Safety Council* 8 (1919): 1088.

59. "Children's Safety Crusade," *Safety*, Apr. 1913, 14; "Safety Campaign Button," *Washington Post*, May 13, 1913.

60. "Safety First Campaign Gives Fine Results," *New York Times*, Oct. 19, 1913.

61. "Lew R. Palmer," n.d., General History Files—Miscellaneous, National Safety Council Library; "L.R. Palmer Dead, Safety Leader, 70," *New York Times*, Mar. 25, 1945.

62. John Sayle Watterson, *College Football: History, Spectacle, Controversy* (Baltimore: Johns Hopkins University Press, 2000), 13–14, 68.

63. "Better Safe than Sorry," *Delineator*, Apr. 1920, 68.

64. For an exposition of this philosophy, see Albert W. Whitney, "Safety for More and Better Adventures," printed copy of a paper read before the Eleventh Annual Congress of the Playground and Recreation Association of America (Apr. 17, 1924), Historical Files, National Safety Council Library, Itasca, IL.

65. Roy Rutherford Bailey, *Sure Pop and the Safety Scouts* (Chicago: World Book Company, 1915), 4.

66. Ibid., 4–5.

67. Ibid., 51–55.

68. Ibid., 66–67.

69. Ibid., 98–99, 112–13.

70. Ibid., 50.

71. Ibid., 71. Sure Pop also draws an explicit parallel to the efficiency movement: "All this talk about efficiency is really part of the same movement, though very few realize it." He advises the children to recruit Chance Carter's older brother, who is studying to be an efficiency engineer, as an advisor to the troop. Ibid., 72.

72. *Proceedings of the National Safety Council* 4 (1915): 12.

73. To his chagrin, this included a representative from the competing Safety First Federation. A Dr. Hoffman from the American Museum of Safety was also included. *Proceedings of the National Safety Council* 5 (1916): 10.

74. "Safety First Merit Badge Requirements Prepared under the Direction of Lew Palmer," *'Be Prepared' for Merit Badge Examinations* (New York: Boy Scouts of America, 1919), 8.

75. Norton, *Fighting Traffic*, 76; *Proceedings of the National Safety Council* 8 (1919): 1011, 1042.

76. "The Rochester Public Safety Campaign: A Report of Organized Safety Activities in the City of Rochester, N.Y.," *Proceedings of the National Safety Council* 8 (1918).

77. *Proceedings of the National Safety Council* 9 (1919): 1012.

78. "Safety First Merit Badge Requirements," *Annual Report of the Boy Scouts of America* (1917), 44. See also "Letter from the Chief Scout," *Annual Report of the Boy Scouts of America* (1917), 33–35.

79. Vivian Lee Richardson, "To Protect the Public Welfare: A History of the Safety and Healthy Movement in Twentieth Century America" (MA thesis, Central Missouri State University, 1992), 64.

80. *Proceedings of the National Safety Council* 7 (1917): 352.

81. E. George Payne, *Education in Accident Prevention* (Chicago: Lyons and Camahan, 1919), 31–32.

82. Ibid., 35.

83. "Prof. E. George Payne Will Teach Here," *New York Times*, Aug. 6, 1922; "Dr. E.G. Payne Dies: Former N.Y.U. Dean," *New York Times*, Jan. 29, 1953. In the 1940s, Payne's interests shifted away from safety toward the cause of promoting racial and cultural tolerance.

84. Richardson, "To Protect the Public Welfare," 64–65.

85. White House Conference on Child Health and Protection, *Safety Education in Schools; Report of the Subcommittee on Safety Education in Schools* (1930), 11, 46, 14, 17. The report's wording suggests that the response was incomplete. It does not give the total number of children from which the percentages are derived. In 1926, the National Society for the Study of Education commissioned a study that surveyed 780 superintendents of schools in cities with a population of ten thousand or more; 288 claimed to teach safety as part of the regular curriculum, and an additional 125 sponsored safety-related school organizations such as junior safety councils or safety patrols. Cited in Robert Allen Dougherty, "A History of Safety Education in the Public Schools of the United States" (PhD diss., University of Chicago, 1938), 41–42.

86. Dougherty, "History of Safety Education," 54. In 1927, the NSC published the National Safety Tests, which Whitney claimed "measure safety information and attitudes with fair reliability." Ibid., 55.

87. E. George Payne, "Contemporary Accidents and Their Nonreduction," *Journal of Educational Psychology* 11 (Sept. 1937): 21. Payne was not alone in this critique. The Education Division was a "pressure group" that pushed for the rapid introduction of safety education because they stood to profit from it. "The Lay Control of Public Schools," *Critical Problems in School Administration*, Twelfth Yearbook of the Department of Superintendence (Washington, DC: Department of Superintendence of the National Education Association, 1934), 103–4, cited in Dougherty, "History of Safety Education," 33.

88. Payne, "Contemporary Accidents and Their Nonreduction," 23.

89. Ibid., 24. Payne is not very clear about what should be done instead, except that educators need to be more "scientific" in designing and implementing safety programs.

Chapter 8: Negotiating Automobile Risk

1. U.S. Census Bureau, *Statistical Abstracts of the United States: 2003* (No. HS-41), Transportation Indicators for Motor Vehicles and Airlines, 1900 to 2001, Mini-Historical Statistics, 77–78, www.census.gov/statab/hist/HS-41.pdf (accessed Mar. 4, 2011); U.S. Census Bureau, *Historical Statistics of the United States, Colonial Times to 1970*, Series B 149–166, "Death Rates for Selected Causes: 1900–1970." On the growing prevalence of automobile accidents in popular culture, see David Blanke, *Hell on Wheels: The Promise and Peril of America's Car Culture, 1900–1940* (Lawrence: University Press of Kansas, 2007); Anedith Jo Bond Nash, "Death on the Highway: The Automobile Wreck in American Culture" (PhD diss., University of Minnesota, 1983).

2. "31,000 More Killed in Our Motor Massacre," *Literary Digest*, June 14, 1930, 11; "A Motor Mind for a Motor Age," *Outlook*, Dec. 31, 1924, 711; "Safety by Selection," *Outlook*, Nov. 15, 1924, 382.

3. Chapters on safety are part of a number of standard works on the early history of automobiles. Particularly useful are James J. Flink, *America Adopts the Automobile, 1895–1910* (Cambridge, MA: MIT Press, 1970); and Clay McShane, *Down the Asphalt Path: The Automobile and the*

American City (New York: Columbia University Press, 1994). Significant recent books include Blanke, *Hell on Wheels*; and Peter D. Norton, *Fighting Traffic: The Dawn of the Motor Age in the American City* (Cambridge, MA; MIT Press, 2008). See also Daniel Albert, "Order out of Chaos: Automobile Safety, Technology, and Society, 1925–1965" (PhD diss., University of Michigan, 1997); and Amy Beth Gangloff, "Medicalizing the Automobile: Public Health, Safety, and American Culture, 1920–1967" (PhD diss., State University of New York at Stony Brook, 2006).

Much of the literature on the history of automobile safety has taken a cue from Ralph Nader in trying to ascertain why automobile companies did not make safer cars sooner. The classic polemic remains: Joel Eastman, *Styling vs. Safety: The American Automobile Industry and the Development of Automotive Safety, 1900–1966* (Lanham, MD: University Press of America, 1984). In the same vein, see Sally Clark, *Trust and Power: Consumers, the Modern Corporation, and the Making of the United States Automarket* (New York and London: Cambridge University Press, 2007).

4. For an overview of the literature on driving and risk perception, see Lennart Sjöberg, Bjørg-Elin Moen, and Turbjørn Rundmo, "Explaining Risk Perception: An Evaluation of the Psychometric Paradigm in Risk Perception Research," University of Trondheim, C. Rotunder Publickasjner, 2004, www.pauladrien.info/backup/LSE/IS%20490/Sjoberg%20Psychometric_paradigm.pdf (accessed Feb. 28, 2011).

5. "Hit and Run Suspect Held for Killing 2," *Philadelphia Evening Bulletin*, Apr. 13, 1926; "Elliott Opens War on Hit-and-Runner," *Philadelphia Evening Bulletin*, Apr. 2, 1926; "Coroner Joins War on 'Hit-Runners,'" *Philadelphia Evening Bulletin*, Apr. 5, 1926.

6. On speed as a theme in advertising, see Eastman, *Styling vs. Safety*, 83.

7. Pennsylvania began requiring examination in 1924. "The Status of Drivers' License Laws," *National Safety News* 11 (Feb. 1925): 28; *Purdon's Pennsylvania Statutes Annotated* (Philadelphia: George T. Bisel, 1931), 161.

8. McShane, *Down the Asphalt Path*, 156–57.

9. Blanke, *Hell on Wheels*, 138.

10. For more on skill and the culture of automobility in this period, see Kevin Borg, *Automobile Mechanics: Technology and Expertise in Twentieth Century America* (Baltimore: Johns Hopkins University Press, 2007); and Kathleen Franz, *Tinkering: Consumers Reinvent the Early Automobile* (Philadelphia: University of Pennsylvania Press, 2005).

11. On early debates about speed, see McShane, *Down the Asphalt Path*, 176–78. The higher limit was prescribed by the National Conference on Street and Highway Safety, "Suggested Model for a Uniform Act Regulating the Operation of Vehicles on Highways" (Washington, DC: Chamber of Commerce of the United States, 1926), 308.

12. Norton, *Fighting Traffic*, 31.

13. "Coroner Joins War on 'Hit-Runners.'"

14. Louis E. Schwartz, *The Trial of Automobile Cases*, 2nd ed. (Albany, NY: Matthew Bender, 1941), 297.

15. Herbert Hoover, *The Memoirs of Herbert Hoover: Years of Adventure* (New York: MacMillan, 1951), 86.

16. George B. Nash, *The Life of Herbert Hoover: Master of Emergencies, 1917–1918* (New York: W. W. Norton, 1996), 429.

17. Ernest Greenwood, "Making Travel Safe," *Independent*, Dec. 13, 1924, 506.

18. The conferences were held in 1924, 1926, 1930, 1935, and 1938. For two interpretations of the conferences' significance, see Blanke, *Hell on Wheels*, 134–36; and Norton, *Fighting Traffic*, 230.

19. See Ellis W. Hawley, "Herbert Hoover, the Commerce Secretariat, and the Vision of an 'Associative State,' 1921–1928," *Journal of American History* 61 (Winter 1974): 116–40.

20. Greenwood, "Making Travel Safe," 506–7; "Hoover Sponsors Public Safety Conference," *National Safety News* 9 (Aug. 1924): 23.

21. Policy mandates attached to Federal Highway funds eventually provided a partial way around this constitutional limitation. The most important legislation was the National Traffic and Motor Vehicle Safety Act of 1966.

22. Clarence R. Snethen, "Los Angeles Simplifies Traffic Laws and Reduces Accidents," *National Safety News* 12 (July 1925): 25.

23. H. M. Brown, "What the Automobile Tourist Should Know about Safety," *National Safety News* 9 (June 1924): 26.

24. Norton, *Fighting Traffic*, 48–49, 50.

25. Ernest N. Smith, "The Outlook for Uniform Traffic Laws," *National Safety News* 14 (Nov. 1926): 25.

26. "Final Text of the Uniform Vehicle Code," *National Conference on Street and Highway Safety* (Washington, DC, Aug. 20, 1926), v.

27. Ibid., 68.

28. Ibid., 69–70.

29. Blanke, *Hell on Wheels*, 135.

30. E. B. Lefferts, "Discipline for the Reckless Motorist," *National Safety News* 16 (July 1927): 32.

31. Albert, "Order out of Chaos," 107.

32. C. T. Fish, "On the Motor-Gypsy Trail," *National Safety News* 12 (July 1925): 6.

33. "A State Police for Illinois," *Orange Judd Farmer*, Mar. 1, 1923, 139; "Knutson Arrested by Virginia Police," *New York Times*, Mar. 11, 1924; "Uniform Traffic Regulation," *New York Times*, May 3, 1930.

34. Daniel M. Albert, "The Nut behind the Wheel: Shifting Responsibility for Traffic Safety since 1865," in *Silent Victories: The History and Practice of Public Health in Twentieth Century America*, ed. John W. Ward and Christian Warren (New York: Oxford University Press, 2007), 369.

35. William J. Novak, *The People's Welfare: Law and Regulation in Nineteenth-Century America* (Chapel Hill: University of North Carolina Press, 1996), 90–91.

36. Flink, *America Adopts the Automobile*, 176–77.

37. "The Status of Drivers' License Laws," *National Safety News* 11 (Feb. 1925): 28.

38. "States Plan Stricter Driver Control," *National Safety News* 35 (Jan. 1937): 56.

39. "Constitutional Law: Police Power: License and Registration of Automobiles," *Michigan Law Review* 8 (Mar. 1910): 417; Robbins B. Stoeckel, "The Case for the Driver's License Law," *National Safety News* 10 (Nov. 1924): 31.

40. "The Status of Drivers' License Laws," *National Safety News* 11 (Feb. 1925): 28.

41. See, for example, "Mrs. Cruikshank #2," *American Life Histories: Manuscripts from the*

Federal Writer's Project, 1936–1940, http://memory.loc.gov/ammem/wpaintro/wpahome.html (accessed July 13, 2009).

42. C. T. Fish, "The Empire State's War on Reckless Driving," *National Safety News* 14 (Nov. 1926): 13.

43. Flink, *America Adopts the Automobile*, 169–70.

44. Fish, "Empire State's War," 13.

45. Ibid., 5.

46. A 1931 study in New Jersey found that only 18% of drivers observed in a large-scale study of 12,000 intersections actually obeyed stop signs. "Eighteen Per Cent Obedience to 'Stop' Signs," *American City*, Sept. 1931, 117. For other examples of bad behavior, see Carl Dreher, "Homicide on Wheels," *Nation*, Aug. 7, 1930, 221.

47. David M. Henkin, *City Reading: Written Words and Public Space in Antebellum New York* (New York: Columbia University Press, 1998). For more on the concept of legibility, see James C. Scott, *Seeing like a State: How Certain Schemes to Improve the Human Condition Have Failed* (New Haven, CT: Yale University Press, 1998).

48. McShane, *Down the Asphalt Path*, 200; Norton, *Fighting Traffic*, 59.

49. Brown, "What the Automobile Tourist Should Know," 27.

50. McShane, *Down the Asphalt Path*, 200–201.

51. "Report of the Committee on Construction and Engineering," *National Conference on Street and Highway Safety* (1924), 21.

52. William H. Connell, "The Highway Business—What Pennsylvania Is Doing," *Annals of the American Academy of Political and Social Science* 116 (Nov. 1924): 124.

53. Ibid.

54. Norton, *Fighting Traffic*, 59.

55. Snethen, "Los Angeles Simplifies Traffic Laws," 26.

56. Morton G. Lloyd, "Traffic Signal Systems in Cities," *National Safety News* 51 (May 1926): 294.

57. Hawley S. Simpson, "Traffic Signals—A Necessity or a Nuisance?," *National Safety News* 15 (May 1927): 11–12.

58. "Traffic Engineering in the City on Seven Hills," *National Safety News* 16 (Aug. 1927): 15.

59. "Road and Highway Guard Tests," *National Safety News* 15 (Apr. 1927): 151.

60. C. R. Weymouth, "Wisconsin's Highway Program," typescript, Series 919, Box 1, State of Wisconsin, Department of Transportation Records, Wisconsin Historical Society.

61. In 1927, the first year that the federal government separated pedestrian fatalities from aggregate vehicle fatalities, pedestrian deaths constituted 41% of the total. "Motor Vehicle Accidents—Number and Deaths, by Type of Accidents," *Historical Statistics*, Series Q224–232, 720. However, studies by local police departments in cities such as Philadelphia estimated that a much higher percentage of automobile accidents involved pedestrians. See Norton, *Fighting Traffic*, 23–24; Earl J. Reeder, "Surveys Show Accident Causes and Suggest Remedies," *National Safety News* 11 (June 1925): 23.

62. See A. Nash, "Death on the Highway." Joseph Furnas's 1934 *Reader's Digest* article ". . . And Sudden Death" and the follow-up volume ". . . And Sudden Death and How to Avoid It" (1935) were particularly influential in shaping popular opinion.

63. Gangloff, "Medicalizing the Automobile," 196; Eastman, *Styling vs. Safety*.

64. "Uniform Vehicle Code—Regulation of Vehicle Operations," *Conference on Street and Highway Safety* (1926), 322.

65. Alan Nevins and Frank Everest Hill, *Ford: Expansion and Challenge, 1915–1933* (New York: Charles Scribner's Sons, 1957), 389–90.

66. Ford Motor Company, *Ford Manual* (Detroit: Ford Motor Company, 1919).

67. Nevins and Hill, *Ford*, 396.

68. Ibid., 461.

69. Albert W. Whitney, *Man and the Motorcar* (New York: National Bureau of Casualty and Surety Underwriters, 1936), 42.

70. Gangloff, "Medicalizing the Automobile," 92.

71. Nevins and Hill, *Ford*, 450. As with four-wheel brakes, safety glass was first offered on a luxury car, the Stutz. Ibid., 416.

72. [Triplex advertisement], *National Safety News* 54 (Oct. 1927): 29.

73. Gangloff, "Medicalizing the Automobile," 83, 94.

74. [Libbey-Owens-Ford advertisement], *National Safety News* 35 (Mar. 1937): 4; [Libbey-Owens-Ford advertisement], *National Safety News* 36 (Sept. 1937): 59.

75. Harry Armand, "Safety and the New Cars," *Safety Engineering* 76 (Dec. 1938): 32.

76. See Gangloff, "Medicalizing the Automobile."

77. Fire insurance described in chapter 1, for instance, covered loss of a particular building but not liability for other buildings burned if the fire spread. Homeowners' insurance, which includes both traditional fire protection and liability coverage for someone injured on the property, is a twentieth-century invention.

78. Kenneth S. Abraham, *The Liability Century: Insurance and Tort Law from the Progressive Era to 9/11* (Cambridge, MA: Harvard University Press, 2008), 71, 69.

79. This is the problem identified by neoclassical economists in the 1960s as "moral hazard," adopting a term that had a somewhat different meaning in the insurance industry. The term was coined in the nineteenth century. See Tom Baker, "On the Genealogy of Moral Hazard," *Texas Law Review* 75 (Dec. 1996): 237–92. Some economists have argued that the availability of no-fault insurance leads to a rise in the number of fatal accidents. See, for instance, J. David Cummins, Richard D. Phillips, and Mary A. Weiss, "The Incentive Effects of No-Fault Automobile Insurance," *Journal of Law and Economics* 44 (Oct. 2001): 427–64.

80. For more on the history of popular risk perception and statistics, see Arwen Mohun, "On the Frontier of the Empire of Chance," *Science in Context* 18 (2005): 337–57.

81. Ambrose Ryder, *Automobile Insurance: A Description of the Various Forms of Coverage, Underwriting Methods and Selling Plans* (Chicago: Spectator, 1924), 215, 217.

82. Georges Dionne, Christian Gouriéroux, and Charles Vanasse, "Testing for Evidence of Adverse Selection in the Automobile Insurance Market: A Comment," *Journal of Political Economy* 109 (Apr. 2001): 444–53.

83. Ibid., 117.

84. Baker, "Genealogy of Moral Hazard," 249.

85. Ibid., 107.

86. Ibid., 20.

87. Ibid., 154.

88. Samuel P. Black Jr. and John Paul Rossi, *Entrepreneurship and Innovation in Automobile Insurance: Samuel P. Black, Jr. and the Rise of Erie Insurance, 1923–1961* (New York: Routledge, 2001), 64, 66.

89. Ibid., 66, 71.

90. Committee of Censors, "Report of the Committee of Censors to the Law Association of Philadelphia: In re: Contingent Fee Accident Litigation," (Philadelphia: Allen, Lane, and Scott, 1929), 8 and passim.

91. Ken Dornstein, *Accidentally, on Purpose: The Making of a Personal Accident Underworld* (New York: St. Martin's Press, 1996), 101.

92. Committee of Censors, "Report," 37–38.

93. The constitutional justification for compelling employers to participate in workmen's compensation was their prior contractual relationship with workers. Typically, no such relationship existed between parties involved in automobile accidents. Ibid., 57.

94. Committee to Study Compensation for Automobile Accidents, *Report by the Committee to Study Compensation*. For a briefer version, see Young B. Smith, "The Problem and Its Solution," *Columbia Law Review* 32 (May 1932): 785–803.

95. Frank P. Grad, "Recent Developments in Automobile Accident Compensation," *Columbia Law Review* 50 (Mar. 1950): 318.

96. Ibid., 319–20. For a lawyer's perspective, see E. W. Sawyer, "Liability Insurance and the Lawyer," *Southern Lawyer* 41 (1937): 41; Adlai H. Rust, "Automobile Liability Insurance Trends," in American Bar Association Section of Insurance, Negligence, and Compensation Law, *Proceedings* 25 (1935–36): 25–33. For an early example of an insurance executive describing proposed measures as socialistic, see Henry Swift Ives, "Compulsory Liability Insurance, with Special Reference to Automobiles," *American Bar Association Journal* 10 (Oct. 1924): 697–702.

97. Grad, "Recent Developments," 305–6. See also Abraham, *Liability Century*, 73.

98. J. David Cummins, Richard D. Phillips, and Mary A. Weiss, "Incentive Effects of No-Fault Insurance," *Journal of Law and Economics* 44 (Oct. 2001): 427.

99. Joseph Laufer, "Insurance against Lack of Insurance? A Dissent from the Uninsured Motorist Endorsement," *Duke Law Journal* 1969 (Apr. 1969): 227–72.

100. Abraham, *Liability Century*, 82.

101. U.S. Census Bureau, *Statistical Abstracts of the United States: 2003* (No. HS-41), Transportation Indicators for Motor Vehicles and Airlines, 1900 to 2001, Mini-Historical Statistics, 77–78, www.census.gov/statab/hist/HS-41.pdf; U.S. Census Bureau, "Death Rates for Selected Causes: 1900–1970," *Historical Statistics of the United States, Colonial Times to 1970*, Series B, 149–66.

Chapter 9: What's a Gun Good For?

1. Bellmore H. Browne, *Guns and Gunning* (Chicopee Falls, MA: J. Stevens Arm and Tool, 1908), 104.

2. Insurance companies and urban coroners such as Peter Hoffman (see chap. 7) collected the first data on shooting deaths. See Louis I. Dublin and Edwin W. Kopf, "An Experiment in the Compilation of Mortality Statistics," *Publications of the American Statistical Association* 13 (Dec. 1913): 643; Peter Hoffman, *What You Must Know for Safety: A Manual of Information and Statisti-*

cal Records Relating to the Conservation of Human Life (Chicago: Public Safety Commission of Chicago and Cook County, 1915), 50, 65–66. The U.S. Census Bureau's 1920 compilation of "principal causes of death" offered a more comprehensive overview: 2,262 died in accidental shootings, 3,169 committed suicide with a firearm, and 4,477 committed homicide using a firearm; 6,304 died in railroad accidents and 9,103 of automobile accidents and injuries in the same year. "Principle Causes of Death in the United States Registration Area, 1920: Census Bureau's Summary of Mortality Statistics," *Public Health Reports* 36 (Nov. 4, 1921): 2724–25. According to the NRA's website, "Firearms are [currently] the second leading cause of traumatic death related to a consumer product in the United States and are the second most frequent cause of death *overall* for Americans ages 15 to 24." NRA Information Center—Gun Violence in America, www.vpc.org/nrainfo/phil.html (accessed Apr. 7, 2011).

3. Lee Kennett and James LaVerne Anderson, *The Gun in America: The Origins of a National Dilemma* (Westport, CT: Greenwood Press, 1975), 36–37; Jan E. Dizard, Robert Merrill Muth, and Stephen P. Andrews Jr., eds., *Guns in America: A Reader* (New York: New York University Press, 1999), 2–3.

4. Michael Bellesiles's controversial book *Arming America: The Origins of a National Gun Culture* (New York: Alfred A. Knopf, 2000) is at the center of the debate. He claimed that before the 1850s guns were far less prevalent than previous historians had suggested and that they were viewed primarily as tools. His critics (and some of his supporters) have questioned his use of evidence, particularly the apparent invention of nonexistent probate inventories. At stake is the question of what the framers of the Constitution meant when they claimed the "right to bear arms" in the Second Amendment. For an evenhanded summary of the controversy, see James Lindgren, "Fall from Grace: *Arming America* and the Bellesiles Scandal," *Yale Law Journal* (May 31, 2002), www.yalelawjournal.org (accessed Mar. 22, 2011).

5. For an overview of the production story, see David Hounshell, *From American System to Mass Production, 1800–1932: The Development of Manufacturing Technology in the United States* (Baltimore: Johns Hopkins University Press, 1984).

6. *Fourteenth Census*, vol. 10, *Manufactures* (1919), 411. Census takers noted that the number of handguns was probably underreported. Military surplus also contributed to civilian supplies. The U.S. government sold or gave away hundreds of thousands of guns after World War I. Reportedly, many people joined the NRA in the early 1920s to get a free military surplus rifle. Kennett and Anderson, *Gun in America*, 205–6.

7. Harrington and Richardson Arms Co., *Catalog Number 6* (Worcester, MA, 1902), 10–11.

8. For more on changes in marketing and advertising in this period, see Roland Marchand, *Advertising the American Dream: Making Way for Modernity, 1920–1940* (Berkeley: University of California Press, 1985); and Susan Strasser, *Satisfaction Guaranteed: The Making of the American Mass Market* (New York: Pantheon Books, 1989).

9. [Ad 211 B P], Book 543, Series 1, N. W. Ayer Advertising Agency Records, Archives Center, National Museum of American History. For more on this gendered imagery, see Laura Browder, *Her Best Shot: Women and Guns in America* (Chapel Hill: University of North Carolina Press, 2006).

10. [advertisement], *Field and Stream* 89 (July 1919): 364.

11. Russell S. Gilmore, "'Another Branch of Manly Sport': American Rifle Games, 1840–1900," in *Guns in America: A Reader*, ed. Jan E. Dizard et al. (New York: New York University

Press, 1999), 105–24. Gilmore says that interest in gun sports declined after the turn of the century. I have not actually seen any evidence to suggest that this is true. It does seem that gun clubs lost their ethnic character with the disappearance of German American *Schutzenbunde* and other similar organizations.

12. *How to Organize a Gun Club* (Wilmington, DE: E.I. du Pont de Nemours, 1912), n.p.

13. Interview with Mr. H. J. Pinkett (Nov. 14, 1938), *American Life Histories: Manuscripts from the Federal Writers' Project, 1936–1940*, http://memory.loc.gov/ammem/wpaintro/wpahome.html.

14. Leon Litwack, *Trouble in Mind: Black Southerners in the Age of Jim Crow* (New York: Knopf, 1998).

15. See Louis S. Warren, *The Hunter's Game: Poachers and Conservationists in Twentieth-Century America* (New Haven, CT: Yale University Press, 1997).

16. For an intriguing analysis of marksmanship in Buffalo Bill's Wild West Show, see Louis S. Warren, *Buffalo Bill's America: William Cody and the Wild West Show* (New York: Knopf, 2005), 242–50; "Annie Oakley Known by Gun," *New York Times*, Nov. 14, 1926, XX13. On imitating the movies, see, for example, Edward McGivern, *Ed McGivern's Book on Fast and Fancy Revolver Shooting and Police Training* (Lewistown, MT: Author, 1938).

17. On telegraph boys, see "Shot by Messenger Boy," *New York Times*, July 5, 1893; on marksmen on trolleys, see "How to Shoot a Revolver," *New York Times*, July 21, 1901; On the 4th of July, see "Pistol Kill Three; Wound 126 Here," *New York Times*, July 5, 1907.

18. "To Arm Brooklyn Mail Men," *New York Times*, Apr. 30, 1921; "Armed Automobiles," *New York Times*, Aug. 14, 1904.

19. "Special Pistols for Women," *New York Times*, Apr. 23, 1922.

20. For a detailed list of nineteenth-century statutes, see Sam B. Warner, "The Uniform Pistol Act," *Journal of Law and Criminology* 29 (Nov.–Dec. 1938): 529.

21. See, for example, Ralph A. Graves, "Lessening the Dangers of Living," *Harper's Weekly* 69 (July 22, 1905): 1046–48; Senate, Committee on Commerce, *To Regulate Commerce in Firearms*, 73rd Congress, 2nd Session, May 28 and 29, 1934, 45.

22. Kennett and Anderson, *Gun in America*, 192.

23. Frederick Hoffman, chief statistician for the Prudential Insurance Company and one of the most influential statisticians of his time, reported to the U.S. Congress that, even in the 1930s, there were still no available statistical analyses describing who was killing whom. He suspected that "brutal crimes to a considerable extent are committed by men and women of education and refinement supposed to be outside the category of persons with criminal intent." Senate, Committee on Commerce, *To Regulate Commerce in Firearms*, 73rd Congress, 2nd Session, May 28 and 29, 1934, 45.

24. "Anti-Firearm Bill Passed," *New York Times*, Mar. 15, 1905. In the 1920s and 1930s, many legal writers traced the Sullivan Law back to a law passed in the 1880s.

25. "Gaynor Shot—X-Ray Shows Bullet Split—His Condition Good—Assailant Shows No Regret," *New York Times*, Aug. 10, 1910.

26. Most notable was the Kennedy assassination. See Kennett and Anderson, *Gun in America*, 231–32.

27. "Carrying Firearms," *New York Times*, Aug. 17, 1910; "Licensing Pistol Carriers," *New York Times*, Aug. 17, 1910. A record of the legislative session has not survived, so there is no direct evidence of the opposition's arguments.

28. "Right to Bear Arms," *New York Times*, Aug. 23, 1910.

29. Surprisingly little is known about Sullivan. The most substantial biography is Daniel Czitrom, "Underworlds and Underdogs: Big Tim Sullivan and Metropolitan Politics in New York, 1889–1913," *Journal of American History* 78 (Sept. 1991): 536–58.

30. For a detailed description of the process of passing the law, see Kennett and Anderson, *Gun in America*, 165–86. Unfortunately, neither Sullivan's papers nor legislative transcript of the hearings survive, and so Kennett and Anderson's account (and mine) is based on newspaper and magazine articles.

31. Kennett and Anderson, *Gun in America*, 181–82.

32. "How 'Big Tim' Sullivan Will Put Down Lawlessness," *New York Times*, Dec. 4, 1910.

33. "To Deprive Police of Clubs," *New York Times*, Jan. 14, 1909.

34. See Magistrates Court, District 1, Docket Books (roll 126 microfilm) v. 42, New York City Archives.

35. Ibid., May 2, 1913.

36. New York County District Attorney, *Record of Cases* (1913), New York City Municipal Archives.

37. See, for example, Case Number 95764, William Wardell, and Case Number 95712, Charles Weller; both were initially charged under Penal Law 1897 but eventually sentenced for assault. New York County District Attorney, *Record of Cases*.

38. "Five Men Arrested for Having Arms," *New York Times*, Sept. 2, 1911; "Doings in Little Old New York," *Aberdeen Daily News*, Dec. 14, 1914, 4; "New York State Weapons Law Responsible for Some Peculiar and Amusing Happenings There," *Evening News*, Oct. 10, 1911.

39. "Streets Perilous as a Battlefield: Coroner Will So Report in Speaking of Violent Deaths in the City," *New York Times*, Feb. 4, 1912; "Has It Occasioned the Arrest of One Notorious Tough?," *New York Times*, Oct. 2, 1911; "War on Gunmen Brings Results for New Regime," *Fort Worth Star-Telegram*, Jan. 31, 1914; "Law Is Effective in Putting Down Gun Numbers," *Evening News* [San Jose, CA], Feb. 18, 1916.

40. "Woman Commended by Court," *Charlotte Daily Observer*, Dec. 10, 1911.

41. See, for instance, "The Sullivan Law," *Forest and Stream* (Sept. 30, 1911): 519; "A Law which Over Reached Itself," *Montgomery Advertiser*, May 19, 1914.

42. Senate, Committee on Commerce, *To Regulate Commerce in Firearms*, 73rd Congress, 2nd Session, May 28 and 29, 1934, 67.

43. "K.T. Frederick; New York Lawyer," *New York Times*, Feb. 13, 1963. Fredericks's obituary makes no mention of his involvement with the U.S. Revolver Association or his later presidency of the NRA. The only surviving papers I could find are about his involvement with wildlife conservation. They are at the University of Colorado.

44. Whether his calling resulted from personal experience or he was initially paid by the U.S. Revolver Association is unclear.

45. For a particularly startling example, see Smedley D. Butler, "Wipe Out the Gangsters!," *Forum and Century*, Oct. 1931, xvii–xviii. Butler, described as a major general in the U.S. Marine Corps, suggested creating an Anti-Crime Legion of heavily armed volunteers who would storm nightclubs to arrest public enemies. "Let's put the law books in cold storage and bring out the high-powered rifles and machine guns," he told his readers.

46. It is also unclear how much help Frederick had. In the 1920s, he implied that he was

only the messenger, but in 1934 he testified in Congress that he had been the sole author of the model law. House of Representatives, Committee on Ways and Means, *National Firearms Act*, 73rd Congress, 2nd Session, Apr. 18, 1934, 51. In 1922, the U.S. Revolver Association did try to get Congress to pass a version of the uniform law for the District of Columbia. The so-called Capper Bill did not pass. Commissioners on Uniform State Laws, *Handbook of the National Conference of Commissioners on Uniform State Laws and Proceedings of the Thirty-Fifth Annual Meeting*, 854 (hereafter Commissioners, *Proceedings*).

47. Commissioners, *Proceedings of the Thirty-Fourth Annual Meeting*, 713.

48. Ibid., 294.

49. Commissioners, *Proceedings of the Thirty-Fifth Annual Meeting, Detroit, Michigan, August 25–31, 1925* (United States: National Conference, 1925), 321.

50. Ibid., 856.

51. Commissioners, *Proceedings of the Thirty-Sixth Annual Meeting*, 571–73.

52. Commissioners, *Proceedings of the Thirty-Seventh Annual Meeting*, 867–68.

53. Ibid., 868.

54. Ibid., 894. Secondary sources offer differing versions of the ABA's position on the uniform law. My understanding is that initially the ABA president endorsed the proposal but the ABA's crime committee did not. Sometime around 1928, a new president of the ABA withdrew his support.

55. Ibid., 871.

56. United States Senate, Committee on the District of Columbia, *Control of Firearms Sales*, 72nd Congress, 1st Session, Feb. 26, 1932.

57. "Mrs. Gifford Pinchot Gets License to Carry Firearms," *New York Times*, Sept. 6, 1931. Three years previously the Pennsylvania Legislature had come within two votes of passing what the gun press called the "notorious Salus Bill"—a measure requiring a license in order to possess a firearm even in one's own home or place of business. See Captain Edward C. Crossman, "The Salus Bill," *Forest and Stream* 67 (Mar. 1928), 166.

58. Kennett and Anderson, *Gun in America*, 196–97.

59. Ibid., 202–3.

60. The Senate Committee on Commerce discussed six different firearms bills in the same hearings. U.S. Senate, Committee on Commerce, *To Regulate Commerce in Firearms*, 73rd Congress, 2nd Session, May 28 and 29, 1934. Both the Senate Committee and the House Committee on Ways and Means held hearings on H.R. 9066, a similar bill and probably the one that was initially considered likely to move forward since it bore the name of the eventual law "The National Firearms Act." For reasons unclear to this author, the bill that eventually went forward was called H.R. 9741 (the number of one of the bills initially discussed in the Senate Committee) even though the text was largely taken from H.R. 9066. U.S. House of Representatives, Committee on Ways and Means, *National Firearms Act*, 73rd Congress, 2nd Session, Apr. 16, 18; May 14–16, 1934.

61. Kennett and Anderson, *Gun in America*, 205.

62. Ibid.; Frederickson's articles reprinted from *American Rifleman*.

63. Committee on Ways and Means, *National Firearms Act*, 42, 43, 48.

64. Ibid., 119.

65. Ibid., 141–42.

66. "Taxation of Manufacturers, Importers, and Dealers in Certain Firearms and Machine Guns," *Congressional Record—House*, vol. 78, pt. 10, 11399. Congress heard testimony about gun violence from a number of women's groups in the 1930s. Legislators were dismissive of women such as Katherine Blake of the Women's International League for Peace and Freedom, who talked about her experience as a teacher with pupils who had been shot. House of Representatives, Subcommittee of the Committee on Interstate and Foreign Commerce, *Firearms*, 71st Congress, 2nd Session, Apr. 11, 1930, 29.

67. Ibid.

68. See Janet L. Malcolm, *The Right to Keep and Bear Arms: The Evolution of an Anglo-American Right* (Cambridge: Cambridge University Press, 1994).

69. Kennett and Anderson, *Gun in America*, 212, 216.

70. "What Are You Going to Do About It?," *Forest and Stream* 64 (June 1924): 361.

71. United States v. Miller et al., 59 S. Ct. 816 (May 15, 1939). In general, the Supreme Court had refused to hear Second Amendment cases before this. *Miller* was only the fourth relevant case between 1876 and 1939. The justices consistently ruled that the Second Amendment applied only to militias and not to individuals. See Peter Squires, *Gun Culture or Gun Control? Firearms, Violence, and Society* (New York: Routledge, 2000), 74.

72. Kennett and Anderson, *Gun in America*, 213.

73. "W.A. Law is Killed in Hunting Mishap," *New York Times*, Jan. 22, 1936.

74. For a detailed list, see "Hunters Told How to Avoid Accidents," *New York Times*, Dec. 2, 1924.

75. See, for example, "Penalties for Carelessness," *New York Times*, May 23, 1894.

76. *The American Shooter's Manual* (Philadelphia: Carey, Lea, and Carey, 1827), 21.

77. Browne, *Guns and Gunning*, 105. See also "Not Accidents, but Crimes," *Field and Stream* 51 (Oct. 1, 1898): 1.

78. "To Protect the Public," *Forest and Stream* 67 (Dec. 22, 1906): 975. Some states now require hunters to take a hunting safety course before obtaining a license, but this is a relatively recent phenomenon.

79. Sporting Arms and Ammunition Manufacturers' Institute, "A Movement for Greater Safety in the Sport of Shooting," typescript, Hagley Museum and Library, 3. It is likely that the driving force behind this initiative was the DuPont Corporation, which had bought a controlling share of Remington Arms in the early 1930s and was a pioneer in the industrial safety movement.

80. Ibid., 5, 12–13.

81. "Looking Back on 1946," *National Safety News* 56 (Aug. 1947): 21; "War Souvenir Hazards," *National Safety News* 56 (Dec. 1947): 82.

82. Anthony Accerano, "Hunting and Danger," *Sports Afield* (Oct. 2001). The actual number of fatal hunting accidents is now surprisingly small—somewhere between 66 and 97 a year, depending on which study one refers to. For one example, see "Hunting Accident Statistics," www3.sympatico.ca/d.rosen/accidentstats.htm. According to the NRA website, they helped start the first hunter education program in New York State in 1949. A tour through various gun enthusiast websites (of which there are thousands) also reveals that the argument that only careless people have accidents is still ubiquitous. See, for example, www.teresi.us/html/writing/gun_control.html#accident_stats (accessed May 21, 2011).

83. Arming for self-defense was a long-standing theme in some African American civil rights circles. See Robert F. Williams, Martin Luther King Jr., and Truman Nelson, *Negroes with Guns* (New York: Marzani and Munsell, 1962).

84. Bobby Seale, *Seize the Time: The Story of the Black Panthers and Huey Newton* (Baltimore: Black Classic Press, 1991), 73. This is a reprint of the 1970 original with a new introduction by the author.

85. Ibid., 73, 75, 115, 124. Both weapons had been obtained from the personal arsenal of "a Third World brother we knew, a Japanese radical cat." Seale, *Seize the Time*, 72.

86. Ibid., 80–81.

87. Ibid., 78.

88. Ibid., 100.

89. Williams, King, and Nelson, *Negroes with Guns*.

90. Philip Sheldon Foner and Claybourn Carson, *The Black Panthers Speak* (Cambridge, MA: Da Capo Press, 2002), xxx.

91. Bernard E. Harcourt, *Guns, Crimes, and Punishment in America* (New York: New York University Press, 2003), 4–5. Gary Kleck says his research shows that guns are used defensively about 2.5 million times a year. Ibid., 6.

Chapter 10: Risk as Entertainment

1. David P. Hahner, *Kennywood* (Charleston, SC: Arcadia, 2004), 44.

2. For more on the history of the industry see Judith A. Adams, *The American Amusement Park Industry: A History of Technology and Thrills* (Boston: Twayne, 1991); Gary S. Cross and John K. Walton, *The Playful Crowd: Pleasure Places in the Twentieth Century* (New York: Columbia University Press, 2005); John F. Kasson, *Amusing the Million: Coney Island at the Turn of the Century* (New York: Hill and Wang, 1978); Arwen P. Mohun, "Designed for Thrills and Safety: Amusement Parks and the Commodification of Risk, 1880–1929,"*Journal of Design History* 14 (2001): 291–306; David Nye, *Electrifying America: Social Meanings of a New Technology* (Cambridge, MA: MIT Press, 1992), 11–12, 122–32. There is also an extensive popular literature on the history of specific parks and types of rides.

3. See "Record Crowd Throngs Coney Island Resort," *New York Times*, July 5, 1905; "A 250,000 Day at Coney," *New York Times*, May 31, 1907.

4. Kasson, *Amusing the Million*, 42–43.

5. Michael Apter, *The Dangerous Edge: The Psychology of Excitement* (New York: Free Press, 1992), 25–29 and passim.

6. On Rarey, see chap. 4. On Sam Patch, see Paul E. Johnson, *Sam Patch, the Famous Jumper* (New York: Hill and Wang, 2003).

7. "Dutton Circus Attractions—Kennywood Park, Pittsburg" (1929), Box 2, Kennywood Park Papers, Historical Society of Western Pennsylvania (hereafter KPP).

8. Mills and Mills to Kennywood Park Corporation, Aug. 8, 1928, Box GST/AJ/05.01, Charles and Betty Jacques Amusement Park Collection, Penn State University Archives, The Pennsylvania State University Libraries (hereafter Jacques Collection).

9. Charles Reginald Sherlock, "Risking Life for Entertainment," *Cosmopolitan* 35 (Oct. 1903): 615, 618, 619, and 623.

10. E. E. Slosson, "The Amusement Business," *Independent* 57 (July 21, 1904): 138.

11. R. L. Hartt, "The Felicities of the Amusement Park," *Atlantic* 99 (May 1907): 672. Even socialist critic Maxim Gorky, visiting Coney Island in 1907, failed to note either the exploitation involved in paying people to risk their lives for entertainment or the lack of class consciousness in crowds eager to see such spectacles. He titled his critique "Boredom" and focused on the ways in which Coney's working-class audiences had been deluded into thinking that the island's rides and attractions were an authentic source of happiness. Maxim Gorky, "Boredom," *Independent* 63 (Aug. 8, 1907): 309–17.

12. Rex P. Billings, Belmont Park, to A. B. "Brady" McSwigan, Kennywood Park, Aug. 14, 1941, Box GST/AJ/01.22; George Hamid, Wirth and Hamid Fair Booking, Inc., to A. B. McSwigan, May 23, 1933, Box GST/AJ/05.11, Jacques Collection.

13. Both the daily papers and the pages of the *Billboard*, the entertainers' trade publication, carried regular testaments to the human cost of the public's taste for risk taking as a form of entertainment. "Hung by His Teeth Half Hour in Air," *New York Times*, Aug. 6, 1908; "Several Accidents," *Billboard*, Aug. 25, 1906, 21; "Motorcyclist Killed" and "Prof. Joe Wilson Killed," *Billboard*, June 14, 1913, 26.

14. Film historians have tied these entertainments to the history of early filmmaking. See Andrea Stulman Dennett and Nina Warnke, "Disaster Spectacles at the Turn of the Century," *Film History* 4 (1990): 101–11.

15. Ibid., 104. Their description is based on a 1905 Biograph Film of the Dreamland version.

16. Edward F. Tilyou, "Why the Schoolma'am Walked into the Sea," *American Magazine* 94 (July 1922): 18.

17. "Kennywood Ticket Report: End of Season, 1924," Box 2, KPP.

18. A. R. Hodge to A. B. McSwigan, Apr. 17, 1925, Box GST/AN/01.01, Jacques Collection.

19. William C. Boyce, "Modern Amusement Parks" [brochure], n.d., n.p.

20. The terms *roller coaster* and *scenic railroad* seemed to have been used somewhat interchangeably through the early 1910s.

21. Robert Cartmell, *Incredible Scream Machine: A History of the Rollercoaster* (Bowling Green, OH: Bowling Green State Press, 1987), 42.

22. "Seventeen Hurt on Scenic Railway," *New York Times*, June 22, 1910. Other similar accidents occurred with some frequency in these years. See "Death on Scenic Railway," *New York Times*, July 3, 1905; "Dreamland Train Derailed," *New York Times*, July 16, 1907; "Needless Danger," *New York Times*, July 28, 1907.

23. "Coney Ride Loses License," *New York Times*, June 23, 1910.

24. Quoted in Cartmell, *Incredible Scream Machine*, 75.

25. "Blame for Coaster Deaths," *New York Times*, Sept. 6, 1911.

26. See Cartmell, *Incredible Scream Machine*, chap. 8.

27. Hahner, *Kennywood*, 44–47.

28. John Miller to R. S. Uzzell, July 25, 1928; John Miller to R. S. Uzzell, Aug. 19, 1925, Box GST/AQ/03.08, Jacques Collection. On Torresdale Park, see www.nephillyhistory.com/1cnephtest/torresdale/torhist.htm (accessed Apr. 29, 2011).

29. John Allen to Will Turner, Oct. 7, 1965, Box GST/AJ/05.11, Jacques Collection.

30. Miller to Uzzell, July 25, 1928, Box GST/AQ/03.08, Jacques Collection.

31. Ibid., 117.

32. Charles Davis, "Renaissance at Coney," *Outing* (Aug. 1906): 519.

33. "Handle Bar Mechanism for Pleasure Railways," United States Patent #1,038,175.

34. The *Cyclone* at Crystal Beach, Ontario, Canada, was particularly notorious for injuring patrons. The park management reputedly stationed a medical team at the ride's exit. Cartmell, *Incredible Scream Machine*, 157.

35. Homer Croy, "There's Money in it, but . . . ," *Popular Mechanics* 49 (Jan. 1928): 59.

36. Davis, "Renaissance at Coney," 674.

37. Andrea Shope, "Dips and the Curve," *Westsylvania*, Summer 1999, 29.

38. George A. Dodge to H. B. Auchy, Nov. 26, 1917, Box GST/AJ/05.01, Jacques Collection.

39. "How to Keep Amusement Rides in Good Condition," *A Reprint of Papers Read at Fourth Annual Convention of National Association of Amusement Parks* (1924), 108.

40. J. P. Hartley, "Some Experiences on Liability Insurance," *A Reprint of Papers Read at Third Annual Convention of National Association of Amusement Parks* (1923), 98.

41. J. P. Hartley, [untitled presentation and discussion], *A Reprint of Papers Read at Fourth Annual Convention of National Association of Amusement Parks* (1924), 127.

42. "Special Public Liability Insurance Plan," in National Association of Amusement Parks, Pools, and Beaches, *Handbook of the Industry* (1943), 99.

43. W. M. Lawson, Smith, Lauerman, Lawson, & Combs, "The Liability Insurance Problem: How Insurance Rates Can Be Materially Reduced through Cooperating of N.A.A.P. Members," *A Reprint of Papers Read at the First Annual Convention* (1920), 22.

44. Between 125 and 135 parks were covered in their entirety. Because of the concession system, it was more common to insure individual rides, especially in big parks like Kennywood. J. P. Hartley, "Some Experiences on Liability Insurance," *A Reprint of Papers Read at Third Annual Convention of National Association of Amusement Parks* (1923), 101.

45. J. P. Hartley, [no title], *A Reprint of Papers Read at Fourth Annual Convention of National Association of Amusement Parks* (1924), 127.

46. John R. Davies, "Liability Insurance for Amusement Parks," *A Reprint of Papers Read at Second Annual Convention of National Association of Amusement Parks* (1922), 138.

47. Hartley, [no title], *A Reprint of Papers Read at Fourth Annual Convention of National Association of Amusement Parks* (1924), 127.

48. Aurel Vaszin to A. B. McSwigan, May 4, 1937, Box GST/AJ/01.21, Jacques Collection.

49. "Cutting Down Accident Frequency," *NAAP Bulletin*, Aug. 1, 1926, 1.

50. Henry Traver to Brady McSwigan, Apr. 15, 1930, Box GST/AN/05.04, Jacques Collection.

51. Hartley, [no title], *A Reprint of Papers Read at Fourth Annual Convention of National Association of Amusement Parks* (1924), 128.

52. Ibid., 129.

53. Hartley, "Some Experiences on Liability Insurance," *A Reprint of Papers Read at Second Annual Convention of National Association of Amusement Parks* (1922), 102–3.

54. Hartley, [no title], *A Reprint of Papers Read at Fourth Annual Convention of National Association of Amusement Parks* (1924), 126.

55. Hartley, "Some Experiences on Liability Insurance," *A Reprint of Papers Read at Third Annual Convention of National Association of Amusement Parks* (1923), 100.

56. "Preparation of Safety Code Started," *Bulletin of the National Association of Amusement Parks* (Mar. 1, 1926), 1.

57. "Program of the Seventh Annual Convention of the National Association of Amusement Parks, Chicago, December 2–4, 1925," n.p., Box GST/AN/01.01, Jacques Collection.

58. "American Engineering Standards Committee," *Transactions of the American Institute of Electrical Engineers* 41 (1922): 877: P. G. Agnew, "The National Safety Code Program," *Annals of the American Academy of Political and Social Science* 123 (Jan. 1926): 51.

59. "Safety Code Committee Reports on Washington Conference," *NAAP Bulletin*, Sept. 1, 1926, 1.

60. "Safety Factors for Amusement Parks (Swimming Pools)," *NAAP Bulletin*, Apr. 1933, 1; "Safety Factors for Amusement Parks (Shooting Galleries)," *NAAP Bulletin*, May 1933, 1.

61. Norval Burch, "Thrills and Chills without Spills: Amusement Parks Give Spine-Tingling Rides but Protect Riders," *National Safety News*, July 1949, 22–24, 87–89, 92–93; H. A. Dever to Brady McSwigan, Jan. 31, 1950, Box GST/AN/05.04, Jacques Collection.

62. Burch, "Thrills and Chills without Spills," 22–23.

63. Ibid., 24.

64. Ibid., 24, 89, 92.

65. J. Gaynor, "Safety in Ride Construction and Operation," *A Reprint of Papers Read at Third Annual Convention of National Association of Amusement Parks* (1923), 95.

66. R. C. Illions to John Allen, May 4, 1955, Box GST/AJ/05.01, Jacques Collection.

67. The law required parks to carry one million dollars in insurance and have their rides inspected monthly. Some industry insiders blamed the failure of a number of Pennsylvania parks on the added expense resulting from the requirement. Jim Futrell, *Amusement Parks of Pennsylvania* (Mechanicsburg, PA: Stackpole Books), 38.

Chapter 11: Consumer Product Safety

1. U.S. Congress, Senate, Committee on Commerce, *Flammable Fabrics Act Amendments, Hearings before a Subcommittee of the Senate Committee on Commerce on S. 1003*, 90th Congress, 1st session, 1967, 57–59. Hackes's story was repeatedly retold by consumer product safety advocates. See Michael Pertschuk, *Revolt against Regulation: The Rise and Pause of the Consumer Movement* (Berkeley: University of California Press, 1982), 37; and Senator Warren G. Magnuson and Jean Carper, *The Dark Side of the Marketplace: The Plight of the American Consumer* (Englewood Cliffs, NJ: Prentice-Hall, 1968), 123–24.

2. *Flammable Fabric Act Amendments, Hearings*, 59.

3. Ibid., 60.

4. Cornell Aeronautical Laboratory, Inc., "Cost to the Consumer of Improving the Safety of Consumer Products," *Report Prepared for the National Commission on Product Safety*, Mar. 1970, 6.

5. Pertschuk, *Revolt against Regulation*, 13.

6. U.S. Congress, House of Representatives, *Safety Devices on Household Refrigerators: Hearing before a Subcommittee of the Committee on Interstate and Foreign Commerce on H.R. 2181*, 84th Congress, 2nd session, May 28, 1956, 14.

7. Ibid. Ironically, Dies had chaired an HUAC investigation of the Consumers Union's possible ties to the Communist Party in 1938. See Norman Isaac Silber, *Test and Protest: The Influence of Consumers Union* (New York: Holmes and Meier, 1983), 28.

8. House of Representatives, *Safety Devices on Household Refrigerators* (1956), 6.

9. Ibid., 38.

10. Ibid., 46.

11. U.S. Congress, House of Representatives, *Safety Devices on Household Refrigerators: Hearing before a Subcommittee of the Committee on Interstate and Foreign Commerce on H.R. 2876 and S. 2891*, 83rd Congress, 2nd Session, Apr. 27, 1954, 28.

12. "A Drive on the Menace of Abandoned Iceboxes," *Life* 35 (Dec. 14, 1953): 57–58.

13. Ibid., 4.

14. Ibid., 58–59.

15. Ibid., 15.

16. See U.S. Department of Health, Education, and Welfare, "Origin and Program of the U.S. National Health Service," *Health Statistics*, May 1958. For a particularly useful retrospective through the eyes of a practitioner, see Julian A. Walker, "Reflections on a Half-Century of Injury Control," *American Journal of Public Health* 84 (Apr. 1994): 664–70. Research grants given out by the U.S. Public Health Service between 1954 and 1965 also provide a sense of this emerging field. See U.S. House of Representatives, Subcommittee of the Committee on Interstate and Foreign Commerce, *National Accident Prevention Center, Hearings on H.R. 133*, 88th Congress, 1st Session, Apr. 9–10, 1963, 47–51.

17. Daniel P. Moynihan, "Epidemic on the Highways," *Reporter* 20 (Apr. 30, 1959): 16–23.

18. Helen L. Roberts, John E. Gordon, and Autino Fiore, "Epidemiological Techniques in Home Accident Prevention," *Public Health Reports* (1896–1970) 67, no. 6 (June 1, 1952): 547.

19. U.S. Public Health Service, *Accident Prevention Program: Accidental Injury Statistics*, June 1958.

20. There are many reasons why mowers are a good example, mostly detailed below. It is also worth noting that every player in the postwar consumer product safety movement (except Ralph Nader) seems to have become involved with the mower problem.

21. Virginia Scott Jenkins, *The Lawn: A History of an American Obsession* (Washington, DC: Smithsonian Institution Press, 1994), 112.

22. For more on the context for this new device, see Kenneth T. Jackson, *Crabgrass Frontier: The Suburbanization of the United States* (New York: Oxford University Press, 1985).

23. The first rotary mower on the market, the Mow-Master, cost $148 in 1947; "Power Lawn Mowers—Preliminary Report," *Consumers' Research Bulletin* 20 (July 1947): 8. By 1957, the average price of a power mower had dropped to $79; "Power Lawn Mowers," *Consumer Reports* 22 (June 1957): 260. In 1960, the Mow-Master could be had for $55.95 (about $400 adjusted for inflation); "Hand-Propelled Power Mowers," *Consumer Reports* 25 (July 1960): 361.

24. Descriptions of these hazards are ubiquitous in the sources used below, but one of the most thorough discussions was written by two University of Iowa epidemiologists who interviewed large numbers of people injured in mower accidents. See William H. McConnell and L. W. Knapp, "Epidemiology of Rotary Power Lawn Mower Injuries," *University of Iowa, Institute of Agricultural Medicine, Bulletin Number 9* (University of Iowa, 1965).

25. Paul W. Kearney, "Cut the Grass—Not Yourself!," *Reader's Digest* 75 (June 1960): 242.

26. A Lexus-Nexus search for lawnmower tort cases that reached state-level appeals courts between 1946, when the first rotary mower came on the market, and 1955, when mower manufacturers turned their attention to mower safety, revealed no cases in which injured people

sued manufacturers. Either cases did not make it to the appeals level, or they were very rare. There were a number of cases in which injured people sued their employers, their neighbors, and the companies that sold them the mowers (claiming misrepresentation).

27. Silber, *Test and Protest*. Safety testing of consumer products had its origins in concerns about fire hazards from electrical appliances. Underwriter's Laboratory tested appliances in the 1910s. See Harry Chase Brearley, *A Symbol of Safety: An Interpretative Study of a Notable Institution Organized for Service—Not Profit* (New York: Doubleday, Page, 1923). The United States Bureau of Standards also did appliance testing. See *Safety for the Household* (Washington, DC: U.S. Government Printing Office, 1918).

28. "Power Lawn Mowers—Preliminary Report," *Consumer's Research Bulletin* 20 (July 1947): 8. Interestingly, the report also noted that the blade hardness did not meet required federal specifications.

29. "Power Mowers," *Consumer Reports* 13 (June 1948): 259.

30. "Big Mower Question: Can a Seal Sell Safety?," *Electrical Merchandising Week* 92 (Oct. 17, 1960): 1. See also "How Will Power Mower Industry's New Safety Standards Work?," *Electrical Merchandising Week* 92 (Dec. 5, 1960): 7.

31. *American Standard Safety Specifications for Power Lawn Mowers* (American Standards Association, June 23, 1960).

32. See, for instance, "Safety with Power Lawn Mowers," *Popular Gardening*, Apr. 1963, 48–51; "How to Use Power Mowers Safely," *Good Housekeeping* 156 (May 1963): 166; Herbert C. Bardes, "Rotary Mowers Feature Safety," *New York Times*, Mar. 8, 1964.

33. "Gasoline-Powered Rotary Mowers: Generally Less Hazardous than in Consumers Union's Last Project, Most Did Good Jobs on Lawns," *Consumer Reports* 28 (July 1963): 333. In the previous tests in 1960, "Consumers Union placed heavy emphasis on safety. . . . The Safety standards applied by Consumers Union were established by Consumers Union engineering staff only after discussions with a number of consultants and safety engineers, combining with extensive research into published material on this subject by such organizations as the National Safety Council, insurance groups, and other experts in the field (including Sweden's National Workers Protection Board)"; "Hand-Propelled Power-Mowers," *Consumer Reports* 25 (July 1960): 356. By 1965, Consumers Union was employing a revised set of ASA standards in their tests, but they still found many mowers wanting; "Electric Lawn Mowers," *Consumer Reports* 30 (June 1965): 286.

34. For more on these legal concepts, see Mark A. Kinzie and Christine F. Hart, *Product Liability Litigation* (Albany, NY: West/Thomson Learning, 2002), 18.

35. "Consumer Interests—Message from the President of the United States," *Congressional Record*, Mar. 15, 1962, 4263.

36. Shelby Scates, *Warren G. Magnuson and the Shaping of Twentieth Century America* (Seattle: University of Washington Press, 1997), 212.

37. Ibid., 213.

38. Dr. Goddard was a physician from the Public Health Services who was part of a movement in epidemiology to make household accidents a public health issue. "Origin and Program of the U.S. National Health Survey," *Health Statistics*, U.S. Department of Health, Education and Welfare (Washington, DC, May 1958).

39. U.S. Senate, *Child Protection Act of 1966: Hearings before the Consumer Subcommittee of the*

Committee on Commerce, United States Senate, 89th Congress, 2nd Session (1966), 20. Goddard also explained that he had looked into the mower problem ten years previously when he was the head of the accident prevention program for the Public Health Service. The mower conversation goes on for several pages.

40. National Commission on Product Safety, *Final Report* (Washington, DC: U.S. Government Printing Office, June 1970).

41. Ibid., 28.

42. Ibid., 29.

43. Ibid., 29.

44. Ibid., 30.

45. "Petition for the Issuance of a Consumer Product Safety Rule," submitted by the OPEI to the Consumer Product Safety Commission, Box 70, RG 424, Consumer Product Safety Commission Records (hereafter CPSC), National Archives. There is no date on the initial petition, but the chronology of the whole process is laid out in "Briefing Paper: Final Standard for Walk-Behind Power Mowers," Jan. 9, 1979, Box 70, CPSC.

46. Richard W. Armstrong to Jon A. Shelton, "Memorandum," Sept. 28, 1973, Box 70, RG 424, CPSC.

47. "Power Lawn Equipment Acceptance of an Offer to Develop Safety Standard and Summary Terms of Offer," *Federal Register* 39 (Oct. 24, 1974). The final goal was to write a rule that "offered adequate protection for consumers at a reasonable cost, without adversely affecting utility, with as little disruption as practicable in the marketplace consistent with the public interest." "Briefing Paper," n.p.

48. See, for instance, Steven Popa to CPSC, May 1977; David Lit to CPSC, May 7, 1977; John Fixer to CPSC, May 14, 1977; May 17, 1977, CPSC, Box 70, RG 424.

49. Susan Wendeln to CPSC, May 9, 1977, CPSC, Box 70, RG 424. Others wondered whether anyone on the Consumer Product Safety Commission had ever actually mowed a lawn or knew what was involved. One wag wrote a letter directly to the president of the United States, Jimmy Carter, suggesting that the commissioners should be forced to mow the White House Lawn until they acquired some common sense. Larry Kelly to President Jimmy Carter, n.d., CPSC, Box 70, RG 424.

50. Donald Roberts to CPSC, May 17, 1977, CPSC, Box 70, RG 424.

51. For an extended discussion of the Consumers Union testing techniques, see Silber, *Test and Protest*.

52. The standard glossed over the technical problems involved by assuming that eventually all mowers would use a "brake-clutch" design in which the engine would keep running while a clutch disengaged the blade from the drive mechanism. Adam M. Ehrlich, "Briefing Paper, Petition on Lawnmowers," Nov. 15, 1973, Box 70, CPSC, RG 424.

53. Southland Mower Company, et al. v. Consumer Products Safety Commission United States Court of Appeal, Fifth Circuit, 619 F.2d 499, 1980.

54. "Consumer Products Safety Commission," *Federal Regulatory Directory, Tenth Edition* (Washington, DC: Congressional Quarterly, 2001), 37.

55. www.celebrationhealth.com/handcenter/lawnmower.htm (accessed Jan. 12, 2009).

56. William A. Niskanen, *Reaganomics: An Insider's Account of Policies and People* (Oxford and New York: Oxford University Press, 1988), 115.

57. Ibid., 129.

58. "Stockman Moves to Kill Consumer Panel," *New York Times*, May 9, 1981.

59. "Exporting Products Recalled in the U.S.," *New York Times*, Apr. 3, 1984.

60. "Nanny 1, Hobgoblins 0," *New York Times*, May 21, 1984.

61. "Amusement Park Safety Stirs a Federal Debate," *New York Times*, Aug. 13, 1984.

62. "Consumer Panel Chief Resigns Her Position," *New York Times*, Dec. 1, 1984.

63. "Accusations Delay Safety Official's Confirmation," *New York Times*, Oct. 20, 1985; "Senator Says Report Clears Safety Nominee," *New York Times*, Dec. 8, 1985.

64. "Sears Warned on Fan Recall," *New York Times*, Mar. 3, 1980.

65. Kenneth S. Abraham, *The Liability Century: Insurance and Tort Law from the Progressive Era to 9/11* (Cambridge, MA: Harvard University Press, 2008), 158.

66. Ibid.

67. Ibid., 164–65.

68. Ibid., 159.

69. John Kolb and Steven S. Ross, *Product Safety and Liability: A Desk Reference* (New York: McGraw-Hill, 1980).

70. Maria Segui-Gomez and Ellen J. Mackenzie, "Measuring the Public Health Impact of Injuries," *Epidemiologic Reviews* 25 (2003): 5. On automobiles, see Ann M. Dellinger, David A. Sleet, and Bruce H. Jones, "Drivers, Wheels, and Roads: Motor Vehicle Safety in the Twentieth Century," in *Silent Victories: The History and Practice of Public Health in Twentieth Century America*, ed. John H. Ward and Christian Warren (Oxford: Oxford University Press, 2007), 345.

71. Ibid., 3–4.

ESSAY ON SOURCES

This is the first large-scale history of risk in American society. It is built from extensive research in original sources coupled with information and analysis gleaned from a wide variety of books and articles. The following bibliographic essay is not exhaustive. Instead, it is intended to guide readers toward works I found particularly useful or that are considered by specialists as definitive with regard to topics in this book.

Theoretical and contemporary studies of risk number in the thousands. Several scholars have created useful primers on the sociology and anthropology of risk, notably Sheldon Krimsky and Dominic Golding, eds., *Social Theories of Risk* (Westport, CT, and London: Praeger, 1992); and Deborah Lupton, *Risk* (*Key Ideas*) (New York: Routledge, 1999). The foundational text of the risk society hypothesis is Ulrich Beck, *Risk Society*, trans. M. Ritter (London: Sage, 1992). Beck has continued to publish a series of volumes restating and elaborating his basic hypothesis. For a complementary take on "reflexive modernity," see Anthony Giddens, *Modernity and Self Identity: Self and Society in the Late-Modern Age* (Cambridge: Polity Press, 1991). Charles Perrow's *Normal Risks: Living with High-Risk Technologies* (New York: Basic Books, 1984) provides a readable guide to the pattern of accidents in complex systems. My book also historicizes insights of some psychologists who study risk perception. An excellent introduction to this literature is Glynis M. Breakwell, *The Psychology of Risk* (Cambridge: Cambridge University Press, 2007). Michael Apter's *The Dangerous Edge: The Psychology of Excitement* (New York: Free Press, 1992) provided a particularly useful framework for explaining elective risk taking.

Since I began this project, the term *risk* and concepts created by social theorists have begun to appear more frequently in works of history. Bill Luckin and Roger Cooter, *Accidents in History: Injuries, Fatalities, and Social Relations* (Amsterdam: Rodopi, 1997), is a pioneering work in this vein. For a particularly good recent example of a risk-oriented history, see John Burnham, *Accident Prone: A History of Technology, Psychology, and the Misfits of the Machine Age* (Chicago: University of Chicago Press, 2009).

This book is also intended as a contribution to the literature on the social history of technology. It examines, in particular, the ways technology and culture have shaped each other over time. For more on the conceptual frameworks of social construction and mutual shaping of technology and society, see Nina E. Lerman, Arwen Palmer Mohun, and Ruth Oldenziel, "Versatile Tools: Gender Analysis and the History of Technology," *Technology and Culture* 38 (Jan. 1997): 3. See also Langdon Winner, "Do Artifacts Have Politics?," *Daedalus* 109 (1980): 121–31; Donald A. MacKenzie and Judy Wajcman, *The Social*

Shaping of Technology: How the Refrigerator Got Its Hum (Philadelphia: Open University Press, 1999); and Wiebe E. Bijker and Trevor J. Pinch, eds., *Shaping Technology / Building Societies: Studies in Socio-Technical Change* (Cambridge, MA: MIT Press, 1994).

The topic of risk in early America has only begun to attract scholarly attention after a very long period of neglect. Carl Bridenbaugh's classic study *Cities in the Wilderness: The First Century of Urban Life in America, 1625–1742* (New York: Ronald Press, 1938) remains a very useful starting point for understanding how colonial city dwellers dealt with fire. No one who writes about fire can ignore Stephen J. Pyne's conceptual work *Fire: A Brief History* (Seattle: University of Washington Press, 2001). Although its heavily illustrated format might suggest otherwise, Donald J. Cannon's *Heritage of Flames: The Illustrated History of Early American Firefighting* (Garden City, NY: Doubleday, 1977) is well researched and very useful. Elizabeth Gray Kogen Spera's "Building for Business: The Impact of Commerce on the City Plan and Architecture of the City of Philadelphia, 1750–1800" (PhD diss., University of Pennsylvania, 1980) has provided a number of historians with insights about the function of fire insurance in shaping eighteenth-century Philadelphia. Nicholas B. Wainwright's *A Philadelphia Story: The Philadelphia Contributionship for the Insurance of Houses from Loss by Fire* (Philadelphia: Contributionship, 1952) is also very informative on the early history of insurance in Philadelphia.

For understanding how different social and political systems shaped firefighting in the Atlantic World, I am grateful to my former student Daniel Winer. His PhD dissertation, "The Development and Meaning of Firefighting, 1650–1850" (University of Delaware, 2009), is now available to other researchers. Jill Lepore's *New York Burning: Liberty, Slavery, and Conspiracy in an Eighteenth-Century Manhattan* (New York: Alfred A. Knopf, 2005) is a thought-provoking study of the cultural power of fire in early America. The theme of insurance and moral hazard is first introduced in the fire chapter. For a clear explanation of the changing meaning of the term see Tom Baker, "On the Genealogy of Moral Hazard," *Texas Law Review* 75 (1996): 237–92.

As with fire, I have also turned to an older literature from the history of medicine to tell the smallpox story, as well as to a handful of new semipopular works. On smallpox, begin with John B. Blake, *Public Health in the Town of Boston, 1630–1822* (Cambridge, MA: Harvard University Press, 1959); and John B. Blake, "The Inoculation Controversy in Boston: 1721–1722," *New England Quarterly* 25 (Dec. 1952). Also useful are two works by Genevieve Miller: *The Adoption of Inoculation for Smallpox in England and France* (Philadelphia: University of Pennsylvania Press, 1957) and "Smallpox Inoculation in England and America: A Reappraisal," *William and Mary Quarterly* 13 (Oct. 1956): 476–92. Recent books on smallpox are more derivative, but still informative. See Ola Elizabeth Winslow, *The Destroying Angel: The Conquest of Smallpox in Colonial Boston* (Boston: Houghton Mifflin, 1974); Ian Glyn and Jenifer Glyn, *The Life and Death of Smallpox* (New York: Cambridge University Press, 2004); Elizabeth Fenn, *Pox Americana: The Great Smallpox Epidemic of 1775–82* (New York: Hill and Wang, 2001); and Donald R. Hopkins, *Princes and Peasants: Smallpox in History* (Chicago: University of Chicago, 1983).

Jan Golinski's *British Weather and the Climate of Enlightenment* (Chicago: University of Chicago Press, 2007) provides an essential guide for understanding how natural philosophers thought about the weather. Also very useful are Peter Eisenstadt, "The Weather and Weather Forecasting in Colonial America" (PhD diss., New York University, 1990); and James Rodger Fleming, *Meteorology in America, 1800–1870* (Baltimore: Johns Hopkins University Press, 1990). Early American weather folklore awaits its historian. In the meantime, I consulted Edward B. Garriot, *Weather Folk-lore and Local Weather Signs* (Washington, DC: U.S. Government Printing Office, 1903). A 2002 conference at the Bakken Institute resulted in the first scholarly collection of articles on lightning rods: Peter Heering et al., *Playing with Fire: Histories of the Lightning Rod* (Philadelphia: American Philosophical Society, 2009). See also an older volume, Basil Schonland, *The Flight of the Thunderbolts* (Oxford: Clarendon Press, 1964). The literature on Franklin and his scientific investigations is huge. Readers might begin with Michael Brian Schiffer, *Drawing the Lightning Down: Benjamin Franklin and Electrical Technology in the Age of Enlightenment* (Berkeley: University of California Press, 2003). J. A. Leo Lemay's *Ebenezer Kinnersley, Franklin's Friend* (Philadelphia: University of Pennsylvania Press, 1964) paints a vivid picture of eighteenth-century electrical demonstrators.

A number of excellent works address the role of the Enlightenment and, more particularly, new scientific ideas in the eighteenth and early nineteenth centuries. Perry Miller's *The New England Mind from Colony to Province* (Cambridge, MA: Harvard University Press, 1953) remains essential for understanding the seeming contradictions of Cotton Mather and other religious leaders' ideas about managing disease. For the antebellum period, see David Jaffee, "The Village Enlightenment in New England, 1760–1820," *William and Mary Quarterly* 47 (July 1990); Donald Zochert, "Science and the Common Man," in *Science in America since 1820*, edited by Nathan Reingold (New York: Science History Publications, 1976), 7–32; Sally Kohlstedt, "Parlors, Primers, and Public Schooling: Education for Science in Nineteenth-Century America," *Isis* 81 (Sept. 1990): 424–45; and Herbert Hovenkamp, *Science and Religion in America, 1800–1860* (Philadelphia: University of Pennsylvania Press, 1978). For an astute analysis of statistical thinking and numeracy, see Patricia Cline Cohen, *A Calculating People: The Spread of Numeracy in Early America* (New York: Routledge, 1999).

The history of human-animal interactions, as specialists like to call it, has only recently gained substantial scholarly attention. A notable and thought-provoking exception is Keith Thomas, *Man and the Natural World: A History of the Modern Sensibility* (New York: Pantheon Books, 1983). For an overview of scientific theories about domestication, see Stephen Budiansky, *The Covenant of the Wild: Why Animals Chose Domestication* (New York: William Morrow, 1992). On domestic animals in colonial America, I relied on Jack Larkin, *The Reshaping of Everyday Life, 1790–1840* (New York: Harper and Row, 1988). The scholarly and popular literature on horses and industrialization is more extensive. See, in particular, Ann Norton Greene, *Horses at Work: Harnessing Power in Industrial America* (Cambridge, MA: Harvard University Press, 2008); Clay McShane and Joel Tarr, *The Horse in the City: Living Machines in the Nineteenth Century* (Baltimore: Johns Hopkins University Press, 2007); and Clay McShane, "Gelded Age Boston," *New England Quarterly* 74 (Spring 2001). On harness racing, see Dwight Akers, *Drivers Up: The Story of American Harness Racing* (New York: G. P. Putnam's Sons, 1947); and Melvin L. Adelman, "The First Modern

Sport in America: Harness Racing in New York City, 1825–1870," *Journal of Sport History* 8 (Spring 1981): 5–32. On the history of horse training, see Clive Richardson, *The Horse Breakers* (London: J. A. Allen, 1998).

Several legal historians have written about legal responses to animal risk. The classic statement is Hendrik Hartog, "Pigs and Positivism," *Wisconsin Law Review* (1985): 900–903. Peter Karsten, "Cows in the Corn, Pigs in the Garden, and 'The Problem of Social Costs': 'High' and 'Low' Legal Cultures of the British Diaspora Lands in the 17th, 18th, and 19th Centuries," *Law and Society Review* 32 (1998), provides a broader, comparative analysis.

Readers interested in industrial risk should begin with Mark Aldrich's indispensible book *Safety First: Technology, Labor, and Business in the Building of American Work Safety, 1870–1939* (Baltimore: Johns Hopkins University Press, 1997). I have utilized his meticulous research and analysis throughout the central section of this book. I have also avoided treating the subject of workplace dangers in detail because I have little to add to what he has already said. Although I have not specifically addressed the topic of mining, Anthony F. C. Wallace's *St. Clair: A Nineteenth-Century Coal Town's Experience with a Disaster-Prone Industry* (New York: Alfred A. Knopf, 1987) provides an analysis of the mentalities present in very dangerous industries which is applicable to railroads and factories.

Aldrich has also written an important work on railroad safety, *Death Rode the Rails: American Railroad Accidents and Safety* (Baltimore: Johns Hopkins University Press, 2006). It should be read alongside a somewhat hidden classic: Charles Hugh Clark, "The Railroad Safety Movement in the United States: Origins and Development, 1869–1893" (PhD diss., University of Illinois, Urbana, 1966). For the early history of the railroad, I have frequently referred to relevant sections of Walter Licht, *Working for the Railroad: The Organization of Work in the Nineteenth Century* (Princeton, NJ: Princeton University Press, 1983); and John F. Stover, *American Railroads* (Chicago: University of Chicago Press, 1961). For the later period, Edward Chase Kirkland, *Men, Cities, and Transportation: A Study in New England History, 1820–1900*, vol. 2 (Cambridge, MA: Harvard University Press, 1948), remains a useful guide. For railroad appliances, Steve Usselman's work is definitive. See, in particular, "Air Brakes for Freight Trains: Technological Innovation in the American Railroad Industry, 1869–1900," *Business History Review* 58 (Spring 1984): 30–50; and *Regulating Railroad Innovations: Business, Technology, and Politics in America, 1840–1920* (Cambridge: Cambridge University Press, 2002). See also John H. White, *The American Railroad Passenger Car* (Baltimore: Johns Hopkins University Press, 1978). On railroad workers' culture of risk taking, see also John Williams-Searle, "Courting Risk: Disability, Masculinity, and Liability on Iowa's Railroads, 1868–1900," *Annals of Iowa* 58 (1999): 27–77. On railroad labor more generally, see Shelton Stromquist, *A Generation of Boomers: The Pattern of Railroad Labor Conflict in Nineteenth-Century America* (Urbana: University of Illinois Press, 1987). On the law, begin with James W. Ely Jr., *Railroads and American Law* (Lawrence: University of Kansas Press, 2001); and Barbara Young Welke, *Recasting American Liberty: Gender, Race, Law, and the Railroad Revolution, 1865–1920* (Cambridge: Cambridge University Press, 2001).

On the factory safety movement, begin with Aldrich, *Safety First*. For a different approach from a historian of medicine, see Christopher C. Seller, *Hazards of the Job: From*

Industrial Disease to Environmental Health Sciences (Chapel Hill: University of North Carolina Press, 1997). On the role of labor statistics, see James Leiby, *Carroll Wright and Labor Reform: The Origins of Labor Statistics* (Cambridge, MA: Harvard University Press, 1960); and William Brock, *Investigation and Responsibility* (Cambridge: Cambridge University Press, 1984). On the history of professionalization, begin with Burton J. Bledstein, *The Culture of Professionalism: The Middle Class and the Development of Higher Education in America* (New York: Norton, 1976); and Andrew Delano Abbott, *The System of Professions: An Essay on the Division of Expert Labor* (Chicago: University of Chicago Press, 1988). Patricia Lengermann and Gillian Niebrugge, "Thrice Told Tales: Narratives of Sociology's Relation to Social Work," in *Sociology in America: A History*, edited by Craig Calhoun (Chicago: University of Chicago Press, 2007), helps put the social work dimension of factory safety in context. William I. Trattner's *Crusade for the Children: A History of the National Child Labor Committee and Child Labor Reform in America* (Chicago: Quandrangle Books, 1970) describes the fight to regulate child labor.

The topic of factory inspection in the United States has received very little scholarly attention. There is no secondary literature that acknowledges the existence of Henry Dorn. More has been written about Florence Kelley. Particularly useful is Sandra D. Harmon, "Florence Kelley in Illinois," *Illinois State Historical Journal* 74 (Autumn 1981): 162–78. A more idealized treatment can be found in Kathryn Kish Sklar, *Florence Kelley and the Nation's Work: The Rise of Women's Political Culture, 1830–1900* (New Haven, CT: Yale University Press, 1995). For one well-written analysis of gender and factory inspection, see Helena Bergman, "That Noble Band? Male and Female Factory Inspectors in the Turn-of-the-Century United States" (MA thesis, SUNY Binghamton, 1995). On safety in the steel industry, David Brody's *Steelworkers in America: The Non-Union Era* (Cambridge, MA: Harvard University Press, 1960) remains extremely useful. The literature on workmen's compensation is vast; for a recent interpretive overview see John Fabian Witt, *The Accidental Republic: Crippled Workingmen, Destitute Widows, and the Remaking of American Law* (Cambridge, MA: Harvard University Press, 2004). Crystal Eastman's *Work-Accidents and the Law* (New York: Charities Publication Committee, 1910) is essential reading, as is John Fitch, *The Steel Workers* (New York: Charities Publication Committee, 1911).

Public safety campaigns have received only passing attention from historians. Two notable exceptions are Peter D. Norton, *Fighting Traffic: The Dawn of the Motor Age in the American City* (Cambridge, MA: MIT Press, 2008); and Barbara Welke, *Recasting American Liberty: Gender, Race, Law, and the Railroad Revolution, 1865*. Journalist Robert Meyer had access to NSC records, before they were destroyed. He wrote two brief but informative articles: "A Look at the First Congress," *Safety and Health*, Feb. 1987, 22; and "One Giant Step for Safety," *Safety and Health*, Mar. 1987, 46. Readers are therefore advised to consult the enormous body of documentation in the *National Safety News* and the NSC's annual conference *Proceedings*. Some local newspapers and a number of national magazines also contain articles about the campaigns.

In contrast, motor vehicle and traffic safety has attracted the attention of numerous historians. Clay McShane, *Down the Asphalt Path: The Automobile in American Life* (New York: Columbia University Press, 1994), puts automobile safety in the larger context of traffic safety. James J. Flink, *America Adopts the Automobile, 1895–1910* (Cambridge, MA:

MIT Press, 1970), remains the best guide to the earliest years of automobility. Daniel Albert, "Order out of Chaos: Automobile Safety, Technology and Society, 1925 to 1965" (PhD diss., University of Michigan, 1997) is particularly strong on driver's education. I have also made extensive use of three recent studies, David Blanke, *Hell on Wheels: The Promise and Peril of America's Car Culture, 1900–1940* (Lawrence: University of Kansas Press, 2007); Peter D. Norton, *Fighting Traffic: The Dawn of the Motor Age in the American City* (Cambridge, MA; MIT Press, 2008); and Amy Beth Gangloff, "Medicalizing the Automobile: Public Health, Safety, and American Culture, 1920–1967" (PhD diss., SUNY Stony Brook, 2006). On automobile insurance fraud, see Ken Dornstein, *Accidentally, On Purpose: The Making of a Personal Accident Underworld* (New York: St. Martin's Press, 1996).

A number of American studies scholars have tackled the subject of amusement parks. In particular, see David Nye, *Electrifying America: Social Meanings of a New Technology* (Cambridge: MIT Press, 1992); John Kasson, *Amusing the Million: Coney Island at the Turn of the Century* (New York: Hill and Wang, 1978); and David Nasaw, *Going Out: The Rise and Fall of Public Amusements* (New York: Basic Books, 1993). For a business history approach, see Judith A. Adams, *The American Amusement Park Industry: A History of Technology and Thrills* (Boston: Twayne, 1991). The definitive history of roller coasters is by art historian Robert Cartmell, *The Incredible Scream Machine: A History of the Roller Coaster* (Fairview Park, OH: Amusement Park Books, 1987). On rides, see also William F. Mangels, *The Outdoor Amusement Industry: From Earliest Times to the Present* (New York: Vantage Press, 1952); and Charles J. Jacques, "The Coasters of the Philadelphia Toboggan Co.," *Amusement Park Journal* 6 (1984). Jacques has also written a number of well-researched books on specific parks. He has recently deposited his research collection at Pennsylvania State University. Many local historians have written about specific parks. On Kennywood, see David P. Hahner, *Images of America: Kennywood* (Charleston, SC: Arcadia, 2004).

There is a large literature on gun use and gun control. Most of it is so polemical and poorly researched that it is useless for my purposes. The best history of gun control remains Lee Kennett and James LaVerne Anderson, *The Gun in America: The Origins of a National Dilemma* (Westport, CT: Greenwood Press, 1975). Despite the controversies surrounding the author (and its polemical tone), Michael Bellesiles's *Arming America: The Origins of a National Gun Culture* (New York: Alfred A. Knopf, 2000) is an important starting point. Also helpful are several of the essays in Stephen P. Andrews Jr., ed., *Guns in America: A Reader* (New York: New York University Press, 1999). Eric H. Monkkonen has produced some of the most important work on crime and violence in American culture. See, in particular, *Murder in New York City* (Berkeley: University of California Press, 2001) for his take on gun cultures and gun violence. Though not a work of history, very useful for thinking about gun cultures is Abigail A. Kohn, *Shooters: Myths and Realities of America's Gun Cultures* (New York: Oxford University Press, 2004).

Beyond books about Ralph Nader, surprisingly little has been written about the history of the consumer products safety movement. Elizabeth Cohen includes one chapter in her influential book *A Consumers' Republic: The Politics of Mass Consumption in Postwar America* (New York: Vintage Books, 2003). On the Consumers Union, see Norman Isaac Silber, *Test and Protest: The Influence of the Consumers Union* (New York: Holmes and Meier, 1983). The regulatory story in chapter 11 is told mostly through the biographies and writings of politicians involved with the process, as well as government documents. Michael Pertschuk, *Revolt against Regulation: The Rise and Pause of the Consumer Movement* (Berkeley: University of California Press, 1982), is an astute analysis of the 1960s and 1970s legislative history. Shelby Scates, *Warren G. Magnuson and the Shaping of Twentieth Century America* (Seattle: University of Washington Press, 1997), describes the senator's role. On the changes of the 1980s, see William A. Niskanen, *Reaganomics: An Insider's Account of Policies and People* (Oxford and New York: Oxford University Press, 1988); and John Kolb and Steven S. Ross, *Product Safety and Liability: A Desk Reference* (New York: McGraw-Hill, 1980). To my knowledge, no one else has written about refrigerators or lawn mowers and safety, but readers interested in the larger context of the lawn mower story should consult Virginia Scott Jenkins, *The Lawn: A History of an American Obsession* (Washington, DC: Smithsonian Institution Press, 1994).

Finally, this is not a legal history in the strictest sense, but much of the function of the law is to manage risk. I have therefore made extensive use of a number of works by legal historians. Lawrence M. Friedman's magisterial study *A History of American Law*, 2nd edition (New York: Simon and Schuster, 1985), is particularly valuable. For the eighteenth and nineteenth centuries, William J. Novak, *The People's Welfare: Law and Regulation in Nineteenth-Century America* (Chapel Hill: University of North Carolina Press, 1996), is an essential guide to the role of the law in early America. Peter Karsten's *Heart versus Head: Judge-Made Law in Nineteenth Century America* (Chapel Hill: University of North Carolina Press, 1997) offers a well-argued critique of the "economic-oriented" approach often identified with James Willard Hurst, *Law and Social Order in the United States* (Ithaca, NY: Cornell University Press, 1977); and Morton J. Horwitz, *The Transformation of American Law, 1780–1860* (Cambridge, MA: Harvard University Press, 1977). On labor law, see John Fabian Witt, *The Accidental Republic: Crippled Workingmen, Destitute Widows, and the Remaking of American Law* (Cambridge, MA: Harvard University Press, 2004). On the intertwined development of insurance and tort, see Kenneth Abraham's insightful treatment in *The Liability Century: Insurance and Tort Law from the Progressive Era to 9/11* (Cambridge, MA: Harvard University Press, 2008). On changes in tort law, see also Randolph Bergstrom, *Courting Danger: Injury and Law in New York City, 1870–1910* (Ithaca, NY: Cornell University Press, 1992). As a nonlawyer, I also used a number of legal primers to better understand the transformation of case law. In particular, see Jerry J. Phillips, *Products Liability in a Nut Shell* (St. Paul, MN: West Group, 1998).

INDEX